Veterinary Medicine

NIPA® GENX ELECTRONIC RESOURCES & SOLUTIONS P. LTD.
New Delhi-110 034

About the Authors

Prof. Dr. Ramesh Chandra Patra, was born on 28th May 1965, and obtained B.V.Sc. and A.H. degree (1989) from Odisha University of Agriculture and Technology (OUAT)-College of Veterinary Science and Animal Husbandry, Bhubaneswar, and M.V.Sc. (1991) and PhD (1998) in Veterinary Medicine from Indian Veterinary Research Institute, Izatnagar. Dr Patra joined Johns Hopkins University, USA for Post-Doctoral Research in Neurotoxicology. Dr Pata started the professional career with joining as Junior Veterinary Officer in 1992, and academic career as Agricultural Research Scientist in 1993 in Indian Council of Agricultural Research. He joined as Professor, Department of Veterinary Medicine in OUAT in the year 2006. Dr Patra was selected to the Academic Management Positions of Faculty Dean (2015) and Dean of Research (2018) in OUAT. Besides, he was deputed as a visiting scientist to the Centre for Environmental Risk Assessment and Remediation (CERAR), University of South Australia for a period of one year in two spells, and to Pirbright Institute, United Kingdom for a period of six months. His areas of research interest include Veterinary Medicine, Pathophysiology of diseases, Oxidative stress and antioxidant, Environmental toxicology and Bioremediation. Odisha University of Agriculture and Technology bagged a process patent for his invention in developing an herbal acaricide of equivalent potency to chemical acaricides. He has published 166 research articles, 08 book chapters, authored 07 books and 03 bulletins, and supervised 10 PhD and 61 Masters Students. Prof Patra is a recipient of DST-BOYSCAST Fellowship (2001), Johns Hopkins University Neuroscience Post-Doctoral Research Fellowship (2002), DBT-Overseas Associateship (2007), Endeavour Research Fellowship by Government of Australia (2009), and DBT-CREST Fellowship (2012). Professor Patra has been conferred with, Sri Ram Lal Agarwal National Gold Medal (2005), and D.C Blood Gold Medal (2003), Award of Honour (2007), P. L. Narayan Rao Gold Medal (2012) by Indian Society for Veterinary Medicine. The Department of Science and Technology, Government of Odisha conferred Dr Patra with Samantha Chandrasekhar Award (2024) for the year 2022 for outstanding contribution in Life Science. He has been recognized as the Fellow of Indian Society of Veterinary Medicine (2007), National Academy of Veterinary Science (2016), Indian Association of Advancement of Veterinary Research (2016), Indian Society for Sheep and Goat Production and Utilization (2024) and Society for Veterinary Science and Biotechnology (2024). Dr Patra is a past Associate Editor of Indian Journal of Veterinary Medicine, Member of Editorial Board of OUAT Research Journal and peer reviewer of many research journals.

Dr. Mirashree Pati is the meritorious Post Graduate Scholars of the Department of Medicine, College of Veterinary Science and Animal Husbandry, Odisha University of Agriculture and Technology Bhubaneswar, with immense practical knowledge on healthcare issues in the grassroot level veterinary institutions. Dr. Pati, currently a PhD Scholar in Veterinary Medicine, is a University Gold Medallist (OUAT), ICAR-Senior Research Fellow.

Dr. Biswa Ranjan Jena was born in Bhubaneswar, Odisha. He completed his B.V.Sc. and M.V.Sc. in 2018 and 2020, respectively, from Odisha University of Agriculture and Technology, Bhubaneswar. He was awarded with University Gold Medal for securing first position in M.V.Sc. course in 2020. Dr Jena was awarded with the Best PG Thesis award 2020 by Veterinary Internal and Preventive Medicine Society. He completed his PhD from Guru Angad Dev Veterinary and Animal Sciences University (GADAVASU), Ludhiana, Punjab in 2023. He was also awarded with the Best PhD Thesis award in 2023 for his valuable research on Bovine Leukosis. Dr. Jena has published 18 research articles in various national and international journals, 7 book chapters, and is the author of one book. He attended more than 15 national and international symposium and conferences, where he presented various research papers.

Veterinary Medicine
General Systemic States Digestive and Urinary Disorder

Ramesh Chandra Patra, PhD
NAVS Fellow (2016), ISVM (2007), IAAVR (2016), ISSGPU (2024) SVSBT (2024) and RAAST (2025)
BOYSCAST Fellow (2001), & JHU Neuroscience Post Doctoral Fellow (2002) in USA
DBT Overseas Associate in Australia (2007) and
Endeavour Research Fellow in Australia (2010)
DBT CREST Fellow in UK (2014)
Govt of Odisha-Samanta Chandrasekhar Awardee in Life Science (2022)
Ex-Senior Scientist, Indian Veterinary Research Institute, Izatnagar 243122
Ex-Professor and Head, Department of Veterinary Medicine
Former Dean, College of Veterinary Science & Animal Husbandry, Bhubaneswar and
Former Dean of Research, Odisha University of Agriculture & Technology
Bhubaneswar-751003, India

Mirashree Pati
University Gold Medallist (OUAT), ICAR-SRF
PhD Scholar (Veterinary Medicine)
Department of Veterinary Clinical Medicine
College of Veterinary Science & Animal Husbandry
Odisha University of Agriculture & Technology
Bhubaneswar-751003, India

Biswa Ranjan Jena
PhD, (Veterinary Medicine)
University Gold Medalist (OUAT), ICAR-SRF
Lal Bahadur Shastri-VIPM Best PG and PhD thesis Awardee

NIPA® GENX ELECTRONIC RESOURCES & SOLUTIONS P. LTD.
New Delhi-110 034

NIPA® GENX ELECTRONIC
RESOURCES & SOLUTIONS P. LTD.

101,103, Vikas Surya Plaza, CU Block
L.S.C.Market, Pitam Pura, New Delhi-110 034
Ph : +91 11 4386 0225, 9717133558, 9540816132
E-mail: newindiapublishingagency@gmail.com
Website: www.niparesources.com

Print ISBN: 978-93-58872-55-2

ebook ISBN: 978-93-58877-36-6

Composed and Designed by NIPA®.

Preface

The book entitled Veterinary Medicine: General Systemic States Digestive and Urinary Disorder, authored by me as the Principal Author, published by NIPA, New Delhi was scheduled for its new edition. Besides, there was a need to reorient the chapters as per the new course curriculum, designed to meet the requirement of the Minimum Standards of Veterinary Education (MSVE), to facilitate degree students for their professional accomplishment. This volume of Veterinary Medicine book with course code VMD 311 has its genesis for the above purpose and is aimed at empowering students to study course specific books.

The Veterinary Council of India (VCI) exercises its authority and wisdom to maintain the MSVE and modifies the syllabus and course contents to make the teaching and learning process dynamic for the students admitted to Bachelor of Veterinary Science and Animal Husbandry degree programme in VCI recognised institutions. As per MSVE 2016, Veterinary Medicine has been taught as a single course with 4+1Credit hours. VCI guidelines in its new MSVE divided the Veterinary Medicine syllabus into six different courses, to be taught during 3rd and 4th year of study programme with six different course codes.

This volume of the Veterinary Medicine book contains chapters on history and scope of Veterinary Medicine, concepts of animal diseases, concepts of diagnosis, differential diagnosis, treatment and prognosis, general systemic states and diseases of digestive and urinary systems. This volume of Veterinary Medicine book covers a total of 41 chapters in the area of veterinary clinical medicine. It has been intended to make the book reader friendly, easy to understand, and to prepare a concise book to deliver better clarity to each chapter. As the author of this book,

I express my sincere thanks to my students, colleagues, friends and family members, for their motivation, encouragement and support in compiling this volume of Veterinary Medicine Book Series successfully.

Authors

Contents

1

History and Scope of Veterinary Medicine

Veterinary Clinical Medicine is a subset of Veterinary Medicine, and refers to study of animal diseases at the patient bed side for diagnosis, monitoring, prognosis and treatment of the disease. The Veterinary Medicine is that branch of veterinary science that deals with diagnosis, treatment, prevention and control of diseases, disorders and injury in domesticated, pet, wild, zoo, captive, free-living animals and birds. Thus, the discipline covers and deals with all aspects of disease in animals including examination of patient, and its environment, laboratory investigations, diagnosis of disease, differential diagnosis from other related diseases, prognosis of disease, treatment, monitoring of disease process, control and prevention of diseases.

The word Medicine has derived from Latin word "Medire" that means to heal. Thus, Medicine is the art and science of healing. Therefore, a Veterinary Medicine specialist engages his knowledge, skill and past experience for the relief of animals' suffering and for control and prevention of animal diseases. He has to interact with the animal owners for the collection of history of the disease, examines the animals and their environment, and provides animal healthcare services for the cure from the disease. Disease diagnosis, herd health management, care of newborns, issues related to veterinary public health, creation of public awareness about zoonotic diseases, epidemiological studies on animal diseases, animal welfare, extension activities in the field of livestock health are some of the important facets of veterinary medicine. An expert in veterinary medicine should have sufficient knowledge on animal physiology, ecology, anatomy, biochemistry, pharmacology and pathology for investigation of the disease and subsequent measures for ameliorating the suffering of the animal and for cure and control of the disease.

In India, the Veterinary Medicine is a clinical discipline of Veterinary Science and Animal Husbandry, where subjects such Veterinary Medicine, Animal welfare, Ethic and Jurisprudence, Wild and pet animal healthcare are being covered. The course curriculum has been extensively designed by the Veterinary Council of India (VCI) with some liberty given to different colleges, and institutions to revise the courses as per the regional needs. Thus, VCI in its regulation (2008) enforces to maintain minimum standard of veterinary education for award of B.V.Sc and AH degree. In western countries, Veterinary Medicine refers to whole gamut of Veterinary Science and most of colleges are named as College of Veterinary Medicine.

History of Veterinary Medicine

The expertise on healing practice for animal dates back to 3000BC. Urlugaledinna in Mesopotamia is considered as an expert in animal healing. Subsequently, there has been reference to veterinary practice and veterinarian in the literature. The practice of veterinary medicine is believed to exist in the Temple Cities of Babylon of Egypt. Archaeological evidences also suggest that the ancient Egyptian nearly over 4 millennia ago had the well-developed veterinary profession. Veterinary and animal husbandry practices are considered to be well developed in ancient India, and there are mentions of animal healing practices in *Rigveda* and *Atharvaveda.* The ancient Indians also had distinction to establish the first state funded well-equipped veterinary hospital during the period of the Great *Ashoka* (238 BC). The treatment of animal diseases has been described in books written by *Charak* (2300 BC), *Salihotra* (1800 BC), and *Palikapaya* (1000 BC). Nakula and Sahadeva, during the period of Pandavas as mentioned in great epic, Mahabharat, had expertise in treatment and management of horses and cattle, respectively. *Shalihotra* is considered to be the first known veterinary doctor in the world who lived around 1800 BC in Northern India. Hippocrates is known as the Father of Medicine. During these periods, veterinary profession has only one facet that is to render animal healing. The involvement of microbes in disease process and identification of many different kinds of bacteria and protozoa by Lewis Pasteur, Robert Koch and many other scientists helped to better understand the disease in animals by the end of the nineteenth century. In 1898, Löffler and Frosch reported that the hoof and mouth disease is caused by "a previously undiscovered agent of disease". The use of the term "virus" evolved gradually, and viruses could be detected by electron microscopes by 1940s.

The veterinary profession can be said to have started with the establishment of Veterinary School in Lyon, France by a veterinarian Claude Bourgelat in 1771. Subsequently, London Veterinary College started in 1791, thus giving scope for creation of dedicated professional group in the area of animal medicine. Treatment of horses remained a focus for many years to meet the need of the army. Then, the interest of the profession spilled to cattle, other livestock and poultry. This was followed by interest in treatment of dogs, cats and other companion and exotic animals.

In India, a veterinary training school started at Pune in the year 1862. Subsequently, Veterinary Colleges were established in different parts of India such as Mumbai

(1884), Madras (1892), Bengal (1893), Bihar (1930), and Orissa (1954). The last one century has witnessed tremendous progress in the field of veterinary science, and the period from 1945 to 1985 is considered as the golden age for veterinary profession. The advances in bio-medical tools for use in clinical veterinary practice such as endoscope, ECG, ultrasound, and modern molecular techniques have assisted in prompt and reliable diagnosis of animal health problems in a non-invasive or semi-invasive way. Development in vaccine production, newer drug discovery, and understanding of disease process added a fillip for better management of diseases in livestock, pet and wild animals. Eradication of once dreaded disease, Rinderpest – the second viral disease to be eradicated from the whole world, and effective control of several infectious diseases including African horse sickness outbreaks in 1960s, Equine influenza in 1980s and recently, bird flu are the notable examples of implementation of advanced animal healthcare practices.

Scope of Veterinary Medicine

The veterinary medicine has wide scope covering all species of domesticated and wild animals and birds. The veterinary medicine has been extensively practised with or without the professional supervision. There has been increasing demands of veterinarians to meet the rising demand for animal health and animal production related sectors such as public health, animal welfare, disease diagnosis, quarantine and research on biomedical and veterinary subjects, more particularly on infectious agents having potential to spread the disease to human beings. Bioterrorism with use of microbes, pathogenic to both human and animals, necessitated the expertise of veterinary professionals to deal with the subjects and to control the spread of pathogens. Strict biosecurity has been enforced in several laboratories and firms with technical support of veterinary professionals. Over the years, the responsibilities of veterinary profession have become multi-dimensional and demand driven. India has made steady progress in livestock sector with index of meat and milk production rising to about 170 in 2005 from the base of 100 in 1991, a sustained annual growth rate of more than 3.5 percent.

India has 15% and 57% of total cattle and buffalo population in the world and ranks number one in milk production, 4 in egg production and 6 in broiler meat production. The country produced about 198.4 million tons of milk in the year 2019, an increase of 6.5% over the previous year. The dairy industry is growing at an annual growth rate of 5%. However, the productivity of cattle is very low because of poor genetic makeup, environmental stress, and poor feeding and healthcare practices. The lower per capita animal production from livestock in India also reveals a great scope of the profession to increase the productivity with implementation of efficient feeding, breeding and healthcare measures. There are about 65 thousand veterinary institutions including hospital, veterinary dispensaries (37,647), and veterinary-aid-centres (28,168) in the year 2019 as per Annual Report (2019-20) of Department of Animal Husbandry and Dairying, Ministry of Fisheries, Animal Husbandry and Dairying, Government of India.

Veterinary Medicine in the 21st Century has several facets besides primary job of providing animal healthcare. The increased demand for food of animal origin, use of animals as sentinels for environmental monitoring, security screening, control measures for existing and emerging diseases, use of animals for biomedical research, drug trials and issues related to animal welfare have broaden the scope of veterinary professionals. The profession of Animal and Veterinary science is now expected to contribute substantially to national growth, to provide food security, to produce safe food for human consumption, and food free from drug residues and contaminants, and to ensure animal welfare.

The world enjoys a negative status for Rinderpest, the second infectious disease to be eradicated. However, Indian livestock and poultry industry is still threatened with economically important diseases like foot and mouth disease, brucellosis and infectious bovine rhinotracheitis (IBR) in bovines; sheep pox, PPR, enterotoxaemia and blue tongue in sheep; goat pox, contagious ecthyma and contagious caprine pleuropneumonia in goats; swine fever, swine pox, gastroenteritis in pigs; herpes virus and *Salmonella Abortus equi* abortions, glanders and strangles in equines, and Ranikhet disease, fowl pox, Gumboro disease, egg drop syndrome and Marek's disease in poultry. Besides these, there are number of wasting diseases such as tuberculosis, John's disease and parasitic infections and infestations, and non-infectious diseases like gastro- intestinal disorders, metabolic and deficiency diseases, and plant and chemical poisoning, which adversely affect the productivity of various livestock species. Thus, there is a need for better animal healthcare support system for livestock, particularly for high yielding animals and crossbred cows.

Veterinary practices do not normally have numerous veterinary technologists, laboratory staff, interns, residents, and highly trained personnel, with various levels of expertise. The differences between human and animal hospitals as regards personnel, equipment, testing capabilities, fees, operating expenses, and the required private investment involved in operating a veterinary facility, are not well understood and appreciated by the public users of veterinary services. So, the veterinary practitioners have the duty to their clients and the public, as well as themselves, to inform their clients and the public, generally and up front, about their facilities and policies. This information should include choices concerning hospitalization, various methods and personnel for patient monitoring, after hours and emergency services, and the use of emergency clinics as per their availability. In the changing scenario of veterinary practice "One world and one health approach" OIE (Office International des Epizooties) has taken some initiatives to practice good veterinary governance. Evaluation of veterinary services through use of Performance of Veterinary Services (PVS) tool, PVS gap analysis, and

PVS follow up evaluation is under progress in different states and OIE Member countries including India. The OIE performance evaluation tool is designed to assist Veterinary Services to establish their current level of performance, to identify gap and weakness regarding their ability to comply with OIE international standards, to form a shared vision with stakeholders and to establish priorities and carry out strategic initiatives.

Despite several challenges for the veterinary students and practitioners, there are many newer options for veterinarians to choose a career within the profession. The veterinarians are also now trained for their roles that perhaps the public are not so familiar with. Some such roles played by the today's veterinarians include ensuring food safety, border control and quarantine, pet and sports animal practice, animal production and health, animal welfare, research and technical services, wildlife health and environmental protection. A postgraduate degree in any discipline of veterinary science is the minimum qualification for teaching and research jobs in India. However, PhD degree in concerned subject is preferred for such jobs. Veterinarian with the B.V.Sc. and A.H qualification can serve as veterinary surgeon in government hospitals and in private run institutions for animal healthcare. Hospitals in cities provide scope to the veterinarians for practising on both the domesticated livestock and pets. A posting in rural areas increases the responsibilities of the veterinarian towards cattle, sheep, goat, and poultry and, in some regions, horses. Veterinarians in urban areas can set up private practice and such practice involving companion animals can be quite profitable, especially in big cities. Dairy sector also employs large number of veterinary doctors for increased healthcare support and breeding of dairy cattle. The spectacular growth of poultry sector in India has also widen the scope of veterinarians to work in poultry farms to maintain health of poultry birds and to develop strains with high egg productivity and increasing poultry production for meat consumption. Veterinarians also work as extension officers at block levels to educate farmers, encourage livestock rearing and production, and to advise on animal farm management and animal health problems.

Veterinary Medicine and its practice must remain relevant to the changing needs of the society. Veterinary medical education is required to meet changing needs of the society through creation of newer areas of professional focus. Academic veterinary medicine should cover existing and anticipated diversity in the society. Veterinary medicine should seek greater collaboration and cooperation with human health in the public health area, with veterinarians playing a lead role in the management of zoonotic diseases, public health and ecosystem health under the umbrella of "One health".

The Veterinary Council of India in exercise of the powers conferred by the Indian Veterinary Council Act, 1984 (Act) framed the regulations to provide for the minimum standards for veterinary education in India. It consists of seven parts:

i. Preliminary that provides for short title, date of commencement and definitions;

ii. Course of Study that lays down the requirements with respect to the degree course of Bachelor of Veterinary Science and Animal Husbandry (course) including the duration of the professional year and procedure to be adopted for imparting training in the veterinary hospitals or institutions and internship

iii. Admission to the Bachelor of Veterinary Science and Animal Husbandry Degree Course

iv. Veterinary Curriculum – Structuring and Organization of Course Curriculum, specifies the subjects to be covered, syllabus, internship, examination and evaluation, and also provides for standards with regards to teachers and examiners, attendance requirements, scrutiny of answer papers and grading;

v. Courses and Course Contents: gives the year-wise distribution of courses and the detailed course contents

vi. Minimum Standards of Veterinary Education: describing the minimum standard requirements for a veterinary college with respect to number of departments, land and infrastructure including a Livestock Farm Complex, minimum staff for each department, equipment and other facilities

vii. Grading Scheme for Veterinary Colleges

2

Concept of Animal Disease

The disease is defined as a deviation from the normal health. Health refers to the very absence of the disease. Therefore, health is a state. where all the body systems function normally and the activities continue within the physiological limits. So, the disease can be defined as the inability to perform physiological functions at normal levels even when nutrition and other environmental requirements are provided at adequate level. The above definition includes both clinically ill animals as well as animals not performing as expected. However, the disease frequencies in epidemiological studies are calculated considering that the disease is binary. That is the animal is either suffering from a disease or it is free from the disease or in good health. But such situation is not always true due to varied sensitivity and specificity of disease tests used for identification of the disease, and manifestations of the disease are not that discrete or binary in all diseases. The concept of the health and disease can be described in following five categories.

Health as normality: The animal should have normal features, appearance and behaviours including posture and gait, rumination, alertness, body coat conditions etc. to be considered as healthy. Under such categorization, health refers to soundness of the body and absence of the disease. This definition of health is not the actual description of health but serves as an operational definition of clinical health.

Health as homeostasis: The concept of homeostasis refers to maintenance of delicate balance within the individual or within the processes of an individual. In holistic term, the word "health" implies to the concept of a mind and body together in harmony with the environment. As per this concept, health is defined as a state where all organs are in a delicate balance within the individual and with the surrounding world.

Health as biological function: The disease is considered as a state, where there are structural and functional alterations in the cells within the body of the living beings. As per this definition, disease is the disturbance in proper performance of the body functions. Functional disturbances are manifested by physical signs which are being detected in a diseased animal.

Health as physical and psychological wellbeing: As per this categorization, health is defined as a state of physical and psychological wellbeing. But there are problems associated with applying this definition to all animals. Domestic animals are more or less subordinate to human interests and this is more obvious in farm animals. Health definition linked to psychological well-being can be misleading in relation to the general purpose of livestock production.

Health as productivity including reproduction: The health in animals is the same as productivity from animals. The animal is considered to be in good health, if it is with normal production and reproduction. A slight modification of health definition related to health has been introduced as positive health. It refers to optimal production with the provision of a complete diet, an environment that is optimal for the animals' physiological needs, comfortable to animal senses, in which animal is secure and is free from fear, and with no undue challenge by pathogenic microorganism or predator. Therefore, the disease is defined as a state in which normal functions are disturbed or altered at cellular, tissue, organ or whole organism level. Alternatively, the homeostatic mechanisms of the body are upset, which leads to malfunction. It can be associated with microorganisms - bacteria and viruses causing infectious disease, parasites; causing clinical and subclinical disease, poor management; especially related to feeding, causing production diseases, alterations to the animal's genetic make-up; causing changes in function at cell level leading to or predisposing disease, toxic agents; causing cell damage and possibly clinical disease.

Classification of the disease: The following parameters or criteria are being used for classification of the disease

Criterion	Disease	Manifestations/ Characteristics
Clinical signs	1 Subclinical	No visible abnormality is recognized/ No apparent signs of illness unless the profiles of milk production, performance or growth rate are monitored regularly, or biochemical or haematological tests are being performed
	2 Clinical	Manifestation of overt and recognizable clinical signs like fever, breathing difficulty, convulsions, lameness etc.
Duration of illness /Course of the disease (period from first appearance of clinical signs till recovery or death)	1 Peracute	The course of the disease is too short and may be in hours. In many instances, animals are found dead without any visible clinical signs being observed by the owner or the veterinarian. Example: Peracute form of anthrax, Haemorrhagic septicaemia, snake bite.
	2 Acute	The course of the disease is in days (<7 days)
	3 Subacute	The course varies from one to several weeks
	4 Chronic	The disease persists for months. Example: Tuberculosis.

Criterion	Disease	Manifestations/ Characteristics
Disease aetiology	1. Infectious	Diseases caused by infectious agents like bacteria, virus, protozoa, helminths etc.
	2. Non-infectious	The disease is caused by factors, other than infectious agents. Example: Metabolic diseases, Deficiency diseases, Diseases caused by allergens and toxicants
Severity	1 Severe	Disease involves high risk with poor prognosis (expected outcome) and treatment may not have positive outcome
	2 Moderate	Moderately severe and the prognosis of the disease is also favourable
	3 Mild	The disease process is slow, the ill effects produced are minimal and the prognosis is favourable.

3

Concept of Diagnosis and Differential Diagnosis

The term "diagnosis" is derived from Greek words *dia* (between) and *gignoskein* (to know). It literally means to recognize a disease and knowing the difference between the two diseases or conditions. Therefore, diagnosis includes determination of specific cause of the disease, abnormality of structure and function produced by the causative agents and clinical manifestations associated with the abnormality. A prompt and reliable diagnosis plays a key role for successful therapeutic attempt and a favourable outcome. Some commonly used terms in the process of making a diagnosis are given below.

Signs: It refers to the manifestation of a particular disorder in an animal which is observed by the clinician, attendant or owner.

Symptoms: Manifestation of disease or disorder that is noticed/ felt by patient himself or herself.

Pathognomonic signs: It is the specific sign observed in a particular disease that provides sufficient clues for making a diagnosis. Roaring sound is produced in laryngeal paralysis.

Pathognomonic lesions: It is the gross or histological alteration that is very specific to a particular disease. E.g. Abomasal ulcers in theileriosis

Syndrome: A group or combination of clinical signs that forms a distinct clinical feature indicating a particular disease or disorder. E.g. Downer Cow Syndrome, Metritis-Mastitis-Agalactia Syndrome in recently whelped pig

Prognosis: It refers to forecasting of outcome of the disease. It is based on clinical observations of the animals and clinic pathological alterations in a disease process and response to continuing treatment.

Differential diagnosis: This is a process of confirming a particular disease from other diseases with almost similar clinical signs. This is achieved by performing specific test or group of tests that identify a particular disease from other diseases. Example: Differentiating lead poisoning from other causes of laryngeal paralysis as seen in rabies.

Making a diagnosis of the disease involves a detailed clinical examination that consists of three different components: examination of the animals including physical examination of the patient and few relevant laboratory investigations, history taking and examination of the environment. History taking or anamnesis comprises of three parts; patient data (name of the animal and owner, address etc), disease history (present and past disease) and management history (floor space, housing, vaccination, feeding habits and breeding etc). Such information is collected through interrogation with the client and makes the entire process easier to arrive at a conclusion. Certain changes in appearance/ attitude are being recorded by the animal owners or non-veterinary personnel and are collected during history taking. Besides, the clinician has to look into the environment for possible cause of the disease. A schematic representation on clinical examination is given below that helps to arrive at diagnosis.

Clinical Examination of patient

Physical and Laboratory Examination	**History taking**	**Examination of the Environment**
General and physical Examination including A. Examination from a distance B. Examination of body parts	Patient data Management history	Outdoor environment, Stocking rate, Feed and water supply, Waste disposal
Examination using palpation, percussion and auscultation Laboratory examination of the body fluids and other biosamples	Disease history including distribution of the disease, past disease, present disease, including nutrition, reproductive management, housing etc	Indoor environment including sanitation, hygiene, floor space and ventilation

Examination of animal includes observation of clinical signs from a distance followed by physical examination at close quarters, collection of biosamples, laboratory investigations etc. Respiration or breathing movements are examined from a distance, preferably with the animal on standing posture. The respiration rate may be counted by observation of nostril movement or by feeling the nasal air movements by placing the palm about 2-3 cm in front of nostrils. During the process of making a diagnosis and more particularly during examination of the patient from a distance, some of the following changes may be noticed by the clinician. Even, the animal attendant or the owner may provide some idea of such changes during history taking. Those are further investigated by the clinician for its correctness.

Behaviour: The changes in behaviour include separation from group or following the group, sluggish, dull or indifferent response, anxiety, restlessness, yawning, hyper-excitation including abnormal movement with vigour, violent licking of own body or eating and licking of other inanimate objects, jumping, and kicking etc.

Voice: Altered voice includes hoarse voice as in rabies, soundless bellowing and continuous lowering of voice.

Appetite: May be partial (inappetance) to complete loss (anorexia) of food intake. This is one of the foremost signs in diseased animals. There may be painful or difficulty in prehension, mastication, swallowing, regurgitation or rumination.

Rumination: Rumination is observed as jaw movement for re-mastication of regurgitated feed. Rumination is reduced or completely absent in certain conditions, mostly in fever or with oesophageal or rumen disorders.

Faecal characteristics: Frequency of defaecation, volume and consistency of faeces provide indications of the possible problem in the diet or diseases of gastrointestinal system.

Urination: Change in colour, volume, frequency of urine gives idea about the disease condition, particularly involving kidney and associated organs. Dribbling of urine is a common sign of partial obstruction of the urethra. Painful urination occurs in cystitis and urethritis.

Posture and Gait: Abnormal posture (when animal is in stationery state), gait (during movement), bodily conditions (obese, thin or emaciated), shape and size of different body regions, etc. point to the unhealthy state of animals.

Skin coat: Skin is a true reflection of health and its abnormalities such as alopecia, discolouration of hair or wool, itching and bleeding can be noticed from a distance.

The body temperature, respiration and heart rate are commonly examined to identify the underlying problem during physical examination of the patient. Increase in body heat coupled with dryness of the muzzle is the common complaints of the animal owner during history taking. As the normal body temperature of most of the domesticated livestock and pets is higher than that of human beings, only touching the body coat may give a misleading conclusion that the animal is having fever. So, accurate recording of body temperature by a semi-skilled person helps in asserting the temperature. Body temperature of domestic animals is generally taken per-rectally using a rectal thermometer. A conventional clinical thermometer for human use may be used for animals. The bulb end of the thermometer is lubricated and gently inserted with a rotatory movement through the anus into rectum. The thermometer is tilted slightly to make contact of the bulb with mucous membrane of the rectum and held for two minutes. Pulse is taken on the middle coccygeal or facial arteries in cattle, facial artery in horses and mules, and femoral artery in sheep and goat. The pulse rate depends on heart rate but may or may not entirely represent heart rate because of pulse deficit, difference between heart rate and pulse rate.

Methods of Diagnosis

***Syndrome / Pattern recognition*:** The diagnosis is made by correlating the exhibited signs in past cases with similar signs and symptoms. The past experience of the clinician helps in making diagnosis. Bloody diarrhoea with typical odour of the faeces may indicate parvo viral infection in dogs. Ascending posterior paralysis is a good indicator of rabies in domesticated animals. But, confirmatory diagnosis of the disease requires laboratory investigation.

***Hypothetico-deductive reasoning*:** List out all the diseases those produce the exhibited clinical signs, and deduct the disease one by one by specific reasoning based on presence or absence of disease specific clinical signs.

***Algorithm/ Arborization*:** Short list the suspected diseases and diagnose the disease through clinico-pathological tests and examination of biosamples one after another

Key abnormality method: The system involved is identified first. Then what are the organs involved and then investigation is carried out to find out the location and type of lesions followed by specific diagnosis.

Data-base method: This is a computer-based method where inference is drawn after feeding the information to the computer programme.

Physical Examination

A careful physical examination of the patient may result in arriving at identification of the problem, making a diagnosis of the disease and institution of the therapeutics for successful management. The physical examinations involving audiovisual inspection such as palpation, auscultation, percussion, ballottement and tactile percussion are conducted following the distant examinations and a close visual inspection of the body, to investigate the abnormalities associated with the disease. The skill of an experienced veterinarian helps to identify the problem with the use of some of these techniques. Others examinations such as temperature, pulse, respiration, state of hydration/ dehydration, and ruminal movement can be monitored by a trained para-veterinary personnel or animal owners. The common methods for physical examination include palpation, percussion, and auscultation.

Palpation

The guiding principle behind this physical examination method is sense of touch. It reveals the information regarding body temperature, pain, oedema, fluid accumulation, consistency and contour of the organ. Palpation includes direct palpation consisting of handling the tissue by means of fingers for which one or both the hands may be used. Indirect palpation is done through probe, catheters, probing (stomach tube) instruments. Palpation of internal organs is very difficult in small and medium sized animals. But in animals like cattle, buffalo and horses, palpation of some internal organs through per rectal examination has got some advantages. Animals should be kept in standing position while undergoing palpation. The clinician should approach the animal very patiently and slowly, otherwise the animal may react before palpation. Palpation of the organ provides the following information.

- Presence of pain, if any, and the intensity of pain
- Shape, size, and consistency of any part of the body
- Cutaneous thermal response: Hypo/ Hyperthermia
- Consistency of different organ such as doughy, hard, emphysematous, fluctuating, or resilient

Doughy: The palpation finding is said to be doughy when the soft tissues retain finger impressions for few seconds during palpation as seen in oedema and impaction of rumen.

Resilient: It is a condition where tissue appears solid-like muscles and moves with palpation. Ex. neoplasm and abscesses.

Hard: The feeling of bone like consistency E.g. ossification of cartilage

Fluctuating: The soft tissue is soft and elastic and undulates on pressure. It doesn't retain finger impression as the case with doughy feeling i.e. hematoma, cyst.

Emphysematous: - The palpation of the affected area produces crepitating and crackling sound on pressure E.g. B.Q. in calves and malignant oedema in bullocks.

Palpation of external and internal body parts

Pharyngeal region: In large animals, palpation of pharyngeal region is usually performed with both the hands. In small animals, a similar technique is applied by using one or two finger to detect pain, swelling and stenosis on pharyngeal region. The internal palpation of this region in cattle can be done using a mouth gag.

Oesophagus and trachea: The cervical region of oesophagus and trachea can be palpated by one hand on one side of neck and feeling along the jugular groove starting from the pharynx downwards. It gives valuable information regarding pain, local swellings deformities, if any. Internal palpation of oesophagus is done by indirect method with the use of probang or stomach tube to diagnose obstruction, stricture or dilatation.

Thorax: Palpation of chest wall corresponding to lungs area has limited value in large animals.

Cardiac area: Palpation is done by pressing the palm of left hand against the cults inside left elbow at the fourth intercostal space. The animal suffering with traumatic pericarditis may elicit sever pain response with deep and strong palpation.

Abdominal organs: Rumen can be palpated through abdominal wall on left side, particularly over the paralumbar fossa. The rumen motility can be detected in the upper part of left flank or at the paralumbar fossa. In case of bloat, the rumen is tense and bulging. In case of impaction, the rumen palpation gives a doughy and firm feeling. External palpation of reticulum is not possible due to its anatomical position on the right side of the rumen. But by applying pressure at the lower end from 6th to 8th intercostal space may produce a reticular grunt in case of TRP. Pressure on lower part of right costal arch or in the area from 7th to 9th intercostal space may produce pain in case of omasal impaction. Normally, it is difficult to palpate abomasum, spleen and liver. However, enlarged liver can be palpated on right dorsal arch.

Lymph node: Mandibular, prescapular, pre-femoral or supra-mammary lymph nodes are normally palpable. The parotid, pharyngeal, cervical and inguinal lymph nodes are palpable when they are enlarged.

Palpation of internal organs through rectum: Per rectal examination of the abdominal and pelvic organs are more helpful in large animal practice, particularly in livestock. Palpation of pelvic cavity organs becomes easy by rectal palpation. The examination through rectum must be carried out carefully to prevent any damage to organs. In cattle, the rumen is distended in bloat and vagus indigestion. In ruminal acidosis, the fluid content of fore stomach can be palpated. In right side abomasal displacement, a large balloon-like structure occupies the right dorsal sac of right peritoneal cavity. In left abdominal displacement, the abomasum can be felt only when it is exceptionally displaced. It will be felt like a tense, oval, balloon on the left of the dorsal sac of rumen. In rectal examination, caecal volvulus or caecal dislocation can be detected. Normally, the peritoneum is smooth and not painful on palpation. In peritonitis, it becomes rough and nodular and produces pain on palpation.

Percussion

This is a physical method of examination of animal wherein audible sound is produced from the underlying organ with the use of finger or instrument. The nature of the sound produced gives a clue about the underlying organ. Depending on whether only fingers (direct/ immediate percussion) or plexor or pleximeter (indirect/mediate percussion) are used, percussion is divided into two categories. Direct percussion is mainly used on thin and small animals using the middle finger to strike on the left hand over the organ to be percussed. This is called as finger-to-finger percussion. The information obtained by this method is very limited in buffalo and fatty animals. Indirect palpation is done by putting an oval or circular plastic plate (pleximeter) directly on the body surface and a stroke is applied with plexor or percussion hammer. However, percussion has limited application and output in animal practice due to presence of excessive subcutaneous fat in pigs and thick wool in sheep, thick bones in large animals. In large animal practice, percussion is done for two reasons. Acoustic percussion is used for areas of lungs, reticulum, rumen, liver, caecum and spleen where the effect extends up to a depth of only 7 cm. However, problem existing deeper to this cannot be identified. Pain seeking percussion is very useful in cattle and buffaloes and is aimed at localization of pain. Deep percussion was used over the reticular area in suspected cases of traumatic reticular peritonitis (TRP) and over abomasal area for torsion or impaction. Very brief strike is applied on the suspected area and the strength is gradually increased. The area of the lesion is identified wherever percussion elicits maximum pain. The best percussion results can be obtained when following points are taken into consideration.

- An empty closed room is suitable for percussion.
- Suitable instrument is very important.
- Animal must be in standing position and at rest.
- The percussion hammer should be held at a high level compared to pleximeter.
- The pleximeter must be placed firmly and flat on the area to be percussed
- Percussion stroke should be perpendicular to pleximeter, must be sharp, clear and decisive and should not be too close.
- Corresponding parts of either side of body should be compared
- The stroke should not be heavier than necessary.

The types of sound produced by percussion can be described as follows.

Tympanic: It is loud sound produced by organs filled with gas under pressure, like sound generated while beating a drum. It is heard in case of ruminal tympany, and left and right displacement of abomasum. Tympanic lung sound is seen in early or late stage of pneumonia, pneumothorax and pulmonary emphysematous condition.

Resonant/ringing sound: This type of sound is produced when large volume of air/gas is present beneath the site of percussion. Eg. Normal lung sound produced during percussion of normal lung area.

Dull sound: This type of sound is produced from on organ not filled with gas or by percussing a solid organ. The dull sound itself indicates that the particular part has tissue with reduced or no gas in it.

Metallic ring sound: It is similar to tympanic sound but its sound character is more pronounced. This sound is very high pitched and resembles with the striking of an empty metallic jar. Example – Ping sound produced in abomasal displacement while tapping at the 9th intercostal space on left side in left side displacement of abomasum.

Empty muscle sound/ absolute damping sound: This type of sound is produced over large muscle, or from liver in cases of hepatitis, heart affected with pericarditis etc.

Tactile percussion: It is a modified form of percussion consisting of both palpation and percussion to know the consistency and boundaries of deeply situated organs. This is accomplished by firm pushing the organ back with the fist of the hand and allowing the organ to rebound on finger tips. It is done normally for ballottement of foetus and fluid ballottement in ascites, exudative peritonitis, rumen acidosis and rupture of urinary bladder to notice the fluid thrill. This method of physical examination also helps in pregnancy diagnosis. In case of dog and cat, the palm of one hand is placed firmly against one side of abdomen and on the opposite side, percussion of various intensities is applied to palpate with the hand placed on the body surface. The impact of fluid wave can be detected by finger tips. In large animals, an assistant is required to generate the wave or fluid thrill by strong percussion.

Percussion area for lungs: The area of percussion is more or less triangular in shape on either side of the chest. In cattle, the superior upper border is formed by an imaginary line drawn from posterior angle of scapula to the tubercoxae.

The anterior border forms the posterior angle of the scapula to the olecranon process of ulna. The left side posterior border is formed starting from the 11th rib or 11th intercostals space goes to the 9th rib in the middle of ends at lower part of 5th intercostal space. So, it is very important to examine the lungs border between 5th and 9th rib. When there is enlarged reticulum, the lungs border will be pushed upward to the 7th or 8th rib. It is better to percuss the lungs at right side. In case of horse, the superior border is formed by a line joining the posterior angle of scapula and last second intercostal space. The anterior border is formed from posterior angle of scapula to olecranon process of ulna. The posterior border is formed by extending upward and backward from olecranon process of ulna to join the dorsal end at the second largest intercostal process. In case of dog and cat, the area available for lung percussion is similar to horse.

Percussion of heart: Heart gives a complete non resonant sound as it is a muscular organ and air free. Heart is located approximately in the lower 2/3rd of thoracic cavity between the 3rd to 6th pair of ribs. Anterior part of the heart is covered by muscles of shoulder area. So, it is difficult to percuss. An increase in the area of cardiac dullness occurs in case of cardiac enlargement i.e. cardiac hypertrophy or dilation, pericarditis and lateral displacement of heart. Reduction of cardiac dullness occurs in case of over distension of lungs i.e. emphysematous condition. Tympanic sound is produced when there is presence of gas in the pericardial sac. However, it is very difficult to reach at a conclusion particularly in large animal practice.

Auscultation

The word auscultation has been derived from Latin word "*auscultare*" which means 'to listen'. Auscultation is the process by which various functional and pathological sounds produced by different organs of the body can be heard. This physical method of examination is used to hear the sounds emitted by internal organs for the purpose of assessing mainly the circulatory, respiratory and gastrointestinal system. Previously it was practised with the help of naked ear. Later stethoscope is used. "Stethos" means chest and "scope" means to examine. In veterinary practice phonendoscope is very common which can amplify even the minute pathological sounds. Auscultation should be made in a calm and quite place to avoid the sounds of surrounding environment. There are two methods of auscultation.

Direct/ immediate auscultation is carried out by putting the ear directly on the body surface to listen the sounds. There are some disadvantages of this method. It is difficult to keep the ear in contact with the animal body where it is restless or vicious. The skin coat of the animal may be soiled or dirty with faeces, secretion and various discharges. Skin may be infected with ectoparasites like fleas, mites, ticks, etc or may be infected with bacterial and fungal infection which may be transmitted to the clinician while keeping the ear in contact with the body.

Indirect auscultation is performed by means of stethoscope in small animals and phonendoscope in large animals. In clinical practice, a flexible stethoscope is used. It consists of sound-detecting device (chest piece), flexible tube or conduction tube and two ear pieces at the other end. The sound-detecting device has a thin plastic sheet on one side and a bell shape with a hole in on the other. There may be a rubber ring in the chest piece to avoid friction with the hairs of the animal. In small animals, auscultation can be performed with chest piece of 5 cm diameter, surrounded by a hard plastic diaphragm.

Heart area of Auscultation: Careful auscultation of heart gives valuable information about the functional status of the heart than any other clinical procedure. The cycle of cardiac activity is divided into two phases, systole and diastole. The cycle is measured from the onset of one ventricular contraction to the beginning of succeeding contraction. Auscultation of heart is performed in large animals on standing position. It is better when the left foreleg is drawn forward. The small animal should be placed on a table of convenient height for clear heart sound and the mouth of the animal may be closed for few minutes. The purpose of auscultation is to determine the character of heart sound and to detect the presence of any abnormal sounds. The best site to auscultate heart is 4th and 5th intercostal spaces.

Heart sounds: Cardiovascular sounds are classified into two groups, transient and murmur. Transient heart sounds are the normal heart sounds. Loud heart sounds are produced in case of cardiac hypertrophy and anaemia. Increased heart sounds are produced in early stage of febrile condition. The heart sounds are reduced in case of cardiac insufficiency and congestive heart failure. The heart sounds are weaker when heart is displaced from thoracic wall or fluid is present in pleural or pericardial cavity like anasarca, hydropericarditis, exudative and proliferative pericarditis, hydrothorax etc. Abnormal heart sound or murmur or adventitious heart sound may originate from the cavities of heart or pericardium. Cardiac murmur may produce hissy, humming, or whirring sounds. Stenosis or narrowing of valvular orifices may also produce humming or whirring sound. Incomplete closure of valves may also produce hissing sound. The functional murmur is produced when there is no structural alternation of the valve of the heart. Pericardial rubbing sound is produced from the pericardium.

Auscultation of lungs area: Normal respiratory sound is consisted of two sounds; 1) Vesicular (lung) sound, 2) Tubular (bronchial) sound. Vesicular sounds are audible when the air rushes through small alveoli and bronchi. These sounds are audible only in case of normal lungs. The vesicular sound resembles the sound produced when the letter "V" is pronounced. This sound is normally heard throughout the respiratory area except in the area where it is masked by the bronchial sound. The vesicular sounds are clear during inspiration. During expiration, it may change its character and resembles the sounds produced when letter "F" is pronounced. In resting large animals, the rate of air movement is reduced and therefore, the respiratory vesicular sound is poorly or hardly audible. In case of dog, cat and other small animals, the vesicular sounds are very clear. Vesicular sound is reduced, when the chest wall is thick and the air content of the lungs is reduced as seen in early stage of pneumonia, passive congestion of lung or the lungs are compressed by the fluid of pleural sac. The tubular/ bronchial sounds are produced while air passes through large bronchi and trachea. It resembles the sound produced when the letter "H" is pronounced. Normally, it is audible over trachea and larynx.

Diseases of bronchi, lungs, pleura and diaphragm result in production of abnormal respiratory sounds, and include different types of rales, emphysematous sounds, friction sounds etc. Moist rales are produced by the presence of fluid or mucous of relatively low viscosity within the alveoli or bronchi or both. It is better audible during inspiration than expiration. These rales are classified as fine or coarse moist rales. Prognosis for presence of fine rales is more unfavourable than coarse rales. Moist rales are found in case of lobar pneumonia, bronchopneumonia, pulmonary oedema, and pulmonary haemorrhage. Dry rales are heard when air is being forced through a bronchial tube which is partially constricted either by dry mucous or by severe swelling of the mucous membrane. Dry rales vary in pitch according to the diameter of the bronchial tube. These produces humming, rattling, whistling, crackling sounds. It occurs in early stage of acute bronchitis, spasm of bronchial mucosa, pulmonary neoplasm and tuberculosis. Crepitating rale is produced by separation of sticky walls of alveoli and bronchioles. This type of sound is produced when the bronchial mucosa is sufficiently swollen or exudation has affected the alveoli so that the opposing wall becomes adherent to one another and separated by incoming air. The sudden separation of the adherent mucous membrane causes a low-pitched crackling sound. It is also called as fine moist rales. It is clear in later part of inspiratory phase. Care should be taken to avoid friction between chest piece of the stethoscope and body coat of the animal that adds confusing sounds.

Auscultation of Abdominal Area

It gives an idea about the functional status of the rumen and reticulum. The contraction of the rumen produces gargling sound and booming gassy sounds due to the movement of the fluid and solid material on which a gas layer is superimposed. Reduction in rumen activity is accompanied by a decrease in frequency and volume of the sounds. A soft fluid sound may be heard at approximately 7 cm behind the point of elbow at the costochondral junction of the left side when ruminal contents move into reticuloruminal fold. A similar type of sound is produced when fluid returns to the reticulum by the contraction of ventral sac of the rumen.

Temperature Recording

Temperature varies in different species of animals. Body surface temperature is usually lesser than the visceral organs. Temperature in animal is generally taken from the rectum by inserting a thermometer through the anus. In case of female animals, vaginal temperature may be considered. But vaginal temperature remains higher during approximately, normally 1oF higher than rectal temperature in healthy animals. The following precautions should be taken for accurate recording of the temperature.

- Animals should be in an isolated area away from herd with minimum excitation.
- It should be restrained properly with minimum disturbances
- It should be kept away from hot exposure prior to recording the temperature.
- It should be given 20–30-minute rest if it has travelled long distance especially in hot summer.
- Temperature should not be recorded after prolonged working, walking, race or ploughing.
- Animal should not be offered hot or cold water at least 20-30 minutes prior to recording the temperature.
- Thermometer should be kept in contact with mucous membrane to obtain correct temp.

Temperature of healthy animals may vary during day time. Morning and evening hour's temperature differs slightly. In a healthy animal, the lowest temperature is recorded in morning and the highest in the evening hours. This type of variation is called diurnal variation. Critical temperature is the highest level beyond which, it is considered as abnormal. The temperature of the animals also varies with respect to age, sex, breed, environment and body weight.

- Age: -Temperature of young animal is 1°F higher than adults of same species.
- Sex: - Temperature of female animal is higher than male animal.
- Breed: -Temperature of purebred & cross breed is little higher than indigenous breed.
- Body weight: - Temperature is inversely proportional to body weight.
- Feeding: - The body temperature, increases during the process of ingestion, mastication, digestion etc.
- Infection: - Any infection or inflammation process causes rise in temperature.
- Environment: - Body temperature goes up in day time during summer and goes down in night time. This variation does not exceed 2-3 oF.

Table 1: Temperature, pulse and respiration rate in domestic animals

Animal Species	**Body temperature**		**Pulse**	**Respiration**
	Normal	**Critical**	**per min**	**(per min)**
Cattle/ buffalo	38°C (100.5°F)	39°C (102)	60-72	20-30
Calf	38.5°C (101.5°F)	39.5°C (103°F)	80-120	24-36
Sheep	39°C (102°F)	40°C (104°F)	70-90	20-30
Goat	39.5°C (103°F)	40°C (104°F)	70-90	20-30
Horse	38°C (100.5°F)	39°C (102°F)	28-46	10-16
Camel	35°C (95°F)	37°C (99°F)	25-32	6-12
Pig	39°C (102°F)	40°C (104°F)	60-90	10-20

Physical Examination of Eye

Examination of the eye from a distance or at close quarters gives valuable clues for diagnosis of the disease. The unilateral discharge from the eye suggests localized infection. Bilateral eye discharge is seen in systemic infection. The type of discharge also provides information regarding the associated disease.

Watery discharge: It can be due to obstruction of lachrymal duct or gland

Serous discharge: Early stages of inflammation

Purulent discharge (Pus discharge): Later stage of inflammation due to pus forming bacteria.

Bloody discharge: Unilateral trauma, injury etc.

Movement of eyelid: The abnormal movement of eyelid is noticed in CNS disorder, and it is painful in acute conjunctivitis, acute keratitis and deranged nervous function. Excess movement of eyelid is noticed in canine distemper, hypomagnesemia, lead poisoning, encephalitis and prolapse of third eyelid in tetanus in horses.

Physical examination of the conjunctiva gives very good indication of the physical state of an animal such as pale colour in anaemia, icteric (yellowish) in jaundice, mud colour in toxaemia, and cyanotic in anoxia, particularly when the oxygen saturation is low.

There are different methods to test blindness. Those include menace reflex, obstacle test etc. Similarly, different organs and body parts are required to be examined one after another for identification of the problem and associated changes in behaviour, posture, gait, reflexes and reaction to physical examination.

Diagnostic Techniques in Veterinary Clinical Medicine

The process of diagnosis of a disease involves general physical examination to determine abnormality of function. Special examination of the suspected organ or system(s) reveals the problem, and locate lesion. Special examination of the lesion specifies the kind of lesion, and examination of specimen determines the cause of lesion. The first two aspects of disease diagnosis are effectively completed by an experienced clinician with the help of routine diagnostic aids. The other steps however, require specific bio-medical tools to perform diagnostic tests such as endoscopy, ECG, EEG, angiography, phonocardiography, ultrasonography, radiography, tomography, image intensification and scintigraphy. These modern techniques permit better evaluation of functional profile of different internal organs. However, all these techniques have limitation in its application particularly while examining the large animal with heavy body mass and thick layer of subcutaneous tissue.

Endoscopy: Endoscopy (Greek - *endon* = within; *skopion* = to look at) is a tubular instrument used to examine interior of body cavity or a hollow organ. The advent of fibreoptic endoscope in 1957 made this technique a valuable tool in veterinary medicine particularly in canine and equine practice. Besides examination of respiratory system *(Bronchoscopy),* the endoscopy is also used to examine oesophagus *(Esophagoscopy) to* determine diverticulum, foreign body, fistula, meso-oesophagus, rupture, stricture and ulceration; stomach *(Gastroscopy)* for gastric ulcers, pyloric stenosis, squamous cell carcinoma, and presence of parasitic (*Gastrophilus, Draschia*) larvae in horses; small intestine (Enteroscopy) for duodenal conditions, ulcers, mucosal lesions; large intestine (Colonoscopy) for the presence of whip worms, brush cytology, foreign body and mucosal lesions and peritoneal cavity (Laparoscopy) for abdominal tumour and abscess, evaluation of peritonitis, chronic colic and abdominal metastatic lesions, location of cryptorchid testes, ovarian and urogenital disorder.

Electrocardiography: This has mainly been used to identify cardiac abnormalities that limit performance of heart and to evaluate heart size. In equine, change in ECG such as absence of P waves, irregular R-R interval and baseline F and S waves are generally indicative of arterial fibrillation, a common cause of poor performance Phonocardiography is used for recording cardiac sounds and murmurs using chest surface microphone. Phonocardiography and telemetric ECG is used to measure cardiac function during exercise in horses.

Ultrasonography: It is a safe non-invasive diagnostic technique, which has wide application in canine and equine practice for identifying cystic structure and to distinguish dense masses within organs. Contractile organs can be evaluated by ultrasonography without using contrast medium. Oscillometry and ultrasonography are considered reasonably accurate techniques for indirect blood pressure measurement. Ultrasonography is also an excellent method for early detection of pregnancy in equines for which portable ultrasound units are available.

Computed Tomography: Computed Tomography (CT Imaging) is also known as CAT (Computed Axial Tomography) scanning. The computed Tomography is based on the X-ray principle. As X-rays pass through the body, they are absorbed or attenuated at different levels creating a matrix or profile of X-ray beam of different strength, which is registered on a banana shaped detector. This measures the X- ray profile. The CAT scan differs from the radiograph in that the former consists of a very thin beam of X-ray, which passes through a cross section of the body in a rotational manner, whereas, the latter is one-dimensional image, where the overlying organs present deep in the area are superimposed.

Magnetic Resonance Imaging (MRI): It is a sophisticated computerized imaging technique, which has been a clinical diagnostic tool since 1980. The distribution of hydrogen nuclei (Protons), found in cellular water, depends on the tissue type and whether or not the tissue is healthy or diseased. The MRI uses radio waves and strong magnetic fields to create images. The MRI machine produces images by creating resonance (vibrations) in the water found in the body. This causes a radio signal to be emitted from the water in the body, which can be interpreted by receptor called a coil. The coil produces a cross-sectional image of body water content found in the tissues. These methodologies can pick up very minute differences in water density that is missed by other methods.

Scintigraphy or Nuclear Medicine: Scintigraphy was introduced in veterinary medicine in 1970s. Its principal use is in orthopaedic conditions. It is highly sensitive and can detect bone damage well before radiological identification. The techniques are also being applied to evaluate vascular perfusion, pulmonary ventilation and muscle damage, to measure cardiac output and in intra-thoracic pathology. The scintigraphy involves the production of images following internal administration of radioactive pharmaceuticals to a patient. The two- dimensional image obtained by sophisticated 'gamma camera' indicates the distribution of radioactive material in the body. Using advanced computer technology, these images can then be analysed to provide both qualitative and quantitative information. In this procedure, the principle of X-ray and radioactivity are combined. The advantage of scintigraphy is that it allows to detect sub clinical signs of damage.

Diagnosis of Chemical Toxicity

Chemical toxicity or poisoning with wild plants, organophosphate, organochlorine compounds and agrochemical and environmental contaminants is not uncommon in livestock and pet practice. These chemicals produce more of less similar kind symptoms. The symptoms and illness are noticed at the same time in a group of animals having access to the toxicant. Diagnosis of chemical toxicity in isolated case is more difficult. It is only the acquaintance with the syndrome produced by the common toxicants and environmental evidence such as distribution of poisonous plant in a particular locality, recent use of pesticides in the agricultural field, particular industrial enterprise in the locality that give clue for diagnosis of the toxicity in animals. So, while investigating the cause of disease in a group of animals, it is highly essential to thoroughly examine the animal's immediate environment in addition to clinical and physical examination of the animals. The confirmatory diagnosis of toxicity in animals also poses a serious problem and, thus, majority poisoning cases are symptomatically treated but not with specific antidote.

5

Prognosis and Treatment

Prognosis has been derived from a Greek word that means “fore-knowing” or “foreseeing”. It is a medical terminology that refers to predicting the likelihood or expected development of a disease, including whether the signs and symptoms will improve or worsen, speed of expected changes or remain stable over time, expectations of quality of life, such as the ability to carry out daily activities, the potential complications and associated health issues, and the likelihood of survival. A prognosis is made on the basis of the normal course of the diagnosed disease, the individual›s physical condition, the available treatments, and additional factors. A complete prognosis includes the expected duration, function, and description of the course of the disease, such as progressive decline, intermittent crisis, or sudden, unpredictable crisis. Therefore, prognosis is defined as the likely outcome or course of a disease and forecasting of the chance of recovery or recurrence. Classically, it denotes to forecast or prediction. Medically, prognosis may be defined as the prospect of recovering from injury or disease, or a prediction or forecast of the course and outcome of a medical condition. It is expressed as poor, favourable, moderate, excellent, fair, or hopeless. The prognosis is applied to diseases or conditions, as well as the outcome expected from an intervention.

Prognostic studies aims to understand the course, determinants, or probability of outcomes in a cohort of patients. Prognostic information in animal healthcare is useful to provide information to patient’s owner, to assign specialized target clinicians for treatment, to target specific prognostic factors for modification throughout treatment, to provide the basis of personalized or risk-based medicine and to design randomized trials. Several primary studies examine prognostic factors, biomarkers, tests, and models. The need to develop systematic reviews of prognosis studies is essential in facilitating the interpretation and usability of prognosis study findings. A prognostic factor is a measurement that has a relationship with a clinical outcome in the absence of therapy or when standard therapy is applied. The routinely collected patient characteristics such as sex, age, body mass index, the presence of other concurent diseases, or symptoms are considered to be the prognostic factors. Besides, the prognostic factors also include biomarkers and molecular variables like imaging, electrophysiological and physiological measurements.

The prognostic estimates can be very accurate, when applied to a large statistical populations. The statement that 45% of patients with severe septic shock, will die within 28 days, can be made with some confidence, because previous research confirmed death in such a proportion of patients. However, this statistical information cannot be applied for the prognosis in an individual patient, as patient-specific factors may significantly influence the expected outcome of the disease. Additional informations are required to determine if the patient belongs to the 45% group, who will die, or to the other 55%, who will survive.

Different categories or values of any particular prognostic factor are associated with either better or worse outcomes for future health. For example, the tumour grade of cancer at the time of histological examination is considered to be a prognostic factor as it is frequently linked to the time to death or disease recurrence. Prognostic scores often provide a more objective or more accurate prognosis than clinical predictions alone.

Prognostic vs Predictive Outcome

Prognostic factors are distinct from predictive factors. A predictive factor is a measurement associated with response or an absence of response to a particular therapy. Commonly, responses are defined using clinical endpoints commonly practised in clinical trials. A predictive factor implies that there is a quantifiable benefit from the therapy that can be appreciated using a predictive biomarker. In short, a prognostic biomarker provides information on overall outcome of the disease, regardless of therapy, while a predictive biomarker gives information about the result of a therapeutic intervention. The prognostic accuracy is highly heterogeneous in palliative care, and clinicians' predictions are more likely to be inaccurate.

The word treatment has originated from the Latin word "tractare", which originally meant "drag about". Later, it came to mean manage, handle, or deal with. Treatment in healthcare practice refers to the act of caring sick animal or human being. Theoretically, there are three classifications of medical treatment; Curative – to cure a patient of an illness, Palliative – to relieve symptoms of an illness, and Preventive – to avoid the onset of an illness. The healthcare providers organize treatment for their patients to control a health problem, lessen its symptoms, or clear it up. Treatments can include medicine, therapy, surgery or other methodologies. Medical treatment refers to management and care of a patient to combat disease or disorder, and includes all treatments using prescription medications, or use of a non-prescription drug at prescription strength. Treatments include medicines, procedures and therapies to improve health and well-being. Cure is a potential outcome of treatment resulting in improvements in health, and relieve from sufferings. Other treatment outcomes may also keep the animal healthy and allow living a better life. Symbol Rx, used for a medical prescription, is derived from the Latin word recipe or recipere, which means to take.

The treatment is classified into different categories depending on the chronology, priority or intensity. Urgent care is required to be handled on the same day, but those are not necessarily emergencies. The urgent care settingmaychange to the emergency care level. Emergency care handles medical emergencies, and is a first point of contact or intake for less serious problems, which can be referred to other appropriate care levels. Intensive care, also known as critical care, is the care for extremely ill or injured patients that require high resource intensity, knowledge, skill and quick decision making. Ambulatory care is the medical service provided on an outpatient basis. Typically, patients can walk into and out of the clinic with their own power, thus called ambulatory. Homecare refers to the healthcare service provided at home, where medical care providers including veterinarians, animal attendant, and home health aides, attend house calls. This is more common in large animal practice, where shifting of animal patients to the veterinary insitutions is vey cumbersome and expensive for the patient owners. Primary care is the main kind of care in general, and ideally a medical home that unifies care across referred providers. Secondary care is the care provided by medical specialists and other health professionals, who generally do not have first contact with patients' owner. A patient reaches secondary care as a next step from the primary care, typically by referral, although it is very uncommon in animal practice. Tertiary care is specialized consultative care, usually for inpatients and on referral from a primary or secondary veterinary professional, in a facility that has personnel and facilities for advanced medical investigation and treatment. Follow-up care is an additional care during or after convalescence. Aftercare is generally synonymous with follow-up care. End-of-life care is care near the death of the animal. It includes palliative care or supportive care, most especially near the end of life. Hospice care is palliative care very near the end of life, when cure is very unlikely, intented to provide comfort to the animal patient.

The treatment decision often follows formal or informal algorithmic guidelines. Treatment options can often be ranked or prioritized into lines of therapy; First-line therapy, Second-line therapy, Third-line therapy, and so on. The First-line therapy is referred sometimes to induction therapy, primary therapy, or front-line therapy. It is the first therapy that is tried. Its priority over other options is usually either formally recommended on the basis of clinical trial evidence for its best-available combination of efficacy, safety, and tolerability or chosen based on the clinical experience of the treating veterinarian. If a first-line therapy either fails to resolve the issue or creates intolerable side effects, additional (second-line) therapies are substituted or added to the treatment regimen, followed by third-line therapies, and so on. Multiple therapies (combination therapy or polytherapy) are tried simultaneously. Combination therapy is advocated to achieve prompt care of a patient to treat mixed infection, to achieve synergistic effect, to prevent emergence of drugs resistance mutants, to minimize toxicity and to achieve broad spectrum effects. Failure of combination therapy results from failure of isolation and identification of specific causes of diseases, lack of proper selection of antimicrobials, resistant organism as the cause of infection, error in sampling, inadequate drug dose, incorrect route of drug administration, poor host defence mechanism, and presence of necrotic material or foreign bodies. Thus, combination chemotherapy is also called polychemotherapy, whereas treatment with one agent at a time is called single-agent therapy or monotherapy. Adjuvant therapy is the treatment given in addition to the primary, main, or initial treatment, but simultaneously, as opposed to second-line therapy. Neoadjuvant therapy is therapy that is initiated before the main therapy. Surgical excision of a tumor is considered as the first-line therapy for a certain type and stage of cancer. However, radiotherapy is used before it as neoadjuvant medication, as it is given first chronologically, but not a primary therapy or main therapy. Premedication is conceptually not far from neoadjuvant therapy, but the words are not interchangeable. The cytotoxic drugs administered to put a tumor on the ropes before surgery are called neoadjuvant chemotherapy, not premedication, whereas things like anesthetics or prophylactic antibiotics before a surgery are termed as premedication. Step therapy or stepladder therapy is a specific type of prioritization by lines of therapy. The step therapy attempts to mix cost containment by someone other than the patient owner (third-party payers) into the algorithm.

Treatment of Infectious Diseases in Animals

Chemotherapy refers to the use of chemicals for therapeutic purposes to alleviate pain and sufferings in medical practice. Similarly, the term antimicrobials are used for the treatment of infectious diseases, so as to destroy the infecting organism without any toxic effect to the host. Chemotherapy Index is the ratio of maximum tolerated dose of a drug to its minimum curative dose. The chemotherapy index has been replaced by therapeutic index, which is a ratio of LD_{50} to $ED_{50.}$ The chemicals used for chemotherapy are divided into two groups. The group attached to the cell receptor is called haptophore, and another group, which causes specific pharmacological effect, is called as toxophore. A drug can produce curative or toxic effect depending on its affinity for the parasites or to the host. A compound with high affinity for the host tissue (organotropic) may be toxic, whereas the compound with high affinity for parasites (parasitotropic) becomes curative with minimal or no toxic effects.

Drug resistance curtails the therapeutic use of antimicrobial agents. The resistance develops due to their short generation time and easy genetic mutation in pathoges. Bacteria can find themselves resisistant against antimicrobials through increased production of antibiotic inactivating enzymes such as beta lactamase, kinase, acetyl transferase, alternation of binding site configuration, development of impermeable cell wall, increased production of metabolic intermediates e.g. paraaminobenzoic acid (PABA), defective production of autolytic enzymes, and alternation of membrane transport system. Development of antibiotic resistance can be controlled by using drugs at therapeutic dose for a short period, use of drug combination, selecting appropriate drug for therapy after antibiotic sensitivity test, use of an effective narrow spectrum drug against the causative organism instead of using broad spectrum antibiotic, and by isolating sick animals as preventive measures to check spread of the disease.

Successful chemotherapy of microbial infection depends on identification of pathogen, drug sensitivity, rapid attainment of effective drug concentration at the site of infection, host defence mechanism, maintenance of drug level for an adequate period, use of supportive drug, and duration of treatment. The therapeutic concentration of drug at the focus of infection depends upon dosage form, dosing rate, route of administration, rate of distribution and elimination of drug, chemical character of drug, and its penetrating power.

Antimicrobial drugs are also classified based on cidal and static action. Bactericidal antibacterials include Penicillin, Cephalosporin, Aminoglycosides, Sulpha and Trimethorprim combination, Bacitracin, Nitrofurans, Metronidazole, Polymixin, Novobyocin, and Fluoroquinolones. Bacteriostatic antibacterials include Tetracycline, Chloramphenicol, Macrolides (azithromycin, clarithromycin and erythromycin), Lincosamide, Sulfa drugs and Trimethorprim.

General Considerations for Success of An Antimicrobial Chemotherapy

- Bactericidal drug that kills the bacteria, sould be used in immunodeficient animals, whereas bacteriostatic drugs, which suppress the growth, should be prescribed in immuno-competent patients.
- Physician should be familiar with the drug sensitivity pattern of the causative agents, and relative prevalence of individual microbial diseases and the differential diagnosis in in the area or locality.
- Broad spectrum drug or a combination of drugs should be used during initial therapy, before having an antibiotic sensitivity report.
- Treatment is initiated with broad spectrum antibiotics, where foci of infection or curative agent are not determined.
- The identification of disease etiology is necessary before the selection of drugs and future course of action in an outbreak.
- Prolonged therapy and oral adminstration of antibiotics may develop resistant organisms.
- The host characterstics such as age, genetic factor, pregnancy and lactation
- Normal host defense mechanism
- Presence of pus in lesions and acidic condition in abscess cavity
- Streptomycin and Tetracyclines are not recommended in pregnant animals.
- Sulfonamide therapy is not advocated in lactating animals.

- Duration of treatment: An effective therapy clears infection within 3 to 5 days or even longer, depending upon the type of infection, acute or chronic.

Incompatibility: Antimicrobial drugs may intract physiochemically to many other agents. Harmful interactions may occur without visible changes. It is advisable not to mix them in the same syringe. Most drugs can be mixed with normal saline with exception of Benzyl penicillin and Sodium penicillin. Similarly, tetracycline should not be mixed with any solution containing calcium or magnesium. Penicillin-G, Chloramphenicol and tetracycline should not be mixed to avoid precipitation.

Mode of action: The term mode of action or mechanism of action (MOA) is used in medicine to describe how a drug or other substance produces an effect in the body. For example, a drug's MOA could be how it affects a specific target in a cell, such as an enzyme, or a cell function, such as cell growth. The knowledge on MOA of a drug may offer information about the safety of the drug, and how it affects the body. It may also help decide the right dose of a drug and which patients are most likely to respond to the treatment.

Drug withdrawal: Drugs should not be present in animal-derived food meant for human consumption. There are specific regulations for the withdrawal of antibiotic drugs used in animals, meant for slaughter. So, specific withdrawal period is allowed for antimicrobials after their administration before slaughter or milk of treated animal used for human consumption.

Adverse effects: The damaging effects of commonly used antimicrobial drugs include direct host toxicity, adverse interaction with other drugs, and interference with protective effect of normal host microflora or disturbance of metabolic function of microflora of gastrointestinal tract of herbivorous. Adverse effects also include tissue necrosis at the site of injection, production of drug residues in animal products, impairment of host immune system and allergic or hypersensitivity reaction.

Antimicrobial Drugs

Antimicrobial therapy is based upon the selective toxicity of the drug on invading organisms rather than mammalian cells. It is important to select a drug to which the organism is sensitive, to maintain an effective tissue concentration above the minimum inhibitory concentration (MIC) until the infection is eliminated, and where the measured MIC is less than the breakpoint concentration. Sensitivity tests using either sensitivity disks or sensitizer micro-well plates are used to estimate the MIC of specific bacteria and then tables are consulted to see if the MIC is below the breakpoint. If the MIC is below the breakpoint, it is predicted that the microbe will be susceptible (S) to therapy.If it is equal to the breakpoint, it is predicted that the microbe will be Intermediate (I), where high therapeutic doses may work.If the MIC is above the breakpoint, it is predicted that the microbe will be resistant (R). The breakpoint concentrations have been determined by groups like the Clinical and Laboratory Standards Institute (CLSI) following review of clinical and laboratory data.

The pharmacokinetic data in data labels of more recently introduced antimicrobials contain breakpoint information. It is pointed out that the sensitivity tests and breakpoints are useful indicators for the clinical outcome. However, additional factors like drug binding, drug distribution, and the stats of an active immune system affect the outcome in the live animals, so that clinical experience is still essential. Extra-label use of specific antimicrobial drugs in food animals is prohibited for reasons of safety or spread of the resistance. These drugs include the fluoroquinolones, chloramphenicol, nitroimidazoles, furazolidone, nitrofurazone and other nitrofurans, and sulfonamide drugs in lactating dairy cattle, except approved use of sufadimethoxine, glycopeptides, and vancomycin. There are six relevant questions that are routinely used for the selection of an antimicrobial agent.

1. Is an antimicrobial agent required? Is there an infection that will respond to the treatment? It should be avoided if the answer is "Just in case".
2. Where is the infection (Location - organ/ tissue)? What are the access problems to be overcome?
3. Which pathogen(s) are usually found at the infection location?
4. Which antimicrobial agent has the necessary pharmacokinetic properties to get to the location? Will the antimicrobial agent get there at a concentration above the minimum inhibitory concentration (MIC) so that the MIC is below the breakpoint?
5. What dose and route of administration is necessary to achieve the desired effect?
6. How long the treatment should be given?

There are four additional factors that influence the selection.

a. Preferrence to a bactericidal compound over bacteriostatic antimicrobials
b. Toxicity and cost of an antimicrobial drug
c. The withdrawal period f specific antimicrobial drugs - Animals must not be slaughtered for meat or dairy animals' milk must not be used during the preslaughter period.
d. The plasma concentration governs the dose intervals on a treatment regimen but the tissue residence times determines the preslaughter withdrawal time in production animals.

Resistance to Antimicrobials

Organisms may produce constitutive or inducible enzymes, which inactivate the drug. The permeability to or uptake of the drug by organisms may be decreased or transport out of the cell may be increased. Alteration of the drug receptor or binding site may result in reduced drug affinity at target loci. The organisms may develop alternate metabolic or synthetic pathways to bypass the effects of the antimicrobial agent.

Antibiotic: An antibiotic is a chemical substance produced by a living organism, generally a microorganism that is detrimental to other microorganisms. Antibiotics are commonly produced by soil microorganisms and probably represent a means by which organisms in a complex environment, such as soil, control the growth of competing microorganisms. Microorganisms that produce antibiotics are useful in preventing or treating disease caused by bacteria and the fungi. Antibiotics came into prominence worldwide with the introduction of penicillin in 1941 as the first antibiotic. Since then, antibiotics have revolutionized the treatment of bacterial infections in humans and other animals. In 1928, Scottish bacteriologist Alexander Fleming noticed that colonies of bacteria growing on a culture plate had been unfavourably affected by a mould, *Penicillium notatum*, which had contaminated the culture. A decade later British biochemist Ernst Chain, Australian pathologist Howard Florey, and others isolated the responsible ingredient as penicillin, and showed that it was highly effective against many serious bacterial infections. Toward the end of the 1950s, scientists experimented with the addition of various chemical groups to the core of the penicillin molecule to generate semisynthetic versions.

The tubercle bacillus (*Mycobacterium tuberculosis*) was conspicuously unaffected by penicillin. However, this organism turned out to be highly sensitive to streptomycin, an antibiotic that was isolated from *Streptomyces griseus* in 1943. Streptomycin also demonstrated activity against many other kinds of bacteria, including the typhoid fever bacillus. Two other early discoveries were gramicidin and tyrocidin, which are produced by bacteria of the genus *Bacillus*. Those were discovered in 1939 by French-born American microbiologist, René Dubos, and were valuable in treating superficial infections but were too toxic for internal use. The researchers discovered the cephalosporin in 1950s. It is related to penicillin and is produced by the mould, *Cephalosporium acremonium*. Scientists discovered a class of antibiotics known as quinolones in the following decade. Quinolones interrupt the replication of DNA, a crucial step in bacterial reproduction, and have proven useful in treating urinary tract infections, infectious diarrhoea, and various other infections involving bones and white blood cells.

Classification of Antibiotic

Antibiotics are broadly divided into two groups on the basis of their effect on microbial cells through two main mechanisms, which are either bactericidal or bacteriostatic. Bactericidal antibiotics kill the bacteria and bacteriostatic antibiotics suppress the growth of bacteria (keep them in the stationary phase of growth). Antibiotics can be categorized by their spectrum of activity—namely, whether they are narrow-, broad-, or extended-spectrum agents. Narrow-spectrum agents (e.g., penicillin G) affect primarily gram-positive bacteria. Broad-spectrum antibiotics, such as tetracycline sand chloramphenicol, affect both gram-positive and some gram-negative bacteria. An extended-spectrum antibiotic is one thataffects additional types of bacteria, usually those that are gram-negative as a result of chemical modification.

The principle governing the use of antibiotics is to ensure that the patient receives one to which the target bacterium is sensitive, at a high enough concentration to be effective but not cause side effects, and for a sufficient length of time to ensure that the infection is totally eliminated. Antibiotics vary in their range of action. Some are highly specific. Others, such as the tetracycline, act against a broad spectrum action on different bacteria. These are particularly useful in combating mixed infections and in treating infections, when there is no time to conduct sensitivity tests. Antibiotics can be given in different ways, including oral – tablets, capsules or liquids that treat most mild to moderate infections, Topical – creams, lotions, sprays or drops that treat skin infections, Injections – given directly into the blood or muscle for more serious infections.

Some common antibiotics are listed below in the table.

Drug class and Generic Name	**Common uses**
Aminoglycosides (inhibit protein synthesis)	
Gentamicin	Infections of the respiratory and urinary tracts, blood, abdominal cavity; pelvic inflammatory disease
Tobramycin	Infections of the respiratory and urinary tracts, blood, abdominal cavity; pelvic inflammatory disease
Cephalosporin (inhibit cell wall synthesis)	
Cefaclor	Infections of the respiratory and urinary tracts and skin; otitis media
Cefamandole	Infections of the respiratory and urinary tracts, skin, bone and joints, and blood; peritonitis
Cefazolin	Infections of the respiratory and genitourinary tracts, skin, bone and joints, and blood; endocarditis
Ceftriaxone	Infections of the respiratory and urinary tracts, skin, blood, abdominal cavity, and bone and joints; pelvic inflammatory disease; gonorrhoea; meningitis
Cefuroxime	Infections of the respiratory and urinary tracts, skin, bone and joints, and blood
Cephalexin	Infections of the respiratory and urinary tracts, skin, and bone; otitis media
Chloramphenicols (inhibit protein synthesis)	

Drug class and Generic Name	Common uses
Chloramphenicol	Infections of the eyes, ears, and skin; cystic fibrosis; prevention of infection in minor wounds
Fluoroquinolones (interfere with DNA synthesis)	
Ciprofloxacin	Infections of the respiratory and urinary tracts, skin, eyes, abdominal cavity, and bone and joints; diarrhoea; gonorrhoea sinusitis; pneumonia; prostatitis; anthrax
Norfloxacin	urinary tract infections, STDs caused by Neisseria gonorrhoea, eye infections, prostatitis
Lincosamides (inhibit protein synthesis)	
Clindamycin	Infections of the respiratory tract, skin, and abdominal cavity; acne; pelvic inflammatory disease
Macrolides (inhibit protein synthesis)	
Azithromycin	Infections of the respiratory tract and skin; STDs; otitis media; chronic obstructive pulmonary disease; pneumonia
Clarithromycin	infections of the respiratory tract and skin; otitis media
Aminoglycosides (inhibit protein synthesis)	
Erythromycin	Infections of the respiratory tract, skin, and eyes; STDs; pertussis; diphtheria; intestinal amoebiasis ; otitis media; acne; Legionnaire disease; prevention of infection in minor wounds
Nitrofurans (inactivate essential cell components)	
nitrofurantoin	Irinary tract infections
Penicillin (inhibit cell wall synthesis)	
Amoxicillin	Various streptococcal and staphylococcal infections
Ampicillin	Infections of the respiratory and urinary tract and blood; meningitis; gonococcal infections; endocarditis
Penicillin G	Streptococcal and staphylococcal infections
Piperacillin	Infections of the respiratory and genitourinary tracts, skin, abdominal cavity, bone and joints, and blood
Ticarcillin	Infections of the respiratory and gastrointestinal tracts; streptococcal and pseudomonas infections; gonorrhoea; tonsillitis; Lyme disease; impetigo; otitis media; meningitis
Tetracycline (inhibit protein synthesis)	
Tetracycline	Rickettsia, pneumonia, chlamydia, intestinal amoebiasis, acne, prevention of infection in minor wounds
Miscellaneous antibiotics	
Aztreonam	Infections of the respiratory and genitourinary tracts, skin, abdominal cavity, and blood
Imipenem-cilastatin	Infections of the respiratory and genitourinary tracts, skin, abdominal cavity, bone and joints, and blood; endocarditis
Isoniazid	Tuberculosis
Metronidazole	Infections of the vagina and gastrointestinal tract
Rifampin	Tuberculosis

Drug class and Generic Name	Common uses
Trimethoprim-sulfamethoxazole	Urinary tract infections, shigellosis, otitis media, bronchitis, traveller's diarrhoea
Vancomycin	Infections resistant to penicillin and cephalosporin

Mechanisms of Action

Antibiotics produce their effects through a variety of mechanisms. A number of antibiotics work by inhibiting bacterial cell wall synthesis. These agents are generally referred to as β-lactam antibiotics. Production of the bacterial cell wall involves the partial assembly of wall components inside the cell, transport of these structures through the cell membrane to the growing wall, assembly into the wall, and finally cross-linking of the strands of wall material. Antibiotics that inhibit the synthesis of the cell wall have a specific effect at one or another phase. The result is an alteration in the cell wall and shape of the organism and eventually the death of the bacterium. Other antibiotics, such as the aminoglycosides, chloramphenicol, erythromycin, and clindamycin, inhibit protein synthesis in bacteria. The basic process by which bacteria and animal cells synthesize proteins is similar, but the proteins involved are different. Those antibiotics that are selectively toxic utilize these differences to bind to or inhibit the function of the proteins of the bacterium, thereby preventing the synthesis of new proteins and new bacterial cells.

Antibiotics such as polymyxin B and polymyxin E (colistin) bind to phospholipids in the cell membrane of the bacterium and interfere with its function as a selective barrier. This allows essential macromolecules in the cell to leak out, resulting in the death of the cell. Because other cells, including human cells, have similar or identical phospholipids, these antibiotics are somewhat toxic.Some antibacterials, such as the sulphonamides, are competitive inhibitors for the synthesis of folic acid (folate), which is an essential preliminary step in the synthesis of nucleic acids. Sulfonamides are able to inhibit folic acid synthesis because they are analogus to an intermediate compound, para-amino benzoic acid (PABA) that is converted by an enzyme to folic acid. The similarity in structure between these compounds results in competition between para-amino benzoic acid and the sulphonamide for the enzyme responsible for converting the intermediate to folic acid. This reaction is reversible by removing the chemical, which results in the inhibition but not the death of the microorganisms. Rifampin interferes with ribonucleic acid (RNA) synthesis in bacteria by binding to a subunit on the bacterial enzyme responsible for duplication of RNA. Since the affinity of rifampin is much stronger for the bacterial enzyme than the human enzyme, the human cells are unaffected at therapeutic doses.Each type of antibiotic has a specific application in medicine and can serve as a useful model for exploring the various mechanisms by which antibiotics exert their effects. The following sections describe antibiotics including penicillin, cephalosporin, imipenem, the anti-tuberculosis antibiotics and the agents like aztreonam, bacitracin, and vancomycin. These agents and groups of agents further illustrate the chemical and functional diversity found among the antibiotics.

Sulphonamides

Sulfonamides are synthetic chemotherapeutic agents those were commonly used as antimicrobial drug prior to the advent of antibiotics. Sulfonamides and its derivatives are popularly known as sulfonamide or sulfa drugs. The clinical use of this antibacterial declined due to frequent development of cross drug resistance in bacteria, isolated from animals. Recently, sulfonamidesare used in combination with other drugsfor synergistic effects and avoidance of emergence of resistant bacteria. Sulfonamides are broad spectrum antimicrobial agents effective against bacteria, clamydia, toxoplasma and coccidia. However, Mycobacteria, Mycoplasma, Rickettsias, Pseudomonas and Spirochaetes are resistant. They are primarily bacteriostatic, but in high concentrations,they may be bactericidal,especially in urinary tract infections. Sulfonamides are used in acute disease condition during rapidly multiplying stage of bacteria. They are less effective in chronic diseases. Therapy using sulfa drugs is continued till complete elimination of infection. More resistant strains of bacteria may emerge, if therapy is discontinued before complete elimination of organism.Sulfonamides therapy is usually recommended for 7 days, if a desired response is obtained within 3 days. Too short duration of sulphanamide therapy is generally avoided due to failure of cure. The initial doseof sulfonamide, termed as loading dose,is kept higher to obtain a theraputic blood concentration that is followed by subsequent smaller doses as maintainance dose.

Sulfonamides inhibit bacterial folic acid synthesis by competitive inhibition of paraaminobenzoic acid (PABA) by competingat the catalytic site of the enzyme, dihydropteriate synthatase, thereby inhibiting conversion of PABA to folic acid in bacteria. However, antibacterial action can be seen by removal of sulfonamide or addition of PABA. Sulfonamide should not be given with procaine pencilin as this combination is antagonistic. Sulfonamide combined with Pyrimethamine is the drug of choice against toxoplasmosis. Calcium ion and antacids taken with sulfonamide can inhibit gastrointestinal absorption of sulfonamide. Urine acidification enhances the risk of crystalluria. The combination of sulpha drug with trimethorprim has synergistic effects.

Pharmacokinetic Properties

Sulfonamides are rapidly absorbed from gastrointestinal tract and distributed to all tissues and fluids. Acetyleated, hydroxylated and conjugated sulphonamides have less antibacterial activity. Acetylated drugs cause renal damage because of their insolubility in urine. Acetylation of sulfonamides does not take place in dogs. Oral route is the common route of administration, in form of tablet, bolus, solution, water soluble powder. Intravenous route (i.v.) is used for the treatment of acute infections, besidesintramuscular (i.m.) or subcutaneous route. Intraperitoneal route (IP) is not acommon route. Alkaline preparation should not be injected. Intrauterine route is used when drug is combined with urea. Topical use is not suitable. However, solution and ointment are generally prescribed for conjunctival infection of eyes.

Adverse reaction: Sulfa drugs maybe allergic or toxic. Nephrotoxicity is a common side effect, characterized by crystalluria, haematuria and obstruction of renal tubules, anorexia, depression, renal colic, frequent urination and rise of blood urea nitrogen (BUN) level. Increased salivation, vomition, diarrhoea, excitement, muscular weakness, ataxia, rigidity of limbs are recorded in dogs, when administered in higher doses. The hemopoietic system is dearranged with longer period use of sulfa drug, resulting in thrombocytopenia and leukopenia.Disturbed liver function results in jaundice. Neuritis and jaundice are observed in cattle. Cyanosis develops in dogs. Decreased egg production, weight loss, production of rough and thin shelled eggs, liver and kidney lesions, agranulocytosis and anaemia are observedin poultry.

Dose and Route of common veterinary sulfa drugs: Sulfathiazole solution is used orally or intravenouslyagainst systemic infections. Sulfamethazine is used orally or intravenously with a long half life @100mg/kg as loading dose followed by 50mg/kg as maintenance dose at 12hrs interval. Sulfamerazine has limited use in veterinary practice. Sulfadiazine is used in combination with trimethorprim. Sulfobromomethazine sodium (Sulfabrom) at single oral dose of 132-198mg/kg is indicated in cattle for calf diphtheria, pneumonia, metritis, foot rot and septic mastitis. This drug is contraindicated during advance pregnancy. Sulfaethoxypyridazine is used orally or intravenously. Sulfadimethoxime is administered as tablet, bolus, powder, andintravenously in dogsand cats @50mg/kg i.v. followed by 25mg/kg at 12 hrs interval. Sulfamethoxypyridazine is indicated in dogsand cats. Sulfachloropyridazine is used for poultry as solution or in drinking water. Sulfasoxazole is usedin dogs and cats in urinary tract infections (UTI) @50mg/ kg body weightat 8 hrs interval. Sulfacetamide solution is available as 30% solution for topical use or 10% ointment for ophthalmic infection. Succinyl sulfathiazole, Phthalyl sulfathiazole and sulfaguanidibe are gut-active sulfa drugsused for gastrointestinal infections.

Sulphonamides in combination with Trimethorprim or Orimethoprim are used for synergistic effects. Trimethorprim or Orimethoprim are synthetic folic acid antagonists derived from diaminopyramidine (DAP). Individually, DAP derivatives and sulpha drugs are bacteriostatic but when combined, their action becomes synergistic and bactericidal. Trimethorprim is combined with sulfamethoxazole or sulfadiazine or sulfadoxine in 1:5 ratios for veterinary practice. The antimicrobial activity of the combination has broad spectrum activity, effective against aerobic and anaerobic Gm+ve and Gm-ve bacteria. Enterobacteria, nocardia chlamydia and protozoa are sensitive to these combination preparations. However, Rickettsia, Leptospira, Mycoplasma and *Pseudonomas auerogenosa* are resistant. Diaminopyramidine interferes with the synthesis of tetrahydrofolic acid from dihydrofolate by combining with enzyme dihydrofolate reductase. The combined preparation can be safely used simultaneously with ampicillin for wide antimicrobial coverage.

Dosage and administration: The combinaion therapy is indicated in urinary tract infection, bacterial prostatitis, enteric infections due to *E. coli*, Salmonella spp, treatment of brucellosis, nocardia infection, acute upper and lower respiratory tract infections in cattle @24-30mg/kg body weight IV or IM at 12 hrs intervals. Initial dose is followed by half dose at 12 hour intervals till 2-3 days after recovery. In ruminants, drug is administered parentrally and is indicated in salmonellosis, diarrhoea, pneumonia, foot rot and urinary tract infection. In acute mastitis, intravenous injection is indicated at every12 hours interval.Sulfa combination is the drug of choice for urinary tract and ear infections, coccidiosis and prostatic infection. Trimethorprim and sulfaquinoxalibe combination is preferred in poultry for treatment of *E. coli*, Pasturella andcoccidiosis @30mg/kg body weight / day.

Nitrofurans

Nitrofurans are broad spectrum synthetic compounds. They are primarily used against Gm+ve bacteria. A few Gm -ve bacteria, mycoplasma, protozoa and fungi are also affected depending upon the concentration used. They may be bacteriostatic or bactericidal. They inhibit the enzymatic oxidative process leading to break in bacterial DNA strands and results in inhibition of DNA synthesis. Nitrofurans are highly toxic when used in high doses for a longer period. They are mutagenic and neurotoxic. They may also cause haemorrhagic diathesis, anemia, anorexia, nausea and vomiting in livestock, and cardiotoxicity in birds. Toxicity limits their use in topical application and for the treatment of enteric and urinary tract infections.Nitrofurans and nalidixic acid are not used simultaneously because of antagonistic effects.

Five types of nitrofurans have clinical significance. Nitrofurazone (Furacin) is used topically for the treatment of wounds and diseases of skin, eye, ear and reproductive track. It is given orally for prevention of coccidiosis in birds. Nitrofurantoin (Furadantin) has a broad antibacterial action against Gm+ve and Gm-ve organisms. It is absorbed from gastrointenstinal tract without disturbing intestinal bacteria, and 40% drug is excreted unchanged in urine. The concentration achieved in the urine is bactericidal, facilitating its use for the treatment of urinary tract infections. The oral dose is 4mg/kgbody weight t.i.d for 5-7 daysfor dogs. It may be given intramuscularly @ 3.3mg/kg body weight twice daily. Furazolidine

(Furoxone) is generally used as feed additives. It has the greatest antibacterial activity followed by nitrofurazone and nitrofurantoin. It interferes in the acetylation of CoA. It is given for prevention of coccidiosis and bacterial enteric infections in poultry and enteric infections in swine caused by *E. coli*and Salmonella spp. As feed additives for swine and poultry, it is mixed @10-200g/900kg of feed. Furaltadoneis indicated for the treatment of bovine mastitis @ 500mg/ quarter. It is added to milk replacer for prevention of colibacillosis in calves. In poultry, it is effective against fowl typhoid caused by *S. typhimurium*, and mycoplasma when used at the dose rate of 0.04% in drinking water of chickens. Nitrifuroquine has been used for the treatment of bovine mastitiscaused by mycoplasma, bacterial mastitis and for dry cow therapy.

Beta Lactum Antibiotics

Penicillins

Penicillin was first discovered from a mold belonging to *Penicillium notatum* that inhibited the bacterial cell wall synthesis. Penicillin G (Benzylpenicillin) is broken down by gastric acid but penicillin V is resistant to gastric acid. The half life is very short (20 min) and can be increased by using with procaine or benzathine penicillin. Procaine penicillin G maintains therapeutic blood concentration for 24 hour. Benzathine penicillin G lasts for 7 days following single intramuscular injection. Benzathine penicillin does not cross blood brain barrier (BBB), placental barrier and intestinal barrier. It is used in mastitis, systemic and local infections in all speciesagainst Streptococcus, Staphylococcus and *Corynobacterium pyogenes,*at the dose rate of 3 lakhs units/ quarter at 24 hrs interval, pyelonephrotis in cattle (*Corynbacterium remake*), Lumpy jaw in cattle (*Actinomyces bovis*), tetanus, pulpy kidney, lamb dysentery (Clostridia spp), Leptospiral nephritis,vibrionic abortion caused by *Leptospira canicola,* and anthrax due to *B. anthracis* at the dose of 1 lakh units /kg body weight at 12 hrs interval. Antibacterial spectrum ofPhenoxymethyl Penicillin (Penicillin -V) is similar to Penicillin G, and is used at the dose of 8 mg/ kg body weight. B-lactamase (Penicillinase) resistant Penicillin are Cloxacillin, Dicloxacillin and Oxacillin. They are active against Penicillin G resistant Staphylococci, stable in gastric acid, when administered orally or parentrally. Benzathiline penicillin is used for treatment and prevention of mastitis in dry period. Nafecillin and Flucloxacillin are administred intramammary. Methicillin is used throughparentral route due to its poor oral absorption.

Adverse effects: There may be allergic skin reactions in small animals following its aminstration. Some bacteria cause opening of beta lactum ring by producing beta lactamase, leading to inactivation of penicillin G. Generally Gm -ve bacilli become resistant.

Broad Spectrum Penicillin: Antibacterial spectrum ofAmpicillin is same as Penicillin G, but it is active against Gm-ve bacteria, stable in acid. Both the oral and parenteral preparations are available. Very high doses are given over a long period without producing toxic reactions. It is used in respiratory tract infection (RTI), urinary tract infections (UTI), meningitis, typhoid fever (*Salmonella typhomurium*), and bacillary dysentery (shigella), and listeria infections. Amoxicillin isa close analogue to ampicillin that is stable in acid. Antibacterial spectrum is similar to ampicillin.

Antipseudomonal penicillin: Carbenicillin has antibacterial spectrum similar to ampicillin, and is administered by parentral route against infectionsdue to *Pseudomonas auerogenosa*. It has good activity against proteus and enteribacter species. Ticarcillin is similar to carbenicillin with greater activity against Pseudomonas, and used in combination with gentamycin. Mezlocillin is more active against klebsiella and pseudomonas than carbenicillin.Piperacillin is similar to ampicillin, and is used parentrally. Antibacterial activity is caused by inhibition of bacterial cell wall synthesis. Benzyl penicillin is used at the dose rate of 10000-20000 IU/kg body weight intramuscularly at 12 hrs interval. Procaine penicillin G is used @10000-20000 IU/kg IM at 24 hrs interval. Benzathine penicillin is adminstered at the dose rate of 500mg/ quarter. Penicillin V is given at the dose rate of 8mg/kg orally. Cloxacillin is given parentrally@4-10mg/kg, intramammary 200mg/ quarter. Oral ampicillin is prescribed at the dose rate of 10mg/kg twice daily, and parentrallyat the dose rate of 7mg/kg body weight twice daily. Amoxicillin is given at the dose rate of 10mg/kg body weight twice dailyorally, or 7-15mg/kg twice daily parentrally.

Cephalosporin

Cephalosporin has originated from species Cephalosporium that inhibits bacterial cell wall synthesis, similar to penicillin. It has Beta lactum ring, thus gets affected by beta lactamase. It is ineffective against bacteria producing cephalosporinase. It is used in upper respiratory tract infections (URTI), urinary tract infections(UTI), skin and soft tissue infections, caused by *Staphylococcus aureus*, and systemic infections caused by sensitive Gm +ve and Gm -ve bacteria and live threatening infections before identification of specific organisms. The side effects and toxicity include rash, eosinophillia, cross reactivity to cephalosporins in those hypersensitiveto penicillin, diarrhoea, elevated liver function test (LFT) parameters, hypoprothrombinemia and nephrotoxicity.

The First generation cephalosporins are highly active against Gm+ve bacteria, except methicillin-resistant Staphylococcus, Enterococci without CSF penetration. It includes Cephalexin that resembles ampicillin and amoxicillin with respect to antibacterial activity. Cephradine and Cephapirin are used asintramammary infusionwith maximum resistantance to beta lactamase. Cephalothin, Cefazolin and Cefadroxil are used at the dose rate of 11-25 mg/kg body weight twice daily intramuscularly. The Secondgenerationcephalosporins include Cefachlor, Cefuroxime, Cefoxifin, Cefamandole, Cefprozil, Cefotitan and Cefmetazole. They are highly active against Gm+ve bacteria, active against anaerobes and less active against Gm-ve bacteria. The Third generationcephalosporinssuch as Cefotaxime, Cefoperazone, Ceftriaxone, Ceftiofur, Cefixime, Cefazidine, Cefpodoxime, Ceftibuten, Cefdinir, Ceftizoxime, Lactamoxel, sodium Moxalactum and Cefepine have very low Gm +ve coverage, and are highly active against Gm-ve bacteria. The Fourth generation cephalosporins are Cefepime and Cefpirole, and Fifth generation includes Ceftibiprole and Ceftaroline. The injectable form of Ceftriaxone is available for its use against respiratory diseases including shipping fever.

Beta lactamase Inhibitors

Clavulanic acid: A combination formulation of Potassium clavulanate with amoxicillin, clavulanate potentiated amoxicillin, is useful in skin and enteric infections, whereClavulanic acidis administered @2.5mg/kg body weight in combination with amoxycillin @10mg/kg body weight twice daily. It has similar spectrum to that of Clavulanic acid but it is a less potent inhibitor.Other beta lactamase antibiotic, Carbapenems (Impipenem) is active against Gm+ve and Gm-ve bacteria including *Pseudomonas aeurogenosa*. Monobactams (Aztreonam) is only effective against Gm -ve enteric bacilli. Carbapenems and Monobactams are other beta lactamase antibioticsthat interfer with bacterial cell wall synthesis.

Aminoglycoside and Aminocyclitol Antibiotics

Aminoglycoside includes Streptomycin, Neomycin, Framycetin, Amikacin, Kanamycin, Tobramycin and Gentamicin. Spectinomycin and Apramycinare included in Aminocyclitol. Those have poor oral bioavailability, and are mainly effective against Gm-ve organisms. They cause ototoxicity and nephrotoxicity and development of resistance. Cross resistance does not develop between the groups. They produce their antibacterial effects by binding with 30’s ribosome subunit resulting in inhibition of protein synthesis.

Streptomycin is used in treatment of enteritis. It is slowly absorped after adminstratiom through intramuscular route, mainly active against Gram -ve organisms. Resistance developed through mutation, decreased cellular uptake and decreased cell permeability to antibiotics. Clinically, it is used against tuberculosis and brucellosis through parentral route at the dose rate of 5-10mg/kg, @20mg/kgorally, and @100mg/ quarter (lactating) intramammaryand 500mg/ quarter fordry cow therapy. It is used in combination with procaine penicillin for synergistic effectsin the treatment of Gram +ve and Gram -ve bacteria.

Gentamicin is slowlyabsorbed after oral route of administration. However, absorption is rapid by intramuscular route. The biological half life is 75 to 110 minutes. Less amount of antibiotic becomes available in CSF, prostate and eye. Gentamicin is effective against *Pseudomonas, E. coli, Salmonella, Klebsiella, Proteus, Staphylococcus*, and*Shigella.* Organisms with resistance to gentamicin are also resistant to neomycin, streptomycin, framycetin and kanamycin. Many strains, resistant to other aminoglycosides, are sensitive to gentamycin. It may cause ototoxicity (impairment of balance), nephrotoxcity andrespiratory paralysis due to neuromuscular blockage, which can be treated by parenteral calcium or anticholine estrase agents. It crosses placental barrier leading to potential foetal damage. Clinically, it is used in urinary tract infections (UTI), respiratory tract infections (RTI), soft tissue infection, metritis and eye infections at the dose rate of 5mg/kg body weight twice on first day followed by once daily.

Neomycin is used inits sulfate form. It is not absorped from the skin. Neomycin is highly effective for the treatment of enteric infectionscaused by Gram negative organisms. It penetrates the blood brain barrier, pleural and peritoneal barrier after intramuscular injection. It may cause an irreversible damage ofloss of hearingdue to damage of 8^{th}cranial nerve, and renal damage. It is effective against *Salmonella, E. coli, Klebsiella, Proteus*, and *Enterobacter*.Clinically, it is used in urinary tract infections (UTI), coliform infections of skin, eye and ear, calf scour, mastitis, for wound dressing and in eyes and for superficial conditions like otitis, corneal ulcer, conjunctivitis, keratitis, abscess and dermatitisin ointment form at the dose rate of 10mg/kg body weight orally, or at 2.5mg/kg body weight parentrally.

Kanamycin: The antibacterial activity similar to that of neomycin. The topical preparation is used for ear infection and oral preparation for enteric infections. It is not absorbed from theintestine. Parentral administration may cause permanent damage to cochlear and vestibular portion of 8th cranial nerve. Clinically, it is usedin necrotic enteritis, pneumonia, diarrhoea, hemorrhegic dysentery (swine), atophic rhinitis in chickenand bovine mastitis at the dose of 15 mg/kg body weight /day in divided dosesintramuscularly.

Tobramycin is related structurally to Kanamycin, and is highly effective against *Pseudomonas* spp. Amikacin is a water soluble acylated derivative of kanamycin. Toxicity includes impairement of hearing and kidney function. Clinically, it is used in infections resistant to other aminoglycosides. It is the drug of choice for burns infected with resistant Pseudomonas, Mycobacterium tuberculosis (99% strain inhibited) andurinary tract infections with Gm–ve bacteria.Netilmicin is the latest aminoglycoside antibiotic, similar to gentamicin and tobramycin in its pharmacokinetic properties and dosage. It is effective against aerobic Gram -ve bacillai. Clinically, it is used in serious infections due to susceptible enterobacteriacae and other aerobic Gm-ve bacilli. Framycetin (Neomycin-B) has properties like neomycin. It is used clinically in enteritis. It is used as a topical antibiotic for treatment of otitis externa in dogs. Spectinomycin has limited Gm+ve and better Gm-ve activity. It is bacteriostatic against most bacteria and mycoplasma and absorbed after administration throughoral and parentral route. Clinically, it isused in Mycoplasmal infection and Colibacillosis in poultry and pigs, and in *E. coli* mastitis in cattle at the dose rate of 11-44mg/kg body weight orally in pigs and 5.5-11 mg/kg body weight twice daily intramuscularly or 22mg/kg twice daily orally in dogs.

Adverse effects of aminoglycoside antibiotic

1. Ototoxicity- Vestibular and auditory dysfunction due to progressive accumulation of drugs in the perilymph and endolymph of the inner ear and irreversible hearing loss.
2. Nephrotoxicity - Accumulation and retention of antibiotic in the proximal tubular cell reults in decreased glomerular filtration, manifested by low urine concentrating ability, proteinuria, hypokalemia and acute renal acidosis.
3. Neuromuscular toxicity includes reduced prejunctional release of acetylcholine, resulting in muscular paralysis. Enlargement of blind spots, and optic neuritis are other side effects.

Tetracyclines

These are a group of broad spectrum antibiotics. Chlortetracycline was the first compound in this group in the name of Aureomycin, because of golden yellow colour of *Streptomyces aureofaciens*colonies. Another tetracycline named oxytetracycline was isolated from *Streptomyces rimoues*. The tetracycles form stable chelate complexes with metals like calcium, magnesium, iron etc that result in removal of essential metal ions from the body particularly complexes of calcium and magnesium and retard the absorption of tetracycline from gastrointenstinal tract.

Classification

1. Natural tetracycline: Chlortetracycline (IV only),Oxytetracycline, Demethylchlortetracycline
2. Semisynthetic tetracycline: Tetracycline, Rolitetracycline, Methicycline, Doxycycline, Mibocycline
3. Others classified according to duration of action
   - Short acting tetracycline: Chlortetracycline, Oxytetracycline and Tetracycline
   - Intermediate acting tetracycline: Demeclocycline, Methacycline
   - Long acting tetracycline: Doxycycline and Minocycline

Activity spectrum: Tetracyclines are broad spectrum antibiotic and are highly effective against Gm +ve but less active against Gm-ve bacteria.Most susceptable bacteria are βhaemolytic Stereptococci, non haemolytic streptococci, clostridia, brucella, haemophilus and klebsiella. Moderately sensitive bacteria are *Corynbacterium species, E. coli, Pasturella, Salmonella*and*Bacillus anthracis. Proteus, Pseudomonas. Shigella, Streptococcus devaluation* and *Staphylococci*are relatively resistant to Tetracyclines.In addition to the bacterial microorganisms, Tetracyclinesare also effective against Mycoplasma, Spirochaetes, Actinomycetes, Rickettsiae and certain large viruses of psittacosia group in animals. In higher concentrations, these are also effective against some protozoa particularly anaplasma. However, these are not effective against yeast, mold andother fungi. Tetracyclines are also known as antibiotic with "blanket activity". Presence of pus, blood, serum or bacterial debris does not impact the antibacterial activity of tetracyclines. The resistance to tetracycline develops slowly in a graded manner. A complete cross resistance is seen against different members of tetracycline except minocycline, which may not show cross resistance with others in the group.

The tetracyclin group of antibiotics inhibits the protein synthesis by binding with 30's ribosome subunit of bacteria. They are principally bacteriostatic but may become bactericidalat at high concentrations. They have the property to chelate polyvalent cations like $Ca^{2+}$and Mg^{2+}. They have minimal adverse effects on host cells.

Route of Administration: Tetracycline can be administered by oral, parentral, topical or intramammary route. The oral route is not indicated in herbivores as they inhibit normal bacterial fermentation of plant fibres. All tetracyclines can be administered by intravenous or intramuscular routes in animals except Chlortetracycline, which is not administered intramuscularly, as it is quite irritant and painful, and results in irregular and incomplete absorption. The intravenous injections should not be adminstered too fast to avoid perivascular leakage of the drug inducing local irritation, thrombophlebitis, and necrosis. The local application of tetracycline as an ophthalmic ointment or buffered aqueous solution for conjunctival membrane in animal is also useful. The intramammary infusion of tetracycline is extensively employed for the treatment of mastitis of cow, buffalo and goat in the veterinary practice.

Pharmacokinetics: All tetracyclines are adequately but incompletely and irregularly absorbed from gastrointestinal tract. Absorption is impaired by the presence of milk and milk product, aluminium hydroxide gel, sodium bicarbonate, calcium and magnesium salts. Single oral dose of tetracycline achieves peak plasma concentration in 2-4 hour, and persist for 6 hour. Hence, 6 hour dosing interval is recommended.The peak plasma concentration is achieved in 1 hour following intramuscular injection, and persists for 12 hours. The peak plasma concentration of tetracycline is achieved immediately after intravenous injection, and persists for 12 hour. The highest drug concentration is seen in kidney, liver, spleen and lungs following tetracycline adminstration.These are also deposited at the ossification sites. These are capable to cross placental barrier and foetal blood concentration.Diffusion of tetracycline into cerebrospinal fluid (CSF) is difficult. Tetracyclines are secreted in high concentration in to milk, equal to or higher than that in the blood. A combined intravenous and local administration has been recommended in case of acute mastitis. There has been limited metabolism of tetracyclines, and aremainly excreted through kidney and milk. Tetracyclines are relatively non toxic, yet produce some adverse effects.

Suprainfection: Gastrointestinal disturbances supervene due to suprainfection by drug resistant bacteria, fungi, yeast etc. They also suppress bacterial fermentation in herbivores resulting in anorexia and diarrhoea.Chlortetracycline is slightly more irritant than other tetracyclines following oral administration, and may result in nausea, vomition and diarrhoea in dog and cat. Intramuscular injections are painful; especially chlortetracycline is extremely irritant and painful. Tetracycline have the property to chelate Ca^{2+} resulting in formation of "tetracycline-calcium orthophosphate complex" aseffects on teeth and bone, which gets deposited in the developing teeth and bones, leading to yellow orbrown discoloration and ill formed teeth and depression of bone growth. Therefore, therapeutic use of tetracycline is contraindicated during pregnancy and in early childhood.Rapid intravenous injection of tetracycline produces hypotension and sudden collapse as an effect on cardiovascular system. This is due to the ability of tetracycline to chelate calcium. Such effects can be avoided by slow infusion of these drug or pre-treatment with intravenous calcium gluconate. Phototoxicity dermatitis maybe observed in some individuals, particularly with administration of Demiclocycline and Doxycycline but, the reaction is rare in animals.The prolonged dosing with Tetracycline may result in fatty infiltration and jaundice, and may precipitate acute hepatitis, necrosisand renal failure. Hypersensitivity is not common with tetracyclines. However, skin reaction including urticaria, eruption, exfoliative dermatitis, anaphylaxis, burning eye, glossitis may also be observed. Cross sensitization is extremely common.

Precautions: Tetracyclinesparticularlychlortetracyclineshouldnotbeadministered through intramuscular route. They should not be used during pregnancy and lactation.Tetracycline should be used with utmost care in patientswith hepatic and renal insufficiency.Tetracycline should not be administered with milk and in patients receiving antacids and also immediately after intake of food.

Therapeutic Indication: Tetracycline may be used for the treatment of mixed infections causing bronchopneumonia, enteritis, UTI, metritis, mastitis, prostitis and pyodermatitis, and in specific disease condition like anaplasmosis, pasturellosis, actibomycosis, actinobacillosis, leptospirosis, nocardiasis, heart worm disease, canine ehrlichiosis, cutaneous streptothricosis, bovine hoof disease, brucellosis, east coast fever, psittacosis / ornithosis, blue comb disease of turkey and chronic respiratory diseases. In addition, tetracyclines arealso employed as feed additives for promoting growth.

Dosage Regimen of Tetracycline in different Species

Tetracyclines	Species	Dose (mg/kg)	Frequency	Route
Tetracycline	Cow, sheep and goat	2.2-4.4	12 hr	IV, IM
	Dog and cat	4.4-11	12hr	IV
Oxytetracycline	Calf	10-20	8-12 hr	Oral
	Dog and cat	25	12 hr	Oral
	Cattle, calf, sheep, goat	5-10	24 hr	Oral, IV, IM
	Dog and cat	7-10	12-24 hr	IV, IM
Chlortetracycline	Calf	10-20	12 hr	Oral
	Dog and cat	25-50	12 hr	Oral
	Dog	5	24 hr	IV
Doxycycline	Dog	5-10	24 hr	Oral
	Poultry	2.5-10	24 hr	Oral

Local infection: 440 mg of tetracycline HCl / quarter is used for the treatment of bovine mastitis. One mg of Tetracycline / g of ointment is recommended for conjunctivitis. Tetracycline in buffered solution is given at the dose rate of5 mg / ml for eye infections. The intrauterine infection is treated with 1 or 2 tablet of 0.5g in cow and buffalo and ½ to 1 tablet for ewes / does.

Chloramphenicol (Chloromycetin)

Chloramphenicol is a highly active broad spectrum antibiotic with excellent pharmacokinetic properties. It is the first antibiotic to be synthesized commercially. Chloramphenical has a bitter taste and is poorly soluble in water. Hence, it is not suitable for oral administration. Chloramphenicol palmitate is tasteless and hence preferred for oral administration.It is a bacteriostatic antibiotic that interferes with protein synthesis by binding with 50's ribosomal subunit of the 70's ribosome. Chloramphenicol is effective against variety of pathogens including Staphylococci, Salmonella, Pasturella, Borderlands, Haemophilus species, coli form organism, Chlamydia, and Rickettsia. It is the second drug of choice for anaerobic bacterial infections.

Pharmocokinetics: Chloramphenicol is inactivated in the rumenof ruminants. The drug can pass the cellular barrier readily and the effective concentration is achieved in most of the tissues and body fluids including CSF and aqueous humour. It can cross the blood brain barrier (BBB) in absence of meningitis, the placental barrier and will not get metabolised in the foetal liver. It also enters into the pleural, peritoneal and synovial cavities. The drug is also administered intramammaryfor the treatment of mastitis, although the drug is excreted in milk. In cat, the drug is not metabolised readily because of failure of glucurorinidation. A large proportion of drug is excreted in urine as an inactive metabolite and hence it is not useful in treating urinary infections.Chloramphenicol should be used with discretion in animals so as to reduce the chances of resistance.

Drug Interaction: Chloramphenicol should not be used concurrently with bactericidal drugs in treating infection when host defences are poor. It should not be used with penicillin G, lincosamines and macrolides. Severe muscle degeneration can occur in broilers, if it is used with Ionophores. Chloramphenicol is a potent inhibitor of microsomal enzymes, so it prolongs the duration of action of many drugs administered along with it like phenobarbital, and NSAIDs. Since the drug suppresses the immune response, it should not be used during vaccination. Chloramphenicol delays the response of anemia to iron, folic acid and vitamin B_{12}. Aqueous solution of chloramphenicol sodium succinate should not be mixed with other preparation before administration because of a high level of incompatibility.

Toxic and Adverse reaction: Chloramphenicol causes bone marrow toxicity,which can be two types; (i) Dose-dependent reversible suppression of bone marrow activity that occurs when the dose is high and (ii) irreversible bone marrow depression with small doses causing aplasia and fatal anaemia. A few cases of anaemia has been reported after exposure to chloramphenicol in the form of ophthalmic ointment or while handling. Hence the use of chloramphenicol has been banned in food producing animals, and in many countries for hunman and animal use. Anaphylaxis, vomiting and diarrhoea have vocationally been reported in dogs and cats.

Macrolide Antibiotics

These antibiotics contain a macrocyclic lactone ring with attached sugars. They are active against many Gm +ve bacteria and some strains of Listeria.

Erythromycin: This antibiotic remains more active in alkaline medium than at neutrality and is unstable in acidic medium. It inhibits protein synthesis by binding with 50's ribosomal subunit. It is effective against *Streptococcus pyogenes, Streptococcus faecalis, Staphylococcus, Corynbacterium renale, Corynbacterium* pyogenes, *Erysipelothrix rhusiopathiae,* and *Brucella suis*. Rickettsiae are moderately sensitive whereas Enterobactericaes, other Gm-ve bacilli and *B.fragilis* are not inhibited. Cross resistance develops against Oleandomycin, Clindamycin and Chloramphenicol.

It is readily absorbed from small intenstine after oral administration. It can cross the placenta and enter into the foetal circulation. Erythromycin is eliminated at a higher concentration through bile and therapeutically effective for biliary tract infection. It is clinicalluy used @ 6.6-8.8 mg/ kg body weight/ day in 3 to 4 divided dosesin non herbivores animalsthrough oral route. Erythromycin lactobionate may be administered parentrally @ 2.2-4.4 mg/kg body weight at an interval of 8-12 hrs. Local or topical use is not preferred because of skin sensitivity.

Oleandomycin and Troleandomycin: Oleandomycin is derived from *Streptomyces antibioticus*.Toleandonycin is a derivative of Oleandomycin, used for oral administration due to its rapid and complete absorption. Antibacterial spectrum is similar to Erythromycin, but these are primarily antistaphylococcal drug, adminstered @ 1.1-2.2 mg/kg 4 times daily orally.

Newer Macrolides

Roxithromycin: It is a semisynthetic long acting acid stable macrolide, whose antibacterial spectrum resembles closely with Erythromycin. It is an alternative to Erythromycin for respiratory, ENT, skin, soft tissue and genital tract infections.

Clarithromycin: The antimicrobial spectrum resembles with Erythromycin. In addition, it is effective against *Mycobacterium avian* complex, other atypical Mycobacteria (*M. leprae*). Erythromycin resistant bacteria are also resistant to Clarithromycin. It is indicated in loweranduppper respiratory tract infections, sinusitis, ototis media, atypical pneumonia, skin infection due to *Streptococcus pyogenes* and *Staphylococcus aureus*.

Azithromycin: It is more active against *H. influenzae*, but less active against Gm+ve cocci than other macrolides. It is highly active against Mycoplasma and other respiratory tract pathogens. Penicillinase producing *Staphylococcus aureus* are inhibited but not effective against methicillin resistant *Staphylococcus aureus*. It is acid stable and indicated in pneumonia, urethritis, genital infection, pharyngitis, tonsillitis, sinusitis, oitis media, *Streptococcus*and Staphylococcal skin infection and soft tissue infection. It is also used in typhoid and toxopasmosis.

Lincosamides: Lincosamides antibiotics are considered with macrolides due to similar antimicrobial spectra, therapeutic application and mechanism of action. It inhibits protein synthesis by binding to 50's ribosome.

Lincomycin and Clindamycin: Lincomycin is used as Lincomycin HCl while Clindamycin is used as Clindamycin HCl hydrate, Clindamycin palmitate and Clindamycin phosphate. Clindamycin is semisynthetically derived from Lincomycin.They are active against Gm+ve pathogens like Penicillinase producing *Staphylococcus, Streptococcus*, *Clostridium tetani*, *Cl.diptheriae*, *Cl. welchi* as well as *Actinomyces*, *Nocardia*and certain strains of Mycoplasma. High activities against anaerobes are the distinct features of Clindamycin. The combination of Lincomycin and Spectinomycin in 1:2 ratios shows excellent activity against *Mycoplasma gallisepticum* and *E. coli* infection in chicks. The combination also shows greater activity in controlling air saculitis of turkey.

Pharmocokinetics: The drug concentration reaches4 to 6 times higher in milk compared to blood plasma concentration following intramuscular injection of Lincomycin (10mg/kg) in dairy cows. Both Lincomycin and Clindamycin can be administered orally, IM or IV route. Cross resistance may occur between Lincomycin and Clindamycin. Lincomycin is used @2-100 gm/ 900 kg of feed for broiler chicken in case of necrotic enteritis by clostridial infection. Lincomycin is used @ 40-100 gm/ 900kg of feed for swine for the control of enzootic pneumonia in pigs. Lincomycin is administered intramuscularly for the treatment of arthritis and pneumonia involving mycoplasma in pigs. Lincomycin is used @ 11-22 mg/ kg IM twice daily for dog and cat. It is used @ 11mg/ kg IM OD for 5-6 daysin pigs. The antibiotics must not be used within 48 hours before slaughtering. Pain, irritation and induration occur following IM injection of both Lincomycin and Clindamycin.

Antituberculosis Agent

Tuberculosis is a chronic granulomatous disease caused by *Mycobacterium bovis* in ruminants (cattle and buffalo). Sheep and horse are rarely affected. It occurs due to *Mycobacterium tuberculosis, Mycobacterium bovis, and occasionally Mycobacterium avium* complex and *Mycobacterium fortuitum*in dog, and by *M. avium*in pigs. Antitubercular drugs are divided into two types.

1. **First line Anti-TB drugs:** They have high antitubercular efficiency and low toxicity, and include Isoniazid, Rifampin, Pyrazinamide, Ethambutol, and Streptomycin.
2. Second line Anti-TB drugs with low antitubercular efficiency and high toxicity. These are Paramino Salicylic Acid (PAS), Ethionamide, Cycloserine, Kanamycin, Amikacin, Cupromycin and Thiacetazone.

Isoniazid: It posses bacteriostatic and bactericidal activity. Isoniazid inhibits synthesis of mycolic acid, the unique fatty acid component of mycobacterial cell wall. It is administered at the dose rate of 4-6 mg/ kg body weight for 2 to 3 months. Peripheral neuritis is the common adverse effects. The neurotoxicity can be prevented prophylactically with administration of Pyridoxine @10mg/kg body weight.

Rifampin: It is a semisynthetic derivative that has bactericidal action on *M. tuberculosis, M. paratuberculosis* and other subpopulation of TB bacilli. It inhibits bacterial DNA dependant RNA synthesis, and is used at the dose rate of 10mg/kg orally in capsule, suspension and syrup form. Hepatitis is the major adverse effect.

Pyrizinamide: It has similar activity as that of Isoniazid. It is used at the dose rate of 25-30mg/kg body weight orally in tablet and syrup form. Hepatotoxicity is the major adverse effects.

Ethambutol: It is used in combination with Isoniazid to hasten the rate of sputum conversion and to prevent development of resistance. It is administered at the dose rate of 15-20mg/kg body weight. The adverse effects are optic neuritis causing loss of vision.

Streptomycin: It is the first clinically useful antitubercular drug.It is tuberculocidal but less effective than Isoniazid or Rifampin. Streptomycin is available in injectible form and given at the dose rate of 15mg/kg body weight. Vestibular and auditory disturbance are the major adverse effects,

Thiacetazone is a tuberculostatic drug, given @ 2.5mg/kg body weight in combination with Isoniazid.Paraamino Salicylic Acid (PAS), available in tablet form, is also a tuberculostatic drug active only on Tuberculosis bacilli at the dose rate of 200mg/ kg body weight. Gastrointenstinal tract irritation and hypersensitization are the adverse effects. Ethionamide has adverse effects of convulsion and hepatotoxicity. Cycloserine is also tuberculostatic. Multiple doses lead to accumulation of drugs. It is given @ 200mg twice daily. The adverse effect includes CNS toxicity.

Fluoroquinolones

These are synthetic broad-spectrum antibacterial with primary activity against Gm-ve bacteria. The first member of quinolone antibiotic was Nalidixic acid. Enorofloxacin and Danofloxacin are the members of fluoroquinolones group developed exclusively for veterinary practices.

Classification of Fluoroquinolones

1st generation Fluoroquinolones: Nalidixic acid, Oxalinic acid, Cinexacin, Piromidic acid, Flumequine.

2nd generation: Norfloxacin, Ciprofloxacin, Ofloxacin, Enorofloxacin, Danofloxacin

3rd generation: Pefloxacin, Marbofloxacin, and Sarafloxacin

New generation fluoroquinolones: Gatifloxacin, Moxifloxacin, Levofloxacin, Premafloxacin, Sparflixacin, Difloxacin, and Orbifloxacin

Pharmacological Features: Fluoroquinolones have broad spectrum activity with a wide safety margin, used in majority of domesticated species of animals. They can be administered by variety of routes, and have efficacy at low therapeutic dose. The absorption is satisfactory after oral administration in non ruminants and poultry. The administration of drugs in a prey produces therapeutic concentration in the predator. The absorption after non vascular route (IM, SC) is complete, but low compared to IV route. These are distributed extensively to all the body tissues, and metabolized through hydroxylation and oxidation. A fluoroquinolone may get biotransformed to other fluoroquinolones molecules such as Enorofloxacin to Norfloxacin, Pefloxacin to Norfloxacin and Difloxacin to Sarafloxacin. The bioavailability is satisfactory after oral and parentral administration. Fluoroquinolones exhibit "Flip - Flop Kinetics". After IV administration, the drug is eliminated faster and hence the levels are maintained for shorter period, whereas parentral administration results in maintenance of levels for a longer period and better bioavailability due to slower absorption.

Fluoroquinolones act as bactericidal agents interfering with nucleic acid synthesis by inhibiting the DNA gyrase enzyme. The MIC values for Gm-ve bacteria are higher than Gm+ve bacteria. They have excellent activity against Enterobactericae, fastidious Gm-ve bacteria (*E. coli, Klebsiella, Proteus* spp, *Salmonella* spp, *Enterobacter* spp.) and *Pseudomonas aeuroginosa* (ciprofloxacin) and good to moderate activity against *Streptococci, Mycobacteria, Chlamydia* and *Mycoplasma*. Fluoroquinolones are more active in alkaline environment for Gm-ve bacteria but susceptibility to Gm+ve bacteria is not affected by pH. Fluoroquinolones have biphasic effect. Their activity is absent below the MIC as well as at concentration above the MIC values. Post antibiotic effect includes persistent suppression of bacterial growth following removal of antibacterial agent. The bacterial action persists even after serum concentration of fluoroquinolones falls below MIC. The higher doses for a shorter period are the practice adopted on many forums.

Resistance to Fluoroquinolones develops via single or multistep mutation in the genes that code for specific bacterial enzyme, DNA gyrase. The use of fluoroquinolones as growth promoter is discouraged owing to the concern of residues in animal products and hence the spread of resistance. Fluoroquinolones are widely used in farm animals, pet animals, birds for treating susceptible infections. Clinically, they are indicated in upper respiratory tract infection, gastrointestinal infection, anthrax, enterotoxaemia, mastitis, uterine infection, meningitis, skin and bone infection, BQ, and wooden tongue in cattle, buffalo, sheep andgoat, infection of skin and soft tissue, oral cavity, urinary tract, prostate, respiratory tract and bone, wound, otitisin dog andcat, Colisepticaemia, Infectious Coryza, salmonellosis, CRD, Fowl cholera, Pasturellosis, Mixed bacterial infection in Poultry.

Formulation and Dosage of Fluoroquinolones

Ciprofloxacin

- Injection 2mg/ml, 40mg/ml, 100mg/ml, tablet 250mg
- Eye ointment 0.3%, eye/ ear drop 0.3%
- Poultry product, ciprofloxacin powder 10%
- Combination products - Ciprofloxacin 20g + Norfloxacin 20g soluble powder for poultry. Ciprofloxacin 250mg+ Tinidazole 300mg on table form

for small animals. Ciprofloxacin 1500mg+ Tinidazole 1800mg as bolus form for large animals. It is given at the dose rate of 4-5 mg/kg bwt OD for 3-5 days in cattle, buffalo, sheep andgoat, and 25mg/kg bwt OD for 3-5 days in dog andcat. Poultry are given @ 50mg/ liter of drinking water.

Enorofloxacin: Injection of 5%, 10%, or 20% concentrations, tablets of 50mg, 150mg are used for treatment at the dose rate of 2.5-5 mg/kg body weight in all species. Poultry are provided with 5% or 10% solution in drinking water. Prophylactic dose is given @ 5mg/ kg body weight, Curative dose @ 10mg/kg bwt.

Flumequine: Ten% powder is used in poultry and large animals. The dose in cattle, buffalo is 6mg/kg bwt, 0.5mg/kg bwt in dog andcat, 50mg/ liter of waterin Poultry chicksand 100mg/ liter of water in adult chicks.

Norfloxacin is available in combination with antiprotozoal drug for the treatment of animals and poultry. They include Norfloxacin 1200mg + Tinidazole 1800mg for large and small animals. Norfloxacin + Metronidazole (each 25mg / 500g powder) is used at prophylactic dose @5mg Norfloxacin /kg body weight, and curative dose @ 10mg Norfloxacin /kg bwt. In dog and cat, the dose is 22mg Norfloxacin /kg body weightb.i.d for 3-5 days. In poultry, 5%, 10% or 20% powder is used through drinking water.

Pefloxacin: Tablet 400mg, Bolus 1g and Infusion 400mg/ 100ml are available for clinical use. It is given @ 10mg/kg body weight bid for 3-5 daysorally in cattle, sheep andgoat, @ 5mg/ kgbody weightintravenously. The dose in dog is 10-40 mg/kgbody weight bid for 3-5 days orally or intravenously. High dose of fluoroquinolones are recommended for short period through subcutaneous route. Enorofloxacin is recommended @ 7.5-12 mg/kg body weightsubcutaneously as a single dose of administration.

Toxicity and Adverse effects: Fluoroquinolones have very high margin of safety, and are well tolerated without reproduction toxicity or teratogenecity. The toxicity like orthopathic effects is observed in immature pups, and dog below 1 year of age, due to damage to the cartilage of weight bearing joints. Dogs between 4 to 21 weeks of age are most susceptible to these toxic effects. Fluoroquinolones cause toxicity to chondrocytes in young dogsand horses, leading to vesicle formation on articular surface. The chelation of magnesium, which is essential for development of cartilage mostly in young and growing animals by the drug leads to its deficiency causing loss of proteoglycans in the cartilage. High dose of fluoroquinolones have shown to cause ocular problems from drug induced changes in the retina. Doses higher than 5mg/kg/day are not recommended in cats to avoid the enorofloxacin induced blindness. Ciprofloxacin may occasionally cause CNS toxicity in the form of restlessness, tremor, dizziness and seizure, which is attributed to inhibition of binding of GABA to its receptors.

Drug Interaction: The plasma concentration of theophylline, caffeine and warfarin are increased if administered along with ciprofloxacin, norfloxacin or pefloxacin. The increase in plasma concentration results from inhibition of metabolism of these drugs which might cause toxic effect. Antacids, iron salt and sucralfate reduce the absorption of fluoroquinolones, if given concomittantly orally. Nonsteroidal antiinflammatory drugs (NSAIDs) are reported to enhance CNS toxicity of fluoroquinolones resulting in seizures.

Miscellaneous Antibacterials

Ionophores, polyether antibiotics, are fermentation products of Streptimyces species. They are Monensin, Salinomycin, Narasin and Lasolacid. Maduramycin and Semudracin are the newer ionophores antibiotics. The antibiotic Monensin alters the cell permeability and its antibacterial spectrum includes Gm+ve bacteria. It is also effective against coccidia and toxoplasma.It is highly toxic to ruminants and other animals when the dose is 5-10 times the recommended dose or when mixed with certain incompatible drugs. Muscle weakness and myoglobinuria are marked features of poisoning in dog, pig and sheep. In cattle, sheep and goat, progressive cardiomyopathy develops and animal die of generalized cardiac failure. In poultry and turkey, there is necrosis of skeletal and cardiac muscle associated with posterior paresis, inability to rise, incoordination, weakness of legs, and reluctant to move. In ruminants, it is used to improve the feed conversion efficiency. In poultry and other species, it is used in control of coccidiosis.

Other ionophores such as Lasolasid, Maduramycin, Narasin, Salinomycin and Semudracin also possess anticoccidial activity. They are also used for their effect in improving the efficiency of rumen fermentation. Salinomycin is used as a growth promoter in swine. Mode of action and clinical use and toxicity are similar to Monensin.

Novobiocin is a semisynthetic bacteriostatic agent effective against Gm+ve bacteria. It is bactericidal in higher concentration. It inactivates the beta subunit of DNA gyrase. Adverse effects includeskin eruption, eosinophilia, thrombocytopenia and leucopenia. It is synergistic with tetracycline. Novobiocin is useful in anthrax, mastitis and cellulitis.In poultry, it is used for staphylococcal infection and fowl cholera. It is combined with Penicillin G for mastitis treatment, especially for dry cow therapy.

Polymixins: It is active against *Pseudomonas aeuroginosa* and also used topically. Out of Polymixin A, B, C, D, E, only B and E are useful because of less toxic effect. Polymixin is also known as Colistin and inactivates the endotoxin of Gm-ve bacteria. Polymixin are used as SO_4 and Colistin is used as methane sulfonate. They are bactericidal, effective against Gm-ve organism like *E. coli, Salmonella* and *Pseudomonas*. Cross resistance occurs among polymixins. They are synergistic with sulfonamide and trimethorprim against many Gm-ve organisms. It also acts synergistically with tetracycline. Synergism is also noticed with chelating agent like EDTA, chlorhexidine. It is used for topical treatment of local infection caused by Pseudomonas. It reduces the activity of endotoxins in body fluids and hence is useful in toxaemia. Systemic use causes nephrotoxicity, neurotoxicity and neuromuscular blocking effects. Polymixins should be avoided for use in animals because of the toxicity. It is used in dogs and catsfor local treatment of otitis externa, superficial eye infection.

Bacitracin is bactericidal to Gm+ve bacteria and inhibits the formation of bacterial cell wall. Bactericidal activity requires presence of divalent cation such as zinc. The spectrum of activity is similar to that of penicillin. The drug is not absorbed orally and is highly toxic exhibiting symptom of nephrotoxicity. It is generally used in treatment of superficial infection of the skin and mucosal surface. Zinc Bacitracin is used as a growth promoter in poultry, cattle and swine and also in the prevention and treatment of enteritis.

Vancomycin: It produces rapid bactericidal effect and inhibits cell wall synthesis. This drug is active against Gm+ve and not against Gm-ve because of non permeability. It is useful parentrally in methicillin resistant *Staphylococcus aureus*.

Virginamycin: The drug is active against Gm+ve aerobic and anaerobic bacteria, and used to promote growth. Mycoplasmas are also often susceptible. It is not absorbed after oral administration in animals. It does have no adverse effects

Methamine (Hexamine): It is available as a salt of mandelic acid or hipuuric acid. Oral administration of enteric coated tablet is well absorbed and excreted unchanged in urine. If urine is strong acidic, it releases formaldehyde which acts as a non specific urinary antiseptic. Urine is rendered acidic by concurrent administration of ascorbic acid or ammonium chloride.

Other ionophores antibacterials include Isoniazid, Rifamycin (Rifamycin SV, Rifampin, Rifamide), Mupirocin, Sodium Fusidate, Tiamulin Fumarate, Carbadox and Arsenicals.

Carbadox is used to promote growth and prevent dysentery in swine. It is more effective under anaerobic condition than aerobic condition. Arsenicals are arsenic containing compounds (arsenilic acid, sodium arsenilate etc.) used to promote growth in pigs and poultry and to treat swine dysentery.

Anthelmintics

Both helminths and protozoa are commonly found intestinal parasites in domestic animals. The transmission from animals to humans constitutes a worldwide potential zoonotic risk. The strict cohabitation between dogs and humans sharing the same living areas increases this zoonotic risk. Among intestinal helminths, both *Ancylostoma* spp. and *Toxocara* spp. affecting dogs are a concern for public health. Rescue shelters for dogs represent a good observatory for the evaluation of the pathogens that circulate among dog populations. Dogs admitted to shelters may carry intestinal parasites that could pose serious risks to other animals, shelter staff, and visitors. However, companion animals placed in rescue shelters remained under the supervision of veterinarians, and are given dewormers to avoid the spread of parasites.

Anthelmintics are drugs or agents that eliminate worms from gastrointestinal tract. The anthelmintics that kill the worm are called vormicides, and those remove the worms from gastrointestinal tract by temporarily paralysing them are called vermifuges. However, there are concerns about the impact that frequent anthelmintic applications may have in terms of increasing resistance selection pressure. Parasitic infestations in animal result in anaemia, weakness, retarded growth of young one, decrease in milk yield in lactating animals. As a whole, the parasitic load makes the animals debilitated and bad looking. The skin and hair texture is not charming in animals, affecteded with worm load. Anthelmintics are classified as Antinematodal, Anticestodal and Antitrematodal.

General Mode of Action of Anthelmintics

1. Neuromuscular blockers - Piperazine for Ascariasis.
2. Cholinomimetics-Livamisole, Methyridine, Morantel, Pyrantel, Bephenium, Thenium
3. Inhibition of glucose transport - Diathiazine (canine whipworm).
4. Disruption of glycogen metabolism- Niridazole (Schistosomicidal drug).
5. Inhibition of glycolysis-Arsenicals (Thiacetarsamide), Antimonials (Potassium antimony tartarate), and Stilbophen.
6. Inhibition of mitochondrial reaction - Benzimidazole, Thiophanate.
7. Uncouplers of electron transport-Salicylanilides (Clioxanide), Niclosamide, Oxyclozanide, Rafoxanide, substituted phenols (Bithionol, dinitrophenol, hexachlorphene, niclofolam, nitroxynil) etc

Properties of Ideal Anthelmintics

1. Efficacy is said to be good, if it removes 95% of gastointestinal nematodes.
2. Wide therapeutic index –It should not produce any toxicity in host.
3. Ease of administration - Most anthelmintics is administered orally in empty stomach. Some anthelmintics are administered parentrally e. g. Ivermectin, ancylol etc.
4. Residue –It should be eliminated from body without any residual problem. The anthelmintics having a short withdrawal time are much safer.

Antinematodal

Drugs that act against round worms are called as antinematodals. These drugs are classified as follows.

1. Simple Heterocyclic compound - Phenothiazine, Piperazine etc.
2. Benzimidazole - Albendazole, fenbendazole, mebendazole, thiabendazole, cambendazole etc.
3. Imidazothiazoles - Livamisole, Butamizole HCl.
4. Tetrahydropyrimides - Pyrantel, Morantel
5. Organophosphorus compounds- Haloxon crufamate, coumaphos, diclorvos, trichlorfon etc.
6. Miscellaneous drugs-Toluene, n-butyl chloride, tetrachloroethane, theniumclorylate, disophenol, phthalotyne, glycobiarsol, avermectins, Hygromycin -B.

Piperazine: It is a drug of choice for ascaris and nodular worm infestation (oesophagostomum) of all species of domesticated animals and is moderately effective in pinworm infection. It doesn't have any effect for other GI tract parasites. It absorbs atmospheric water, therefore when stored, the container should be tightly closed and protect from light. It forms hexahydrate of piperazine with moisture.Salts of Piperazine include Piperazine adepate, Piperazine citrate, Piperazinephosphate, Piperazine sulfate, Piperazine tartarate and Piperazine hydrocloride.

It is active against ascarid and nodular worm infection in ruminants. It doesn't possess activity against abomasal and small intestine nematodes. It is 100% effective in ascaris infection indog andcats. It is also effective against whip worm and tape worms. It possesses high efficacy against *Ascaridia galii.* The caecal worm, Hetrakis gallinarum, is not susceptible. It acts through neuromuscular block in susceptible worms. It is convenient to give tablet of piperazine in dogs and cats. In other animals it is given in bran mashes. In poultry, the powder or suspension is given in drinking water. It is given at the dose rate of 100-300mg/kg in cattle, 45-65 mg/kg in dog and 32 mg/kg in cat. Piperazine posseses good margin of safety in all animals. Large doses produce vomition, diarrhoea, incoordination and head pressing in dog and cat. Dose of 4 times more than the therapeutic dose can produce transitory diarrhoea and anorexia in calves. It is contraindicated in animal suffering from hepatic and renal disorders.

Phenothiazine: It is an old drug, marketed as powder containing 95% pure phenothiazine (phenovis). It is most effective against large stomach worm (Haemonchus), Nodular worm (oesophagostomum) in ruminants, less effective against stomach worms (ostertagia, tricostrongylus) and hook worm (bunostomum). It has limited activity against round worms of small intestine (cooperia, nematodirus and tricostrongylus). It is not effective against larvae or immature round worms as well as flukes and tape worms. The commercial preparation that contains piperazine and carbon disulfide along with phenothiazine are effective against ascarids, bots and strongylus in a single dose treatment. The red colouration of urine should not be confused with haematuria.

Horses are more susceptible to phenothiazine toxicity than cattle, sheep and goat but birds are resistant. It should not be used in dogs and cats because of higher susceptibility to phenothiazine. Debilitated and anaemic cattle are more susceptible to poisoning as it causes haemolysis of RBC, which may lead to anaemia and death. Photosensitization may occur especially in calves and may also be observed in sheep, goat and fouls. Its use is contraindicated in weak, anaemic and emaciated animals. It should not be given with organophosphorus as it potentiates the toxicity of organophosphorus. It should not be given in advance pregnancy. Other disadvantages include pink coloration of milk and urine, hair, and permanently discolouredwool. It is administered through oral route as single dose treatment@10g/ 45kg in cattle. The total dose should not be more than 70g. The minimum dose in calves is 10g. Sheep andgoat are given @ 25-30g/kg body weight. Chicken are given @0.5g/ bird and Turkey @2g/bird.

Benzimidazoles: Albendazole, Fenbendazole, Mebendazole, Thiabendazole, Cambendazole, Oxfendazole, Oxbendazole, Parbendazole etc. come under this group.

Thiabendazole: It is commercially available as powder, suspension and tablet form. It is a broad spectrum anthelmintic. It is both larvicidal and ovicidal. It has excellent efficacy against all major GI nematodes except whip worm, lung worm and filarial parasites in ruminants (cattle, sheep, goat). It is also effective against haemonchus, tricostrongylus, bunostomum, oesophagostomum and strongyloids. It is given at the dose of 50-110 mg/kg and the treatment is repeated after 21 days. Ostertagia, Cooperia and Nematodirus are less susceptible. In poultry, it is used against gape worms @ 0.1% for 2-3 weeks in drinking water. It is very safer anthelmintics and ruminants can tolerate 20 times of its therapeutic dose. Pregnant ewes do not tolerate the high dose. When given at the dose of 100mg/kg in piglets, it can reduce glycogen level, ascorbic acid and increase succinate dehydrogenase.

Mebendazole: It posses broad spectrum activity andit is mostly used in equines. In poultry, it is used against *Syngamus trachae*. The dose in horse is 5-10 mg/kg body weight. Indog andcat, it is given at the dose rate of 5-10 mg/kg body weight. It is a safer compound andthe therapeutic index is higher than 40.The drug should not be used during pregnancy.

Albendazole: It is better absorbed from GI tract and its active metabolites albendazole sulfoxide is concentrated 100 times more than mebendazole in blood.

Fenbendazole: It is effective against round worms of cattle, sheep, goat and pigs. It also eliminates lung worm and tape worm and kill all mature and immature worms. It is given at the dose rate of 5mg/kg body weight in cattle, buffalo, sheep, goat and pigs.

Levamisole hydrochloride: It is a broad spectrum anthelmintic and an immunomodulator and used in animals where immune system is depressed. It is mainly active against adult and larvae of haemonchus, ostertagia, tricostrongylus, cooperia, metastrongylus, ascaris, hyostrongylus, trichuris and dictyocaulus (lung worm). It is administered at the dose rate of 7.5mg/kg body weight orally in cattle, sheep, goatand pig, and 8-36mg/kg in poultry. It is injected subcutaneously at the dose rate of 2ml/50kg body weight. Cattle treated orally should not be slaughtered within 48hrs and within 7 days after parentral treatment. It is not used in animals of breeding age.

Tetramisole Hydrocloride: It is active against Dictyocaulus (lung worm) and effective against both immature and mature worms. In pig, it is used against ascaris, hyostrongylus, esophagostomum and metastrongylus. In cat, it is used against lung worm (Aelurostrongylus). It is given at the dose of 15 mg/kg body weight in cattle, sheep, goat and pig, and mg/kg daily in cat for 6 days. Overdoses in cattle and sheep causes head shaking, salivation and licking of lips.

Butamisole Hydrocloride: It is an injectible anthelmintic used in dogs for the treatment of whipworm and hookworms (*Ancylostoma caninum*) infestations at the dose rate of 2.4mg/kg body weight subcutaneouslyas a single dose. The intramuscular injection is not admissible due to pain. Toxicity includes vomition, ataxia, tremor, lateral recumbency and conclusion. It is ontraindicatied in debilitated and diseased animals, renal and liver disorder, heart worm positive dogs. It is not used within 24 hrs following cestocidal drug bunamidine.

Pyrantel: It is a broad spectrum anthelmintic against GI parasites of cattle, dogs and pigs. The salts available are tartarate, pamoate and embonate.Pyrantel tartarate is readily absorbed from GI tract of pigs and dogs and less in ruminants. Pyrantel was the first anthelmintic to be used massively in tablets to treat intestinal nematodes in dogs and might be, therefore, the first drug to fail as a result of anthelmintic resistance

Morantel: It is more effective and safer anthelmintic in ruminant than pyrantel. Pyrantel and Morantel are primarily effective against adult gut worms and immature larval stage that remains in the lumen or the mucosal surface. Pyrantel pamoate is given at the dose of 7.5mg/kg orally in pupsat 2, 4, 6 and 8 weeks of age. Morantel is given @ 8.8mg/kg orally in Cattle, and 10mg/kg orally in sheep.

Febantel: It is abroad spectrum anthelmintic for use in horses. Pregnant mare can be safely treated by febantel.

Miscellaneous Antinematodals

N-Butylchloride: It is effective against ascarid and hookworms infection, and can be administered in gelatin capsule to dog and cat. It does not have efficacy against parasites of other domestic animals. Overnight fasting of dogs and cats and use of purgatives enhance the worm expulsion. It is given at the dose rate of 125mg bid orally in pups, and 500mg orallyin adults.

Disophenol (Ancylol): It is an injectible antihookworm compound used in dogs and cats and administered through subcutaneous route. It is effective against canine hookworm and adult *Spirocerca lupi*. The treatment is repeated after 21 days to remove the worm, which were in larval from at the first dose. Disophenol is not effective against ascarids, whipworm and lung worm (*Paragonimus*) of dogs and cats. It is effectively used to treat wild animals (leopard, lion, panthers, cats). It is also used to treat Gapeworm (*Syngamus trachea*) of poultry and turkey. Disophenol is administered as powder form through oral or subcutaneous route. Intramuscular injection is painful. It is given @10mg/kg subcutaneously in dogs and 6.6mg/kg body weight in leopard, loin, and blackpanther. It is safer for dogs and cats as it possesses three fold margin of safety. Even 2 days puppy can be given the drug without any adverse effect.

Phthalotyne: It is a whipicidal (*Trichuris vulpis*) drug in dogs, and can be administered orally in tablet form or intravenously at the dose rate of 200mg/kg orallyin dogs.

Glycobiarsol: It is also used in whipworm infection at the dose rate of 220mg/kg daily for 5 daysin dogs.

Hygromycin – B: Hygromycin is an antibiotic, added in poultry and swine feed for consumption over a period of several weeks.It is effective against *Ascaridia galli*, *Capillaria obsiquota* and *Heterakis gallinarum* of poultry and *Ascaris suum*and*Oesophagostomum* of swineat the dose rate of 12g/900kg of feed inswine and 8g/900kg of feed in chicken.

Ivermectin: Avermectins are produced by fermentation of an actinomycete. Avermectins consist of complex of 08 components. They are A1a, A2a, B1a, B2a, A1b, A2b, B1b, and B2b. Each component has anthelmintic activity. These are not antibacterial or fungicidal. Ivermectin is the combination of B1a and B1b in the ratio of 80% and 20% respectively. Less than 1 mg/kg body weight of avermectin is required for anthelmintic activity by either oral or parentral route of administration. It is active against many nematodes and many arthropods including all gastrointenstinal nematodes, lung nematodes, certain ectoparasites of cattle, sheep, goat and swine and sarcoptic mange of dogs. Heart worms are effectively eliminated. It possesses 10 fold margin of safety in cattle, swine and dogs. It produces mydriasis after 10 fold oral dose in dogs. Acute toxicity symptoms include CNS depression, listlessness and ataxia. It is safer for pregnant animals. The therapeutic dose is 0.2mg/kg subcutaneouslyforcattle, orally in sheep and goat, and 0.3mg/kg subcutaneously in pigs.

Drugs Acting Against Heart Worms

Thiacetarsamide sodium: Adult worm die within 5-7 days after 4 doses of thiacetarsamide. Complete rest is essential during the first 14 days. Only limited exercise is allowed. The rise in body temperature accompanied by coughing following treatment indicates a pulmonary embolism. Thiacetarsamide sodium is nephrotoxic and hepatotoxic. The perivascular leakage during intravenous injectoin can cause huge swelling and tissue sloughing. Dogs are allowed their feed one hour before the injection. The treatment should be discontinued, if vomition persists or discolouration of urine is noticed after the first injection. The toxicity produced by thiacetarsamide is treated with BAL (Dimercaprol) at dose of 8.8 mg/kg/ day intravenously in four divided doses. The dose of thiacetarsamide is 2.2 mg/kg intravenously twice daily for 2 days.

Diathiazine: It is a cyanine dye, effective against the heartworm of dogs at the dose rate of 6.6-11 mg/kg /day orally with feed once daily for 7 days. The lowest dose of 6.6 mg/kg is given to avoid toxicity. The spiliting of daily doses into two parts (3.3 mg/kg twice daily) can minimize the occurrence of toxicity. High doses produce vomition, diarrhoea, anorexia and asthenia. Death may occur in dogs as it is nephrotoxic and hepatotoxic.

Diethylcarbamazine Citrate: It is a piperazine derivative that acts as filaricide and available as a syrup, powder or tablet. The combination of diethylcarbamazine with oxbendazole protects the dogs against hookworm infection. The use of the drug in microfilariae positive dog is contraindicated. It is used at the dose rate of 6.6 mg/kg daily for prophylactic schedule.

Anticestodal Drugs

Anticestodal drugs are known as antitapeworm or taeniacidal drugs. They are referred to as taeniacidal when the drugs cause the death of tapeworms, and drugs causing the expulsion of worms by paralysing them are referred as taeniafuses.

Earlier Anticestodal Agents used in past:

1. Natural Organic Compound (originated from plants)

Pumpkin seeds have active principle cucurbitine with low efficacy against cestodes.

Male fern (Dryoptins Felicia's) has active principle, filicic acid that has ability to paralyse the muscles of tapeworm.

Kamala - Acts against cestodes of cats (*Dipylidium caninum*and *Taenia taeniaformis*).

Nicotine - Nicotine along with $CuSO_4$ is extensively used in ruminants for elimination of tapeworm. It is less efficacious.

Arecoline - It is obtained from betel nut. Its salts are arecoline hydrobromide, arecoline acetarsol and arecoline carboxy phenyl stilbonate. Arecoline hydrobromide is effective against infection of *Taenia* and *Echinococcus granulosus* in dogs. It paralyses the worm musculature. The dog should be fasted for 12 hours and administered at the dose of 1 mg/kg orally as tablet with meat through oral route. Purgatives are advised by the clinician. Toxicity includes discomfort, vomition and unconsciousness. Atropine sulphate is advised in case of toxicity @ 0.44mg/kg. Arecoline is not advised in cats due to excessive bronchial secretion that may produce suffocation. Arecoline acetarsolis available in tablet form and given promptly due to its unstability solution. It is given against Taenia and Dipylidium infection in dogs and cats. It does not possess activity against Echinococcus. Toxicitiesare not well tolerated and vomition may occur. Acute obstruction of upper respiratory tract by the vomitus can lead to death of cats. Salivation, restlessness, ataxia and difficulty in respiration have been reported. Puppies less than 3 months of age and cat less than 1 year of age should not be treated with arecoline acetarsol. It is contraindicated in febrile animals, gastrointenstinaland cardiac disorders. In case of toxicity, atropine sulphate @ 0.44 mg/kg proves to be an effective antidote. Arecoline carboxyphenylstilbonate is available in tablet form and effective against Taenia and Dipylidium in dogs. It is given at the dose of 10.3 mg/kg orally. Toxicity signs include depression, nausea, vomition and colic.

2. Inorganic compound

Tin oxide or chloride - It is obsoletednowadays. Lead - arsenic salt of lead has been used to treat Monenzia infection in calves, lambs and kids.

3. Synthetic Organic Compound

Bunamidine hydrochloride: It has efficacy against all tapeworms of dogs and cats in a single dose. It is active against *E. granulosus, Taenia, Mesocestodes, D caninum*. However, it is not used for treatment of ascaris because of its low efficacy. Tablets, administered orally. should not be crushed or dissolved before oral administration as it is an irritant to mucous membranes of oral cavity. Bunamidine causes death of dog when given intravenously at the dose of 5 mg/kg body weight. It is safe to all stages of pregnancy in bitches. It is given @ 25 mg/kg orally in dog and cat. The diet is given 3-4 hours after bunamidine treatment.

Bunamidine hydroxynaphthoate: It is effective against *Monenzia expansa*and*M. benedeni* in sheep and goat. It is ineffective in dogs, if given in empty stomach. It is given in food daily for 4 days. It acts well against*Echinococcus*and*Taenia* infestation. Side effects include vomition.

Niclosamide: It is exclusively used for treatment of tapeworm infestation, and effective against *D. caninum*and Taenia spp. in dogs and cats, and Monenzia infestation in cattle, sheep and goat. Niclosamide is highly effective against Thysanosoma (fringed tapeworm) of cattle, sheep and deer. Tablets of Niclosamide are administered orally in dogs and cats and as a drench in cattle, sheep and goat @ 100 mg/kg orally, in dog and cat @ 50-100 mg/kg orallyand @ 100 mg/kg orally rabbit, chicken, and monkey. The above doses should be given after overnight fasting in all species. It is a safer drug. It does not produce signs of toxicity in sheep and cattleif the dose exceeds 40 times than the recommended dose. The limit of safer drug dosage has not been established in dogs and cats.

Dichlorophen: It possesses bactericidal and fungicidal activity, and is used taeniacidal agent in animal practice. Its properties are similar to Niclosamide. It is effective against in treatment of Taenia and Dipylidium of dogs and cats. It has less activity in Monenzia expansa infestations in sheep. Niclosamide is administered at the dose rate of 300 mg/kg body weight in dog, 100-200 mg/kg body weightin cat, and 500 mg/kg body weightin sheep. Overnight fasting in all species is required before its administration. It is not a drug of choice for *Echinococcus* in dogs.

Resorantel: It is highly (95-100%) effective against Monenzia infection in cattle and sheep. It is also used against rumen flukes, especially paramphistomiasis.It is 90% effective against rumen flukes in cattle and sheep for both adult and immature forms. It is given at the dose rate of 65 mg/kg as a drench in cattle and sheep. Slight diarrhoea is seen as the side effect.

Bithionol: It is bacteriostatic, fungicidal as well as anthelmintic. It is used against tapeworm infestation of dogs, cats and poultry and also has activity against tapeworms andrumen flukes of cattle, sheep and goats. It is highly effective against Taenia in dogs and cats but not used against *D. caninum*. Besides poultry, most tapeworms of geese and quail are susceptible to this drug. The therapeutic dose incattle, sheep, goat, dog and cat is 200 mg/kg orally. It is 200 mg/kg (2 doses) at 4 days interval in feed in chicken, 600 mg/kg in a single treatment in Geese and 200 mg/kg as a single dose in quail. Bithionolshould not be givenwithCarbon tetrachloride, sodium antimony tartarate, and hexachloroethane.

Praziquintel: It is highly effective against all species of mature and immature tapeworms. Anticestodal efficacy covers *Taenia pisiformis*, *T. taeniaformis* (hydatigena), *T. ovis*, *T. multiceps* and *D. caninum*, *Echinococcus granulosus* and *E. multilocularis* are susceptible at higher dosages. It is effective against Lung flukes (Paragonimus) in dog, Tapeworm infestations in ruminants, poultry and snakes, *Monenzia*, *Stilesia*and*Avitellina* of sheep and goats, intestinal flukes of swine (*Fasciolopsis buski*), ovine cenuriasis but not effective against in larval stage of Echinococcus sp (hydatid cyst). However, it is highly effective against adult and juvenile form of *Echinococcus* in canines. It is administred through oral, subcutaneous or intramuscular routes. Subcutaneous route is not preferred due to least efficacy. The recommended dose is 5 mg/kg indog, 1 mg/kg in cat, and 10-15 mg/kg in ruminants. Bithionol has wide margin of safety. However, it increases ALP, dermal and eye problem, and has teratogenic effect at higher dosage.

Benzimidazole: They are primarily used against nematode infection. They are mebendazole, fenbendazole, albendazole, cambendazole. Doses vary according to species and nature of cestodes. Mebendazole is given @20 mg/kg in dog and cat, 20-50 mg/kg in ruminants.The dose of albendazole is 50 mg/kg in dog, and that of Fenbendazole is 50 mg/kg for dogs and 15 mg/kg for ruminants.

Antitrematodal Drugs

A. Antitrematodal Drug Affecting Adult Flukes

Carbon tetrachloride: It is the first effective drug for the treatment of *Fasciola hepatica*. It is also effective against ascaris infection in dogs and chicken. It has activity against blood sucking nematodes including *Ancylostoma* in dogs and cats, *Haemonchus*and*Bunostomum* in ruminants. The Presence of fat and oil in the stomach increases its absorption. It is administered orally as a drench with rice gruel in ruminants. It is given in gelatin capsule in dogs. It is administered to animals before feeding. Toxicities include liver damage, hyperaesthesia, convulsion followed by coma. Death may occur due to hepatic failure. Toxicity is treated by calcium and dextrose administration by intravenous route. Protein feed is withheld from animals suffering from toxicity.

Hexachloroethane: It is less toxic than carbon tetrachloride in cattle and used against *Fasciola*. It is effective against *Haemonchus* and *Tricostrongylus*. However, it has not effective in intestinal nematodes of ruminants. The recommended dose is 15-100g in cattle, 8-15g Sheep orally, and repeatedessentially after one month. It may cause hepatotoxicity

Hexachlorphene: It is effective against liver flukes of ruminants and cestodes in canine. It is highly effective against *F. hepatica and F. gigantica* in ruminants. Immature flukes present in bile are susceptible and but those present in liver parenchyma are not affected as the drug is less available to immature flukes due to its plasma protein binding. It is administered @ 25 mg/kg incattle. It has a narrow margin of safety. Accurately calculated dose is very essential in cattle. Higher doses produce excitability or depression, followed by death.

Bithionol sulfoxide: It acts against liver flukes and has anticestodal properties. The recommended dose is 60 mg/kg in ruminants in feed or as boluses only.

Oxyclozanide: It is effective against liver flukes and has poor action against rumen flukes. It is used in duck farm for elimination of fluke, *Notocotylus attenuates*. The recommended dose for ruminants and duck is 15 mg/kg orally. It is a safer drug as therapeutic index is 4.

Niclofolan: It is used against both the mature and immature liver flukes of ruminants, generally administered by oral route in tablet form and also parentrally through subcutaneous and intramuscular route in cattle at the dose rate of 4 mg/kg for mature and 16-20 mg/kg for immature flukes or 0. 0.6-1 mg/kg SC or IM. It is given@6 mg/kg in sheep, and 3-5 mg/kg SC in pigds. Fever, tachypnoea and sweating are common side effects.

Rafoxanide: It is available as bolus or suspension for clinical use. Principally, it is used for liver flukes of ruminants and also effective against haemonchus, bunostomum and nasal bot of sheep at the dose 7.5 mg/kg orallyin ruminants. It has a good margin of safety in recommended dose. High dose may cause diarrhoea and blindness. It is contraindicated in lactating animals should not be treated whose milk and milk products are consumed by human within 28 days.

B. Antitrematodal drug used against immature Flukes

Diamfenetide: It has high efficacy against immature *F. hepatica* and is mainly used prophylactically against liver flukes in sheep. Milk and edible tissues of animals are not harmful if consumed by human beings7 days after its dosing. It is 100% effective against flukes of 1 day to 9 weeks of age. The recommended dose in sheep and goat is 100 mg/kg orally. Toxicities include temporary impairment of vision and loss of wool.

C. Drugs acting against Paramphistomiasis

Drugs used against liver flukes and cestodes in ruminants give good result in paramphistomiasis. These drugs are Tetrachloroethane, Niclofolan, Niclosamide, Resosantel, Bithionol or Bithionol sulfoxide.

D. Drugs acting against Paragonimiasis

Paragonimiasis causes lung infection in dogs and cats. The drugs used for this are Bithionol, Praziquintel, Albendazole and Fenbendazole.

Antifungal Agents

Fungal infections are termed as mycoses and can be divided into superficial infections, affecting skin, nail, scalp or mucous membrane, and systemic infections affecting deeper tissues and organs. In animals, *Aspergillus fumigates* causes brooder pneumonia in chicken and mycotic abortion in cattle. Superficial fungal infections are classified into dermatomycosis and candidiasis. Ringworm (dermatomycosis) is most common fungal skin disease in animals. The dermatophytes that cause ringworm in animals, birds belong to genera *Trichophyton* and *Microsporum*. *Trichophyton verucosum* var discoides causes outbreak in cattle and is transmissible to human. Occasionally, it affects dog, goat and horses. *Microsporum canis* is the major cause of ringworm in dogs and cats.

Drugs used against Fungal Infection: Antifungal or antimycotic agents are used against fungal infection. These agents are either fungicides or fungistates. Majority of antimycotic agents are fungistates.

A. Topical Agents

Benzoic acid and Salicylic acid: It has bacteriostatic and fungistatic properties. Benzoic acid (6%) is combined with Salicylic acid (3%) to make Whitfield's ointment for use in ringworm infection by *Trichophyton* in cattle.

Undecylenic acid: It has fungistatic properties and long exposure with high concentration may have fungicidal effect. Zn and Cu salts of acid are often used to minimize the irritant effect.It is effective against *Microsporum*.

Natamycin: It is an antibiotic, effective against dermatophytes, bovine yeast mastitis and mycotic keratitis. Twent ml of 2.5% solution of natamycin is infused into the affected teat in cases of mastitis, caused by *Candidia spp*. once daily for 3 days. It is used as a suspension containing 0.1 mg/ml to treat ringworm in animals. It should be treated indoor or in the evening to avoid inactivation by sunlight.

Nystatin: It is an antibiotic and not absorbed either from the mucous membrane or from skin. It is used to treat bovine mastitis, avian crop mycosis, equine metritis, canine otitis. It is not effective against dermatophytes.

Dichlorophen: Besides taeniacides, it is useful as fungicide against *Trichophyton* and *Microsporum*.The main advantage of *Dichlorophen* is low number of dressings required. Only two dressings can clear up bovine ringworm.

Monosulfiram: It is an insecticide (miticide) and included in some ear preparation for fungicidal and miticidal activity.

Copper sulfate: $CuSO_4$ is strongly fungicidal and has astringent and caustic properties. It is used against ringworm as 5% ointment or 1-2% aqueous solution.

Crystal Violate: Crystal violate is an effective fungistat against *Trichophyton* and *Microsporum* spp. Gention violate is a mixture of methyl alcohol and crystal violate. Crystal violate is used as 1% alcoholic solution on the ringworm lesion, and recovery occurs within 2-3 weeks.

Azoles: Cotrimazole, Econazole, Triconazole and Sulconazole are azole antifungal agents only for topical application.

B. Systemic Agents

Antifungal Antibiotics

Amphotericin: Amphotericin is poorly absorbed after its oral administration. Therefore, it is used by this route for fungal infection of gastrointenstinal tract. It can also be used topically. Amphotericin is complexed with lipids or with p-cyclodextrin or encapsulated in liposomes or nanospheres for intravenous infusion. It can also penetrate blood brain barrier (BBB) only when the meninges are inflamed. Therefore, intravenous amphotericin aling with flucystosine is effective in cryptococcal meningitis. The most commonunwanted effects include renal toxicity, besides hypokalemia, hypomagnesemia and anemia. Intravenous infusion sometimes causes thrombophlebitis. Intrathecal injection causes neurotoxicity. Topical application also causes skin rashes. Liposome encapsulated and lipid complexed preparations have fewer side effects, but they are considerably more expensive.

Griseofulvin: It is a narrow spectrum fungistatic agent, used to treat dermatophytes infection of skin, nail, but requires a prolonged treatment. Usually, it is given 6-9 months orally. The unwanted effects are gastrointestinal upset, headache, and photosensitivity.

C. Synthetic Antifungal Agents

Azoles - This group of broad-spectrum fungistatic agents include Fluconazole, Ketoconazole, Miconazole and Econazole.

Ketoconazole: It is used orally to treat systemic fungal infections. However, toxicityand relapse is common even after successful treatment. Main hazard is liver toxicity and it may prove fatal. Liver toxicity may occur without clinical evidence, and may progress even after discontinuation of the drug. Other signs are gastrointenstinal disturbance and pruritus.Rifampicin, histamine H2 receptor antagonist and antacids decrease the absorption of ketoconazole.

Fluconazole: It is the drug of first choice for most type of fungal meningitis as it reaches high concentration in the CSF and ocular fluids. Vaginal tissue, saliva, skin and nail also achieve fungicidal concentration of fluconazole.It can be given orally or intravenously. The unwanted effects include mild nausea, headache, and abdominal pain. Sometimes, hepatitis is also reported.

Itraconazole: It is given orally. Encapsulated in p-cyclodextrin is available for intravenous route. It dose not penetrate to CSF. The unwanted effects include gastrointenstinal disturbances, headache, and dizziness. Rarely hepatitis, hypokalemia can also occur.

Miconazole: Miconazole is given orally for infection of gastrointenstinal tract. It can also be given topically. The unwanted effects are gastrointenstinal disturbances, pruritus, blood dyscrasias and hyponatraemia.

Flucystosine: It is a synthetic antifungal agent. Drug resistance occurs during treatment when given alone. It is therefore combined with amphotericin for severe infection such as cryptococcal meningitis.

Terbinafine: It is a fungicidal compound active against a wide range of skin pathogens. It is highly lipophillic and keratophillic. It is used both orally and topically to treat fungal infection of nails.

Capsofungin: It is effective against candidiasis and invasive aspergillosis that are refractory to amphotericin. Oral absorption is poor and it gets extensively protein bound.

Micafungin (FK463): It is effective against *Aspergillus* and *Candidia* species.

Amorolfine: It is effective against fungal infection of nails. It inhibits the fungal sterol synthesis following oral adminstration.

Antiviral Agents

Antiviral drugs are a class of medication used for treating viral infections. Most antivirals target specific viruses, while a broad-spectrum antiviral is effective against a wide range of viruses. Antiviral medications can help your immune system fight back by helping in stopping virus multiplying in the the body. This lowers the amount of virus in with a goal of having less severe symptoms and quick recovery.

Types of antiviral drugs: Direct virus-targeting antiviral drugs include attachment inhibitors, entry inhibitors, uncoating inhibitors, protease inhibitors, polymerase inhibitors, nucleoside and nucleotide reverse transcriptase inhibitors, nonnucleoside reverse-transcriptase inhibitors, and integrase inhibitors. Some antiviral drugs is given below.

Antiviral Drug	Use	Type
Acyclovir (Aciclovir)	Herpes Simplex, chickenpox, varicella zoster virus	guanosine analogue RTI
Adefovir	Hepatitis B	RTI
Amantadine	Influenza	Influenza A virus M2 proton channel antagonist
Ampligen	Avian Influenza	Immunomodulatory double-stranded RNA

Amantadine: A viral membrane protein known as M2, forms an ion-channel for H+ influx from the endosome into the virion, prior to fusion of the viral membrane with the endosomal membrane during replication of influenza viruses within the host cell. Amantadine binds to M2 protein and blocks its ion channel activity and thus inhibits viral uncoating and replication. In addition to its antiviral activity, amantadine antagonizes the N-methyl-Daspartate (NMDA) receptor in the CNS. NMDA receptors are important in pain sensation, especially chronic pain. Amantadine combined with other analgesics such as opiates or NSAIDs alleviates chronic pain.

Acyclovir: Acyclovir is metabolized to the monophosphate by thymidine kinase, which is more active in the virus than in the host cell. The host cell then converts monophosphate to triphosphate that inhibits the viral DNA polymerase, ending the nucleotide chain prematurely.

Zidovudine (AZT): Zidovudine is used in cats to treat FIV infection, where it produces temporary alleviation of the clinical signs and increases quality of life and survival time in most cats, particularly when clinical signs of immunodeficiency are evident. It does not inhibit the viremia. Clinical improvement occurs 14 days after the start of treatment. Zidovudine is not effective against feline leukemia virus at nontoxic doses.

Cat omega interferon: Feline omega interferon can be used to treat cat viral infections, including Calici virus, FeLF, FIV, and other feline viral infections as well as canine parvovirus.

Difference between Anti-Microbial and Anti-Bacterial: Antimicrobial is an agent that kills microorganisms or stops their growth. Antimicrobial medicines can be grouped according to the microorganisms they primarily act against. For example, antibiotics are used against bacteria, and antifungals are used against fungi. Antibacterial is a substance that destroys bacteria or suppresses their growth or reproduction.

6

Control and Prevention of Infectious Diseases

Prevention is always better than cure. It is more appropriate in animal healthcare practice, as there is unavailability of adequate healthcare infrastructure in remote areas and the patient owners fails to afford expensive and cumbersome treatments. The control of infectious diseases essentially require proper sanitary conditions, ensuring contamination-free environment, avoiding transmission of the disease pathogens, prompt control of outbreak, breaking the life cycle of the parasites, enhancing nonspecific immunity, and immunizing animals against specific pathogens through vaccination. This chapter describes the steps to be taken to reduce the disease incidence.

Prevention of Environmental Contamination: The premises such as sheds, stables, and kennels and pastures should be kept free from contamination. The removal of parasites from the host is the most appropriate step, achieved by use of parasite-killing agents, thus preventing pasture contamination. Destruction of adult parasites in hosts prevents passage of eggs or larvae, thus excluding contamination of the environment. Ovicidal drugs should preferably be used to destroy the eggs, thereby avoiding hatching of eggs and development of infective stages on the pasture. Anthelmintic treatments prior to rainy seasons using larvicidal drugs prevent contamination of pastures when conditions are becoming favourable for egg and larval development. Proper disposal of the litter gives satisfactory control over fecal-oral transmission of monoxenous parasite with a direct life cycle, or that spends most of its adult life in a single host, where parasite's progeny is transmitted from one host to another during the life cycle. Faeces or litter may be heaped to destroy the eggs or larvae of parasites. The optimum stocking density in pens and pastures significantly reduces the parasite burden in animals, and the associated problem of contamination of sheds and pastures.

Control of Intermediate Host, Vectors and Reservoirs

The perfection in management practices by limiting the contact between intermediate and final hosts can prevent infectious disease pathogens to pass through several hosts to complete the life cycle. Direct action may be taken to reduce or eradicate intermediate host populations. The population reduction of the intermediate hosts like snails, by use of chemical (Molluscicides) or through biological control by allowing grazing of snail-eating ducks incompletes the life cycles of the trematode. The reduction of snail population can be achieved by proper drainage, fencing and other management practices. The number of insect and tick vectors can be reduced by use of chemicals (insecticides/acaricides), biological control (hymenopterous insects, entomopathogenic fungi and *Bacillus thuringiensis*) and genetic control (sterile male technique, chromosomal translocation). The use of vaccines at appropriate time also controls the vector population. Destruction of reservoir hosts is important in controlling certain parasites, such as rodents for Leishmania and Antelopes for African trypanosomes.

Control of Internal Parasites

The control internal parasites is a challenging task in grazing livestock even by introducing multi-label and multi-directional approach. It is almost impossible to suggest general recommendations to control parasitic diseases in varied geo-climatic conditions and methods adopted for rearing the livestock in the country like India. Besides, there is an urgent need to design sustainable parasite control strategy in view of increasing incidence of anti-parasitic drug resistance in animals. Periodical deworming against internal parasites in the definite host prevents development of infective stage in the premises and pastures, thus keeping the susceptible hosts free from infective stage of parasites.

Control of Arthropod Pests

Manure, filth, damp and dark corners, stagnant water etc. are favourite breeding places of insects. So, these places should be cleaned periodically. Eggs of ticks and mites deposited in cracks and crevices on the walls, floors and wood work of the animal house should be removed or destroyed periodically. Periodical dipping or spraying of animals (once in April-June and once in July-September) with suitable insecticides does not allow lice, flies, fleas, mites and ticks halting on the skin, thus preventing localization of the infective pathogen. The insides of animal sheds should be scrubbed and cleaned daily to remove debris. Areas around the animal sheds should be kept dry and clean. Interior of animal sheds like roof, walls and corners should be regularly cleared of cobwebs and spider webs and sprayed with insecticides, at least once in a month. Dusting of animals with DDT, Lorexane, gammaxene or some patent preparations available in the market, can be used to control cattle warble flies. Individual animals can be dusted by hand in small herds. A gunny bag or any other bag having sufficiently large pores through which dusting powder can escape out is filled and hung at a convenient place at a opportune height in case of larger herds. As the animals pass under the bag they rub their backs against the bag, and get a dusting in the process. The bags are hanged at the entrance to stanchion barn, hay or straw feeding bunk, gates leading out on to the pasture etc. Organophosphate insecticides like Malathion, Parathion, and Neguvon etc. are very destructive to insects but are quite toxic to animals as well. Newer generation synthetic pyrethroids like Deltamethrin, Cypermethrin etc. are available in the market for the control of ticks and mites. Precautions should be taken during the use of chemicals and manufacturer's instructions should be scrupulously followed during their usage.

Control of Outbreak

Entry and exit of the farm animals, animal products, vehicles and persons should be stopped in case of an outbreak. The containment vaccination must be followed as per veterinarian's advice. Grazing in a common place should be avoided to prevent transmission by contact. All visitors to the farm should be banned. Foot dips containing antiseptics must be provided at the entry of the farm and sanitation and hygiene should be geared up.

Isolation of Sick Animals

Sick animals should be isolated. Isolation refers to segregation of animals, which are known to be or suspected to be affected with a contagious disease, from the apparently healthy ones. Segregated animals should be housed in a separate isolation ward situated far away from the normal animal houses. The isolation ward should never be at a higher level than that of the healthy shed. The segregated animals should be placed at one end of normal animals' buildings as far away from healthy stock as practicable, if a separate accommodation is not available. The isolated animals should be brought back into the herd only when the outbreak ends and they are fully recovered. Attendants working on sick animals and equipment such as buckets, shovels etc. used for them should not be used for healthy stock. The sick animals should be attended daily, only after the healthy stock. The equipment should be thoroughly disinfected before they are used on healthy stock next day. The attendant should wash his hands and feet in antiseptics, and discard the clothes in which he worked before working for healthy animals.

Quarantine for Newly Purchased Animals

Quarantine is the segregation of apparently healthy animals, especially those are brought into the herd for the first time, which might have been exposed to the risk of infection in the past places of housing from infected animals. The idea is to allow sufficient time for any contagious disease that the quarantined animal may be harbouring, to become active and obvious. Hence, the quarantine period depends on the incubation period of a disease. But in practice, a quarantine period of 30 days is allowed as it covers incubation period of almost all diseases. The quarantine period should be about six months for rabies. Animals should be thoroughly screened for parasitic infestation by faecal examination during the quarantine period, and deworming should be carried out on the 23rd/24th day, if needed. The animals should also be subjected to dipping or spraying on the 25th/26th day for removing ectoparasites.

Vaccination of Farm animals

Vaccination is a practice of building up immunity artificially in the animal's body against specific infectious diseases by injecting biological agents called vaccines. The term vaccine is used to denote an antigen (substance from an organism), may be a live, attenuated or dead bacterium, virus or fungus and used for the production of active immunity in animals. The vaccine may also include substances like toxins, toxoids or any other metabolites etc. produced by microbes and used for vaccination. The young farm animals should be vaccinated at regular intervals at appropriate time as per the schedule against a particular disease.

Deworming of Animals

The regular deworming of livestock and companion animals, preferably during the early stages of infestation is essential to keep the herd and premises worm-free. The most suitable time of deworming is, when the worm load is less. In adult animals, deworming is done following faecal sample examination. It is good to deworm adult females after parturition. Young animals should preferably be dewormed every month using a suitable anthelmintic. Older stock can be dewormed at 4-6 months intervals. Deworming schedule is very crucial for buffalo calves as mortality due to worms is very high. It is advisable to deworm heifers twice a year up to two years of age in areas of heavy endoparasite infestations, particularly in hot-humid regions. Adult stock can be drenched twice a year, once before monsoon season (May-June) and next one during monsoon (August-September).

Elimination of Carriers

The tissues of a recovered animal, apparently in good health may continue to harbour the disease-causing organisms. Such pathogen carrying animals are known as 'carriers'. The carrier state may remain for years, and the animal becomes a potential danger to susceptible animal population. Common diseases for which carriers have been observed in farm animals are Tuberculosis, Leptospirosis, Brucellosis and Foot and Mouth Disease. Carriers of diseases in a herd should be diagnosed and eliminated to make the herd completely free from disease pathogens. Certain diagnostic screening tests can be used for spotting out carrier animals in a herd. These tests are periodically employed on all animals in a herd for diagnosis and culling of carriers. Some of the commonly used screening tests are tuberculin test, Johnin test, agglutination test and test for detection of subclinical mastitis.

Tuberculin Test: This is quite a reliable test for diagnosing non-overt cases of tuberculosis in all species of farm animals. Tuberculin test should be carried out in animal farms once every six months in the initial stages and later on, depending on the herd health status, the test can be conducted annually. January is the ideal month for conducting tuberculin test under Indian conditions. The important methods of the test are intradermal, subcutaneous and ophthalmic administration of tuberculin, the purified protein derivatives (PPD) of *Mycobacterium tuberculosis* (tubercle bacteria) into infected animal. The intradermal route of tuberculin administration in all bovines is the most practicable, reliable and popular. The best site is the side of the neck. In bovines, it can also be done in one of the folds of the skin by raising the tail, or on the vulva. The suitable site in the neck, is the middle third of the neck, as sites near the shoulder or mandible give less pronounced reactions. A small area of skin is clipped and cleansed with spirit. The allergic responses are set up following intradermal injection of tuberculin that constitutes a 'reaction'. On the contrary, tuberculin even in large doses, gives no reaction in healthy animals. Purified Protein Derivative (0.1 ml) is injected intradermally. The reaction creates a bead-like swelling detectable by the finger, and is measured by slide calipers. The positive carrier animals should be culled and destroyed from the herd.

Johnin Test: Johnin is the purified protein derivative of *Mycobacterium paratuberculosis* (Johne's bacterium) used for the diagnosis of Johne's disease in cattle and buffaloes. Johnin test is performed in the similar way like single intradermal test done for tuberculosis. A painful indurated skin swelling with an increase in skin thickness more than 4 mm is taken as positive. All positive animals are culled and destroyed following the local regulations.

Agglutination Test for Brucellosis: This serological test is a based on the principle of reaction between antigen, the dead bacteria, and antibody, agglutinins present in the body fluids, mainly the serum of the infected animals, resulting in agglutination following mixing of the agglutinins present in the serum and other body fluids of animals suffering from brucellosis or carriers with a suspension of killed culture of *Brucella abortus*. The latter will cluster together and the reaction is known as agglutination. The non-infected animals lacking agglutinins do not show such agglutinations. Rapid plate agglutination test can be done at the site of the animal. Standard Tube Agglutination Test is done in the laboratory. Agglutination can be conducted using whole blood, serum, milk, whey, semen, etc. Stockman can only attempt to collect sterile samples of blood (from jugular vein) or milk of the animals periodically, maybe once in a year, and get them tested in the nearest laboratory. All positive reactors should promptly be eliminated from the herd.

Disposal of Carcass

Proper disposal of carcasses of animals died of infectious disease, help preventing the spread of disease pathogens to other animals and humans. Carcasses should never be disposed off by depositing them in or near a stream of flowing water, as this may carry infections to places downstream. An animal died of an infectious disease should not be allowed to remain longer in sheds as biting insects, rodents, etc. can reach it. It is not safe to open carcasses of animals that have died of an infectious disease, particularly by an anaerobic organism. All carcasses should be disposed of properly either by burying or by burning.

Burials of carcass: The most common method of carcass disposal is burial. This is reasonably a safe method, if done deeply enough and in soil from which there is no drainage to neighboring places. Deep burial is necessary to prevent worms carrying bacterial spores to the surface as well as to prevent carnivorous animals from digging up the carcass. The carcass should be carried to the burial place in a trolley but never by dragging it over the ground. The burial pit should be got ready before the carcass is taken there. The pit should be so dug that the highest part of the carcass must be at least 1.5 m below the level of the land surface. Bedding used for the dead animals, its excreta, left over feed and the top 5 cm soil form where the dead animals was lying, if the floor is not cemented, should also be buried along with the carcass. Drainage of water out of the burial place can be checked by seeing to it that the burial place is an area where the general water level is at least 2.5 m below the ground. The carcass is covered with a thick layer of freshly burnt quicklime and then filled with dirt and topped with some rocks, to further circumvent marauders.

Burning of Carcass

Burning is the most sanitary method of destroying carcass. It is preferably done close to the site of the death without dragging the carcass. The trench should be dug after deciding the site for burning. The trench should be at least 0.5m deep, shallower towards the ends, and comparing in width and length to the carcass's size. General direction of the trench should be that of the prevailing wind direction. The trench is first filled with wood, some iron bars placed across it and the carcass placed thereon. By firing the wood, the carcass will be completely consumed and, with it, all infectious materials. The carcass utilization or carcass frying or rendering plants are usually available in towns and cities, for industrial utilization of animal carcasses. In such situation, the skin is removed with due regards for the dangers of disease dissemination. The removed skins are usually disinfected by immersion in a disinfecting solution and the remainder of the carcass 'fried out' for its fat, the latter being used in soap manufacture. Farmers can inform these plants whenever there is a carcass so that these plants can collect the carcass.

Disinfection of Animal Houses

Under ordinary conditions, daily scrubbing and washing of houses and the sunlight falling in the houses, are sufficient enough to keep them moderately germ-free. However, disinfection is highly required in cases of disease outbreak, and should be carried out scrupulously. All floors, walls up to a height of 1.5 meters, interiors of mangers, water troughs and other fittings and equipment coming in contact with animals are disinfected. The first step for disinfection of animal houses is removal of all filth, as the power of disinfectants is greatly reduced in the presence of organic matter. Floors, walls up to a height of 1.5 m, interior of water troughs and mangers should be well scrubbed and all dung, litter etc. should be removed and stacked separately, where animals cannot reach. The dung, litter etc. should first be disinfected *in situ* in case of an outbreak of anthrax by sprinkling suitable disinfectant. The top 10cm soil of earthen floor should be removed and disposed of along with the litter. The place should be scrubbed and washed with 4 percent hot washing soda solution after removal of filth, prepared by mixing 4 kg washing soda in 100 liters of boiling water. The approved disinfectant solution should then be coated liberally over the place by sprinkling or preferably by spraying and left as such for 24 hours for its action. The animal house should again be washed with clean water and left to dry by wind and sunlight. The interior of water troughs and mangers should be whitewashed. This can be done even routinely at fortnightly intervals.

Disinfection of Pastures

Any obvious infective material, like carcass, aborted foetus, dung etc. over the pasture should be removed promptly. Animals should be prevented from grazing on the suspected pasture for at least three to four months. The pasture can be ploughed up and left fallow for about six months to destroy pathogens by direct sunlight. The followings are the commonly used disinfectants.

Bleaching powder (Chloride of lime): It can be used for disinfection of animal houses, when a contagious disease has occurred and for sterilization of water supplies. It should not be used in milking barns as its strong odour may taint milk. The concentration should not be less than 30 per cent available chlorine. It is applied as dusting. Bleaching powder must be stored in airtight bins as damp surroundings, exposure to light and air causes it to decompose rapidly.

Boric acid: It can be used as an udder wash. It is a weak antiseptic and is likely to harm nervous system, if absorbed into body in large quantities. It is used as wash for eyes and other sensitive parts of body. It is used as 6 per cent solution and applied by splashing. Nowadays, antibiotic solutions are replacing boric acid as eyewash solution.

Caustic soda (Sodium hydroxide): Caustic soda is a very effective disinfectant for general use in farm buildings and animal houses, as it is an excellent cleaning agent as well as a powerful germicide. It is highly destructive to virus of Foot and Mouth disease, hog cholera etc. However, it is not effective against tuberculosis and Johne's diseaseorganisms. The concentration of 2 percent solution is required for general use and 5 percent solution is used against spores of anthrax and black quarter. It is applied by splashing. Rubber gloves, goggles and protective clothing should always be worn when caustic soda solution is being used as it burns skin and damages fabrics.

Cresols: The cresols are only slightly soluble in water and are generally emulsified with soap. It is effective against a wide range of organisms including acid fast

Mycobacteria responsible for tuberculosis and Johne's disease. However, it is not effective against viruses and bacterial spores. It is a good disinfectant for use on floors, walls, equipment etc. but not the milking barns, because of its phenolic odour. Concentration of 2-3 percent is required for disinfectant purpose. It is applied by splashing. Only soft water is used for preparing solutions, as hard water precipitates soap. Lysol is a solution of cresol with soap.

Lime (Calcium Oxide, quick lime): It is also known as slacked lime. It is a deodorant as well as a disinfectant. It can be used for sprinkling on manure and animal discharges, on floors or as a whitewash or milk of lime. It is applied by splashing, scrubbing or sometimes dusting. Only freshly prepared lime is used.

Phenol (Carbolic acid): It is effective against several types of bacteria. However, it is not so effective on spores or viruses. The presence of organic matter but not oil or alcohol does not reduce its disinfectant activity. It is very toxic, corrosive and irritant. Concentration of 1-2 percent is required for its disinfectant action. It is applied by splashing. Great care should be taken while using phenol to protect eyes, skin and clothing.

Quaternary Ammonium Compounds (QAC): These are cationic detergents. They have no effect on spores and viruses. They can be used to disinfect dairy utensils, udders, milkers' hands and towels used for wiping udders. Cetrimide, a white powder, is an example for QAC. Concentration of 0.1 percent solution is required for its disinfectant purpose, and 0.5 percent cream is applied on teats and hands to prevent mastitis. Wiping of udder with clothes wetted in 0.1 percent solution and washing of milkers' hands are the mode of application. Utensils should be scrubbed with boiling water before rinsing with QAC.

Soap: Itis an anionic detergent and a very weak germicide. But it has great usefulness in cleaning various surfaces including skin. It can be used as a preparatory to the application of a disinfectant. It is applied by scrubbing. It should preferably be used only as surface-sanitizing agent.

Sodium hypochlorite: It is a chlorine compound. It is an excellent disinfectant but it is not effective against tuberculosis bacteria. Its effectiveness is reduced in presence of organic matter. Concentration required for washing utensils is 200 parts per million of available chlorine, about 300 ml sodium hypochlorite and about 200g of washing soda in 100 liters of hot water. About 60 ml of Sodium hypochlorite in 10 liters of clean water is used for udder washing. It is used for rinsing of utensils, wiping of udder. It should be stored in air-tight containers as hypochlorites deteriorate rapidly when exposed to air.

Washing Soda (Soda ash, Sodium Carbonate)

It is good for disinfection of barn premises after an outbreak of viral disease like Foot-and-Mouth disease. It is a good detergent at the concentration of 4% solution.

It is applied by scrubbing. Lye is better against Foot-and-mouth disease virus than soda ash. It should be used as a hot solution.

Vaccination and Immunity

Vaccine is a substance that prevents the disease caused by certain pathogens upon its introduction into the body. Vaccine consists of dead pathogens or live attenuated organism, or a part thereof. The vaccine induces immunity against a specific pathogen either by production of antibodies or by activation of T lymphocytes. Vaccination is the process of antigen administration to stimulate protective immune response against an infectious agent.

Types of Vaccines

1. **Whole pathogen vaccine:** The most well-known method of vaccine preparation is to use the whole disease-causing pathogen to elicit an immune response similar to that seen during natural infection. Example - Live and inactivated vaccine
2. **Subunit vaccine:** These vaccines typically contain one or more specific antigens from the surface of the pathogen. Adjuvants are often added to subunit vaccines that help to strengthen and lengthen the immune response to the vaccine. Example - Toxoid, protein antigen etc.
3. **Recombinant vectored vaccine:** Recombinant vector vaccines are experimental vaccines similar to DNA vaccines, but they use an attenuated virus or bacterium to introduce microbial DNA to cells of the body. "Vector" refers to the virus or bacterium used as the carrier. Example - Bacterial and viral vector
4. **Nucleic acid vaccine:** Nucleic acid vaccines work in a different way to other vaccines in that they do not supply the protein antigen to the body. Instead, they provide the genetic instructions of the antigen to cells and in turn, the cells produce the antigen, which stimulates an immune response. Nucleic acid vaccines are quick and easy to develop, and provide significant promise for the development of vaccines in the future. Example – RNA and DNA, DNA Plasmid vaccine

Major Group of Vaccines

Live-attenuated (weakened) vaccines: These vaccines contain modified strains of a pathogen (bacteria or viruses) that have been weakened with respect to virulence but are able to multiply within the body and remain antigenic enough to induce a strong immune response. The canine distemper, blue tongue, Rota virus vaccine and some influenza virus vaccines are some examples of this type of vaccine.

Heterologous vaccines: Heterologous vaccines are a sub-group of live attenuated vaccines produced from strains that are non-pathogenic. The vaccine confers protective immunity against a pathogen that shares cross-reacting antigens with the microorganisms in the vaccine.

Killed-inactivated vaccines: Bacteria or viruses are killed or inactivated by a chemical treatment or heat to produce this type of vaccines. This group includes the most influenza virus vaccine, plague vaccine, rabies vaccine, or hepatitis A virus vaccine.

Subunit vaccines: Subunit vaccines contain only the antigenic component of the pathogen instead of using the entire microbe to stimulate the immune system. In some cases, these vaccines use only epitopes, the very specific parts of the antigen that is recognized by antibodies or T cells and binds to epitopes. The chances of adverse reactions to the vaccine are lower as subunit vaccines contain only the essential antigens and not all the other molecules that make up the microbe. Example -Lyme disease vaccine

DNA Vaccine: Some cells of the vaccinated animal take up the genetic material introduced into the body, as antigens of microbes. The DNA then instructs those cells to synthesize the antigen molecules. The cells secrete the antigens and display them on their surfaces. In other words, the body's own cells become vaccine-making factories, creating the antigens necessary to stimulate the immune system. Example -West Nile fever virus in horse

Recombinant vector vaccines: Recombinant vector vaccines are experimental vaccines, similar to DNA vaccines, but they use an attenuated virus or bacterium to introduce microbial DNA to cells of the body. Vector refers to the virus or bacterium that is used as the carrier. Example - RD and IBD vaccine

Toxoid: These vaccines are used when a bacterial toxin is the main cause of illness. The immune system learns how to fight off the natural toxin when it receives a vaccine containing the harmless toxoid. The immune system produces antibodies that block the action of the toxin. Example -Vaccines against diphtheria and tetanus

Gene deleted vaccines: These are genetically engineered vaccines which involve removal or mutation of the virulence gene of the pathogen. Example - Pseudorabies vaccine

Peptide vaccine: These are the subunit vaccine prepared by chemical synthesis of short immunogenic peptides.

Route of Vaccine Administration

- Deep subcutaneous or intramuscular route (most vaccines)
- Oral route (with feed)

- Drinking water and spray in poultry
- Intranasal route
- Eye drop and beak dipping in birds

Vaccination failure: A vaccination failure occurs when the animals/birds do not develop adequate antibodies titre levels, and the vaccinated animal/bird continues to remain susceptible to a field disease outbreak. The failure may occur for the following reasons.

At the vaccinated animal / bird level: The vaccination failure may occur due to reasons attributed to vaccinated animal or bird.

1. **Maternal antibodies:** A high level of maternal antibodies in the young chicken may interfere with the multiplication of live vaccines, reducing the immunity conferred. The chick coming from a breeder hen typically has a high level of antibody against Gumboro (Infectious Bursal Disease) disease, the chick will typically have high levels of antibodies from the mother (maternal) for several weeks. Some of the vaccine virus is inactivated if vaccination is performed in the presence of these antibodies.
2. **Stress:** Stress reduces the animal's ability to mount a robust immune response. Environmental extremes of temperature and relative humidity, inadequate nutrition, parasitic load, and other diseases are incriminated to induce stress in animals and birds. So, the vaccination should be avoided during stressful conditions and should be resumed, once they become healthy.
3. **Animals incubating the disease pathogen:** The animals or birds to be vaccinated may be incubating the disease at the time of vaccination. Despite proper vaccination, the vaccinated animals or birds become diseased as around one week time is required for antibody production to begin and reach protective levels. Immunoglobulin G, M, and A are first detected approximately 4 to 5 days following exposure to a live virus-type vaccine. Additional days are required for titres to reach protective levels.
4. **Immunosuppression:** The term immunosuppression refers to circumstances where the non-cellular (antibody) and cellular components of the immune system are not functioning properly. This may result in the development of only limited protection from the vaccination and an excessive vaccine reaction with morbidity and mortality. The infection with IBD virus or Marek's disease virus, and consumption of feed contaminated with high levels of mycotoxins make the birds immunosuppressed, resulting in vaccination failures.
5. **Strong Field Challenge:** No vaccine can withstand strong challenge in the field conditions.

6. **Weaning of Vaccine Immunity:** Antibodies are protein in nature and are thus subjected to metabolic breakdown. The vaccinated animal may suffer from the disease, if the antibody titre is low.

At the vaccine level: The vaccination failure may occur because of reasons at the vaccine level such as inactivation of Live vaccines, inappropriate strain of the pathogen used in the vaccine, poor distribution of live vaccine, and poor-quality vaccine

1. **Inactivation of Live Vaccines:** Live vaccines may be inactivated due to improper handling or administration. Therefore, such vaccines are checked for the lot numbers and expiry dates on the vials before administration of live vaccines, and the manufacturer instructions are strictly followed for storage and handling of vaccine. Once a vaccine is reconstituted, it must be used for vaccination as per instruction. Infectious bronchitis vaccine loses 50 percent of its potency in warm conditions in about 1 hour.
2. **Improper Vaccine Strains:** The vaccine may not contain the proper strains or serotypes of organism required to stimulate protective immunity. Although the vaccine is administered properly and uniform and adequate antibody titres are present, the chickens still suffer with the disease, particularly with infectious bronchitis and infectious bursal disease.
3. **Poor distribution of Live Vaccine:** - Poor distribution of live vaccine administered by the water or spray route may result in chickens being "missed" in parts of the house. Relying on transmission of the vaccine virus from bird to bird is risky, and can result in excessive rolling-type reactions of long duration and delayed immunity in the flock. Chickens that missed the dose of the killed vaccines do not build protection, as killed vaccines will not spread from bird to bird.
4. **Poor Quality Vaccine:** Vaccine may be of poor quality such as low vaccine titre, contaminated, etc. The vaccine manufacturing industry is highly regulated and has extensive internal quality control. So, vaccination failure due to problems with the vaccine are rare.

Schedule and route of vaccination in commercial layers

Age	Vaccine	Route
1 day	Marek's Disease	Intranasal/Intra ocular
4-10 days	Ranikhet Disease (F-strain)	Intranasal/Intra ocular
18-21 days	Infectious Bursal Disease	Intra ocular/Drinking water
24-48 days	Infectious Bronchitis	Intra ocular/Drinking water
5-6 weeks	Ranikhet Disease (F-strain), Booster	Intra ocular/Drinking water
8-9 weeks	Ranikhet Disease (R_2B-strain)	S/C
10-11 weeks	Fowl pox	Scratching
14-16 weeks	Infectious Bronchitis (Booster)	Intra ocular/Drinking water

Age	Vaccine	Route
16-17 weeks	Fowl pox (Booster)	Scratching/Intra muscular
18-19 weeks	Ranikhet Disease (Killed)	I/M or S/C
40 weeks & above	Ranikhet Disease&Infectious Bronchitis combined vaccine	Drinking water
45-50 weeks	Infectious Bursal Disease	I/M or S/C

Schedule and route of vaccination in broilers

Age	Vaccine	Route
1 day	Marek's Disease (at hatchery)	s/c at neck
5-7 days	Ranikhet Disease Virus (F1)	I/O or I/N
14 days	Infectious Bursal Disease	I/O or I/N
21 days	Ranikhet DiseaseVirus Lasota	Drinking water
28 days	Infectious Bursal Disease (Booster)	Drinking water

Vaccination schedules for cattle and buffaloes

Sl. No.	Name of Disease	Name of vaccine	Age at first dose	Dose	Booster dose	Subsequent dose
1	Foot and Mouth Disease (FMD)	FMD Vaccine	3 months and above	2 ml IM	3, 9, 12 months	Biannual
2	Haemorrhagic Septicaemia (HS)	H.S. Alum Adjuvant	All ages > 6 months	5ml SC (<273Kg), 10ml SC (>273Kg)	-	Biannual
3	Black Quarter (BQ)	Polyvalent Alum Adjuvant	All ages > 6 months	5ml SC, 10ml SC (>273 Kg)	10th day	Biannual
4	Anthrax	Anthrax Spore Vaccine	All ages > 4 months	1ml SC	-	Annually in endemic areas.
5	Brucellosis	Brucella S19 vaccine	4-8 months of age (Only female calves)	2ml IM	-	Once in a lifetime
6	Theileriosis	Rakshavac T	3 months of age and above	3ml SC	-	Once in a lifetime. Only required for crossbred and exotic cattle.
7	IBR	-	All ages > 3 months	-	1 month after first dose	Six monthly (vaccine presently not produced in India)

Sl. No.	Name of Disease	Name of vaccine	Age at first dose	Dose	Booster dose	Subsequent dose
8	Rabies (Post bite therapy only)	-	Immediately after suspected bite.	1ml SC	4th day	7,14,28 and 90 (optional) days after **first** dose.

Immunity

Immunity is the ability of the body to protect against all types of foreign body like bacteria, virus, toxic substances etc., which enters the body. Immunity is conferred by immune system through a complex network of lymphoid organs such as bone marrow, thymus, spleen etc.

Types of Immunity

Innate immunity: It is conferred by anatomical barrier, phagocytic cells, blood protein and cytokines. It is antigen independent, no lag time, no antigen specific immunity and no immunologic memory. It is contrary to antigen-dependent adaptive immunity with a lag period, antigen specific with development of memory cell.

Acquired immunity: The defence mechanism gained by the body against the pathogens is called acquired immunity. It is divided into passive immunity and active immunity based on whether it is acquired passively without any processing or actively, respectively. Passive Immunity is short lived, built without participation of host, with no memory, no Lag period and high titre is achieved instantly. Active immunity is long lived gained by participation of hosts immune system with memory and Lag period. The titre is time dependent. Either of the active or passive immunity is further classified as natural or artificial. Humoral Immunity is B-cell dependent, where antibodies are effecter molecules and are effective against extracellular parasites.

Cell mediated Immunity is T-cell dependent, CTL (Cytotoxic Lymphocytes), NK (Natural Killer) cells, Macrophages are effecters and it is effective against intracellular parasites, tumour cells, and tissue grafts.

Antigen: Antigen is a substance that stimulates the formation of specific neutralizing substances or antibody, when introduced into the tissues of a susceptible animal, and it reacts specifically in some observable way or produce lymphokines or both antibody and lymphokines.

Factors Determining Antigenicity

- Molecular size
- Complexity

- Shape
- Stability
- Degradability
- Foreignness
- Genotype of the recipient animal
- Dosage and route of administration
- Adjuvants

Types of Antigens

According to Antigenic Specificity

- **Species-specific antigen:** An antigen is specific in relation to a specific species of host only.
- **Organ specific antigen:** The antigen is specific in relation to a particular organ in a single species. For example, kidney protein is different from lung protein.
- **Heterophile antigens:** -They are antigens of identical nature present in the cells of some bacterial species and also in the tissues of different animals. Antibody formed against one antigen cross-react with other antigens. There is a phenomenon of antigenic sharing. Examples - Forssmann antigens are a group of related heterophile antigens, they are present in the cell wall of many bacterial species (glycoprotein with carbohydrate side chain) belong to the genera streptococcus, shigella, salmonella and clostridium and also on the surface of red blood cells (glycoprotein) of horses, sheep, cats and mice.

Alloantigen (isoantigen): These are antigens present in one individual and are antigenic to some individuals of the same species. The corresponding antibodies are called alloantibodies. Example - Blood group antigen, Transplantation antigen etc.

Syngeneic antigen: The antigens present in the individuals of the same genetic makeup. Example- antigens present in monozygotic twins.

Autoantigen: -In certain circumstances, own body tissues develop antigenic properties, and antibody are designed against the antigen. The auto antigens are sequestrated without contact with the lympho-reticular system but the release of antigens by any mishap provides an opportunity to produce autoantibody. Example- sperm, lens protein etc.

Superantigen: -A group of molecules that do not have to be processed by Antigen Presenting Cells (APC) in order to activate T cells. Examples - Bacterial antigens (Streptococcal A antigen).

According to T Cell Response

T dependent antigen: Complex antigens such as RBC, serum proteins etc. are rapidly metabolized, require T cell help and produce memory and class switching.

T independent antigen: LPS of *E coli*, pneumococcal polysaccharide, flagellin, Lectin, Con A, PHA, Pokeweed mitogen etc., are composed of repeating units. They are slowly metabolized, with no memory or class switching.

Microbial and non-microbial antigens

Microbial antigens: It includes bacterial antigens, viral antigens and other microbial antigens.

Non-microbial antigens: They are certain dust particles, pollen grain, snake venom, etc.

Antibody: Antibody or immunoglobin is a specialised glycoprotein produced from activated B cell (Plasma cell) in response to an antigen that triggered its production.

Class of Antibodies

Immunoglobulin G (IgG): It constitutes 80% of the serum antibodies, and has half-life of 7-23 days. IgG crosses placenta, provides natural immunity to foetus and neonates at birth and activates complement by classical pathway. It renders immunity to new born, and neutralizes toxin. IgG3 binds to Fc receptor by phagocytosis.

Immunoglobulin M (IgM): It constitutes 5-10% of total serum protein. The half-life is about 5 days. It is the first immunoglobulin produced in a primary response to antigen. It activates the classical pathway, and provides defence against multivalent antigens. It acts as opsonin.

Immunoglobulin A (IgA): It has 6% share of the serum antibodies. It is present in body secretion such as saliva, mucous, tears, colostrum and milk. Half Life is 6-8 days. It activates complement by the alternative pathway. It is a secretory antibody, provides local immunity and promotes phagocytosis and intracellular killing of microorganism.

Immunoglobulin D (IgD): It constitutes0.2 % of the serum antibodies. It's half-life is 3 days. IgD together with IgM. is major membrane bound immunoglobulin on unstimulated B-lymphocyte that acts as recognition receptor for antigen. It functions for 8-cell activation, acts as a receptor for antigen binding.

Immunoglobulin E (Ig E): They are mostly present extra cellularly. Half-life is 2 days. The do not cross the placental barrier. It contains small percentage of lymphocyte. The function of this Immunoglobulin includes immediate hypersensitivity, binding to Fc receptor on Basophils and mast cells. release of substance like histamine, vasoactive mediators. It does not activate complement but plays a role in immunity against helminthic parasites.

Cells of Innate Immune System

A. Myeloid Innate immune cells

1. **Neutrophils:** Neutrophils, martyrs of immune system, are the major cell type of myeloid system, and are considered as the first line of defence against pathogenic microorganism. They are classified as granulocyte, and phagocyte. The primary functions are antimicrobial effectors, particularly in acute bacterial infection. The mechanism of actions is phagocytosis, degranulation, neutrophil extracellular trap formation.
2. **Eosinophils:** They are classified as granulocytes that have characteristic eosinophilic granules that stain red with eosin. The primary functions include antiparasitic effector, particularly in helminth infestations, some antiviral actions, and roles in allergy. Mechanism of actions are degranulation and limited phagocytosis.
3. **Basophils:** They are granulocytes of myeloid type cell. They are mediators of inflammation. Mechanism of action is degranulation. They have characteristic appearance of blue-purple basophilic granules with basic dye.
4. **Monocytes:** Monocytes are circulating precursors of macrophages among other cell types, and dendritic cell. They have kidney shaped nucleus. In general, monocytes, macrophages and dendritic cell progeny serve three main functions in the immune system, such as phagocytosis, antigen presentation and cytokines production.
5. **Mast cells:** Granules of mast cells have large amounts of histamine, heparin, TNF-alpha and other inflammatory mediators. Mast cell plays an important role in the coordination of chronic inflammation and tissue repair.

B. Lymphoid Innate Immune Cell

1. Natural Killer cells (NK cells): NK cells earn their name because of their ability to innately recognize and kill abnormal host cells without previous sensitization. NK cells don't express either T-cell or B-cell receptors, which distinguishes them from NK T lymphocytes. They are lymphoid cells with round nucleus, azurophilic cytoplasmic granules. Destruction of virus infected or abnormal host cells (including tumour cells) is the primary function. Recognition of virus-infected or abnormal host cells and targeted release of cytotoxic granules is the mechanism of action. Both the innate and adaptive immune system work together as a functional unit. Innate immune system is the first line of defence against invading pathogen in vertebrate animals.

The complement system and phagocytes are especially important aspect of innate immunity that deal with extracellular pathogens. NK cells of innate response help to eliminate infected cells. During adaptive immune response, antibodies and some T-helper cells directly help protecting against extracellular infections and other T-helper cells and Tc cells are most effective against intracellular infection.

The production of antibodies against antigen in animals is a complex and highly regulated process, known as immune response that involves the following steps

1. **Recognition of antigen:** The immune system identifies an antigen (Bacteria, Virus or Foreign particles) by Antigen Processing Cells (APC) such as dendritic cells and macrophages.
2. **Antigen presentation:** APCs engulf the antigen and break it down into smaller fragments, and display on their cell surface using a protein called major histocompatibility (MHC) molecule.
3. **Activation of T cells:** T-helper (CD4 T cells) cells recognize these Ag-MHC complex and become activated. They release signalling molecules called cytokines which coordinate the immune response.
4. **B-cell activation:** They have antibodies on their surface that can bind to antigen. When a B-cell encounters an antigen that matches its antibody receptor, it becomes activated
5. **Clonal expansion**: The activated B-cell undergoes clonal expansion dividing into large number of identical cells.
6. **Antibody production:** Plasma cells are derived from activated B-cells and start producing antibodies specific to the antigen.
7. **Antibody Function:** Antibodies circulate in the bloodstream and lymphatic system that binds with antigen when they encounter it, making it processed for destruction. This process can involve the activation of the complement system, which enhances the immune response.
8. **Memory cells:** Some of the B-cells produced during clonal expansion becomes memory B-cells.
9. **Regulation:** The immune response is tightly regulated to prevent excessive inflammation and immune reactions.
10. **Resolution:** Once the antigen is cleared, the immune response is dampened and excess immune cells eliminated through apoptosis (programmed cell death.)

Immune response: The collective and coordinated response to the foreign substances in an individual is mediated by the cells and molecules of immune system.

1. **The primary immune response**: The response elicited after antigen comes in contact to the immune system for the first time. During this time the immune system has to learn to recognize antigen and how to make antibody against it and eventually produce memory lymphocytes.
2. **The secondary immune response:** It occurs when the animal is exposed to the same antigen on subsequent occasions (2nd, 3rd, 4th etc.). At this point, immunological memory has been established and the immune system can start making antibodies immediately.

Primary immune response	Secondary immune response
Occurs as a result of primary contact with an antigen	occurs as result of second or subsequent exposure of same antigen.
Responding cells - B-cell and T-cell	Responding cell-memory cell
Lag phase (Latent period) is often longer	Lag phase is shorter due to presence of memory cells.
Level of Ab reaches peak in 7 to 10 days.	Level of Ab reaches peak in 3 to 5 days.
The first produced antibodies are mainly IgM.	Mainly, IgG antibodies are produced.
The primary response appears mainly in the lymph nodes and spleen.	The secondary response appears in the bone marrow, followed by spleen and Lymph nodes
Both thymus-dependent and thymus-independent antigen give primary immune response	Only, thymus dependent antigens give secondary immune response.

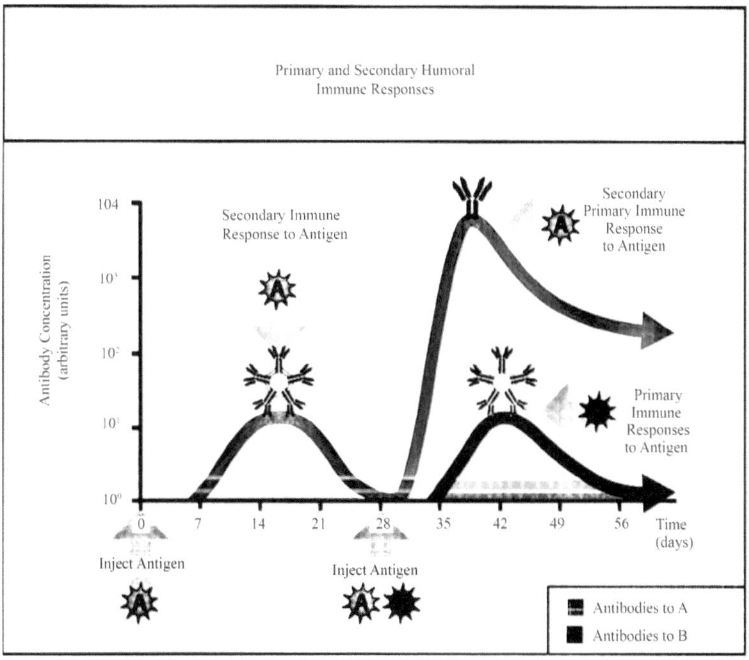

Primary and Secondary Humoral Immune Response

After injection of antigen A on day 0, an animal produces a primary antibody response to antigen A. This response lasts for 30–40 days and is dominated by IgM (pentamers) antibody. The exposure to antigen A plus a new antigen B, the animal produces a robust secondary antibody response to antigen A but a primary response to antigen B.

Vaccination in Poultry

Effective and proper vaccination programme in poultry is necessary to prevent mortality and losses from many dreadful poultry diseases. Vaccination programmes are available against the major poultry diseases. All birds reared in the farm either for your own stock or for sale to the public must be properly vaccinated. The schedule of vaccination should be as follows:

1. Newcastle vaccine should be administered at the time of hatch or within first 7 days of age. One drop of vaccine in the eyes or one drop in the nostril is given. Care should be taken not to release the chick until it has inhaled the nasal drop and/or blinked its eye.
2. Newcastle and Fowl Pox Vaccination: This vaccination must be given when birds reach 6 weeks of age. Fowl Pox and Newcastle vaccine are suspended in 50% glycerine saline solution and the birds' inoculated by the web stick method in the web of the wing.

(b) It is essential that all four vaccinations are given and in a proper sequence. A sufficient stock of properly refrigerated vaccine must always be kept on hand and all vaccinations are to be given according to manufacturer's instructions. It is equally important to observe birds after vaccination to make certain that the vaccine has been effectively given.

(c) Precautions: Water soluble antibiotic/ electrolyte can be given in the water 4 days prior to and 5 days after the vaccination to reduce stress.

Vaccination Calendar

The vaccination schedule is a general guide. Each farm and area will require some changes in the schedule. Following table can be used as a general guideline.

Layers

Name of Vaccine	Route	Age of birds
La Sota or F vaccine Newcastle	Intranasal drop	3 to 7 days
Marek's vaccine (in Hatchery)	Intramuscular	1 day
Infectious Bronchitis (1st dose)	Eye drops	2 - 3 weeks
La Sota Newcastle	Drinking water	5 - 6 weeks
Fowl Pox (1st dose)	Wing Web	7 - 8 weeks

Name of Vaccine	Route	Age of birds
R2B Newcastle	Sub cut or Intramuscular	9 - 10 weeks
Infectious Bronchitis	Eye drop or drinking water	16 weeks
Fowl Pox (2nd dose)	Skin Scarification	18 weeks
La Sota (if necessary) Newcastle	Drinking Water	20 weeks
La Sota (if necessary) Newcastle	Drinking Water	40 weeks

It is necessary to keep proper records on date of vaccination and on vaccines used including type, brand, serial number, date of purchase and date of use of vaccine.

Broilers

Age	Name of Vaccine	Route
3-5th day	Lasota	Intra ocular or intra nasal
7-9th day	Infectious Bronchitis Disease.	Intra Ocular or Drinking Water
16-18th day	Infectious Bronchitis Disease.	Drinking Water
24-26th day	Lasota (booster)	Drinking Water

7

General Systemic States

The general systemic states are generalized in nature and involve whole body systems. They are common to many diseases and contribute to the pathogenesis of the disease. Pathogenesis is defined as the changes in functioning at the cellular level and establishment of disease process. Hyperthermia, hypothermia, fever, toxaemia, septicaemia, bacteraemia, shock, dehydration, acid-base imbalance, hyponatremia, hypokalemia etc. are some examples of general systemic states.

Hyperthermia

Hyperthermia is defined as elevation of body temperature beyond the critical temperature due to high environmental temperature, in absence of circulating toxin in the blood vascular system, and results from failure of thermoregulation by the hypothalamus. Hyperthermia differs from fever in the mechanism that fever is associated with change in the body's temperature set-point in the hypothalamus. Thus, body tries to gain heat and reduce heat dissipation to increase the body temperature to an elevated set-point level.

Neurons in both the preoptic anterior hypothalamus and the posterior hypothalamus receive two kinds of signals. The peripheral nerves transmit signals from warmth/cold receptors in the skin. The temperature signals are also received from blood-bathing regions. These two types of signals are integrated by the thermoregulatory centre of the hypothalamus to maintain normal temperature. However, the body temperature rises above the set point due to failure of the thermoregulation and is associated with production of excess heat in the body through metabolism or absorption of more heat from the environment and failure to dissipate it. The common examples are heat stroke and adverse drug reaction. In heat stroke, the regulating mechanisms eventually become overwhelmed and unable to effectively deal with the absorbed heat when the body is exposed to excessive heat in a humid climate for a prolonged period causing the body temperature to climb uncontrollably. Hyperthermia is relatively a rare side effect of many drugs, particularly those that affect the central nervous system. Malignant hyperthermia is a rare complication of some types of general anaesthesia. Severe muscular exhaustion, especially when humidity is high, heavy hair coat of fatty animals and confinement in an inadequately ventilated area are some of the factors for hyperthermia. Hyperthermia can be created artificially by drugs or medical devices. Hyperthermia therapy may be used to treat some kinds of cancer and other conditions, most commonly in conjunction with radiotherapy.

Hyperthermia is a medical emergency and requires immediate therapeutic measures to prevent disability and death when the body temperature is sufficiently higher than the critical temperature. Hyperthermia is associated with increased thirst due to dryness of the mouth. The heart rate increases due to rise in blood temperature but the blood pressure falls due to peripheral vasodilatation. The respiratory rate and depth increase due to the effect of increased temperature on respiratory centre. The formation of urine and urinary output decrease due to reduced renal blood flow. High rise in body temperature may lead to depression of nervous system activities and that of respiratory centres leading to death of the animal. Circulatory failure may also occur due to myocardial weakness, fast and irregular heart rate. The body endogenous metabolism increases and extensive degenerative changes are set in the different parts of the body in prolonged hyperthermia.

Clinical findings: An increased body temperature above the critical limit is the primary sign observed in affected animal. The heart and respiration rate increases, and the pulse is weak. Sweating and salivation are recorded initially as a compensatory mechanism of heat dissipation, but not later when the sweating is almost absent. In the early stage, there is increased thirst and animal search for cool places and often prefer to lie in water. But, when the body temperature goes above 106oF, the respiration becomes laboured and general distress is evident. Beyond this point, respiration becomes shallow and irregular with rapid and weak pulse, and this is accompanied by convulsion, coma and collapse. The death occurs in most species when the temperature reaches 106oF to 108°F.

Hyperthermia for a longer period may be associated with high incidence of embryonic mortality in 3 to 6 weeks pregnant animals, and abortion. Breeding efficiency is adversely affected by prolonged heat stress and number of litters born to an ewe may be reduced. The conception rate reduces and there is increased incidence of anoestrus in adult animals.

Necropsy finding: There will be peripheral vasodilatation, blood clotting is slow, rigor mortis and putrefaction sets early.

Diagnosis: Examination of the environment gives an ideal impression for hyperthermia in animals. As rise in body temperature above the critical level is the primary manifestation in other systemic states, it has to be differentiated from fever, toxaemia and septicaemia. However, temperature in fever seldom exceeds 106oF. In toxaemia, fever is not a significant clinical picture. In septicaemia, petechial haemorrhage of mucous member and skin can be observed and blood culture for microorganism may come positive.

Treatment: Slow intravenous administration of fluid, cold application including cold pack, immersion and spraying may help in reducing the body temperature.

Adequate amount of glucose and protein may be given to compensate increased utilization and deficient intake. Animal should be provided with adequate drinking water and air movement inside the shed should be ensured. Shelter alone is a most important factor in maintaining comfort of livestock. If animals are to be confined to environment of high temperature and humidity, tranquilizing drugs such as chlorpromazine may be administered to reduce body activity.

Hypothermia

Hypothermia occurs when the body cannot maintain a normal temperature and it drops below the level that is required for normal metabolism. Hypothermia is caused by prolonged exposure to low temperatures and is considered as a medical emergency requiring immediate treatment. It results from excessive loss of heat or the body heat is insufficiently produced. Decreased muscular tone as seen in parturient paresis, acute ruminal impaction, during anaesthesia and sedation, profuse diarrhoea, peripheral vasodilatation in shock, reduction of metabolic activities in terminal stage of many diseases result in hypothermia. It may be a significant cause of post-shearing mortality in sheep due to loss of hair coat that gives protection against inclement weather, particularly in cold weather.

During a protracted period of hypothermia because of very low environmental temperature, the digestibility is reduced, and animal consumes more food to achieve a similar digestible energy intake. The thermoregulatory mechanism of neonates plays an important role, especially in piglets, lambs and calves, as they fail to maintain the body temperature in low environmental temperature, and neonatal mortality rate increases in an organized herd. Cold region pigs have lower blood glucose level and low glycogen content of the liver and muscles than those maintained in hot region. Cold region pigs huddle more and neglect their food.

Signs and symptoms of hypothermia vary on the severity and some of those include gradual loss of mental and physical abilities, shivering, cold extremities, listlessness, lethargy, drowsiness, decreased rectal temperature, slow and shallow breathing and reduced heart beat. Severely hypothermic patients may collapse, and goes into coma leading to death. The following protocol should be used to examine a hypothermic animal.

1. Attention to ABCD: A. Airway; B. Breathing; C. Circulation, D-Degrees. One should make sure that the animal has an open airway, is breathing, and has a heartbeat.
2. Brief history on circumstances such as duration of exposure to cool environment where animal is traced and the environmental temperature during exposure should be collected.
3. Brief physical examinations including a) feel of body temperature, and rectal temperature, b) level of consciousness and neurological examination.

c) cardio-pulmonary examination, d) associated trauma and e) weight of animal should be assessed. Depending upon the availability of staff and equipment, chest x-ray, urinalysis, complete blood examination, and arterial blood gas analysis are also recommended.

Factors responsible for increased susceptibility to hypothermia

- Poor body condition
- Very young or aged animals
- Calves with lower birth weight
- Ewes/ sow with higher litter number
- Young ones delivered prematurely
- Breed susceptibility: Body coat condition, length of the wool or hair
- Climatic conditions, particularly windy, low environmental temperature associated with rain fall
- Food intake, particularly milk intake during the early post natal life

Treatment: Early detection and treatment of new born can improve the condition and survival rate. Provision of warm and dry environment and injection of 20% glucose solution through intra-peritoneal route increases the body temperature through conservation of heat and heat production by elevation in body metabolism. When temperature comes to near normal the animal should be provided with warm food and young ones with mother's milk or colostrum. The primary objectives of the treatment and handling of a hypothermic animal are to keep the animal alive by warming, to avoid any further exposure to cold, and to transport the animal to a site of complete veterinary care.

The treatment of a hypothermic animal should be undertaken with observation of some thumb rules. The hypothermic animal may appear to be beyond help because of skin and membrane colours, dilatation of pupils and depressed vital signs, but a positive attitude works. The direct application of hot objects or excessive pressure should be avoided. It is to be ensured that items such as oxygen and fluids coming into contact with the animal are warmed. Severely hypothermic animal should not be kept in a shower or given bath. The use of drug is not beneficial in treating severe hypothermic animals since the cold heart will not respond as per expectation. The administered drugs are not metabolized normally by the liver and kidneys. Rather, these drugs accumulate in the body and become active as it warms. Lactated Ringers solution should not be used as the hypothermic liver may not be capable of metabolizing the lactate. The administration of cold fluid must be avoided.

Fever

Fever refers to an increase in internal body temperature to levels above critical body temperature and is most accurately characterized as a temporary elevation in the body's thermoregulatory set-point, usually above the normal body temperature. It is also known as pyrexia, and is derived from the Greek word *pyretos* meaning fire, or a febrile response, from the Latin word *febris*, meaning fever. This is a general systemic state of the body associated with feeling of cold despite an increased body temperature, and increases in heart rate, muscle tone and shivering, all of which are caused by the body's attempts to counteract the newly perceived hypothermia and to reach the new thermoregulatory set-point. Although hyperthermia, associated with increased body temperature, is not associated with elevated set point in thermoregulatory centre, unlike fever, but results from excessive heat production or insufficient thermoregulation, or both.

Fever is considered as body's immune mechanism to counter perceived threat inside the body, be it bacterial, viral or endogenous pyrogen. So, increased body temperature (hyperthermia) associated with toxaemia is termed as fever. Fevers are of two types (septic & aseptic). Septic fever, a more common type, is associated with infection with bacteria, virus, protozoa or fungi resulting from generalized or localized infection. On the contrary, aseptic fever is caused due to non-infectious reasons such as injection of foreign protein such as vaccine, some chemicals like dinitrophenol, break down of tissue as occurs in surgical procedures, tissue necrosis and immune reactions due to anaphylaxis.

Pathogenesis: Pyrogens are the substance responsible for inducing fever. These can be either internal (endogenous) or external (exogenous) to the body. The lipopolysaccharide (LPS) present in the cell wall of all Gm -ve bacteria is an example of an exogenous pyrogen. Pyrogenicity can vary, some bacterial pyrogens are known as superantigens that can cause rapid and dangerous fever.

Endogenous pyrogen includes cytokines, especially interleukin 1, which is a part of the innate immune system of the body and are produced by phagocytic cells. Other examples of endogenous pyrogens are interleukin 6 (IL-6), and tumour necrosis factor-alpha (TNF α). These cytokines are released into general circulation, from where they migrate to the brain and bind with endothelial receptors on vessel walls, or interact with local microglial cells. Interaction of cytokine with the microglial cells leads to activation of the arachidonic acid pathway and production of PGE2. Exogenous pyrogen like LPS binds with an immunological protein called lipopolysaccharide-binding protein (LBP). The LBP–LPS complex then binds to the CD14 receptor of a nearby macrophage. Such binding results in synthesis and release of various endogenous cytokines such as IL-1, IL-6 and TNF α. In other words, exogenous factors cause release of endogenous factors, which, in turn, activate the arachidonic acid pathway.

PGE2 release comes from the arachidonic acid pathway. This pathway is mediated by the enzymes phospholipase A2 (PLA2), cyclooxygenase-2 (COX- 2), and prostaglandin E2 synthase. These enzymes ultimately mediate the synthesis and release of PGE2, the ultimate mediator of the febrile response. The temperature set-point of the body will remain elevated until PGE2 is no longer present. The brain ultimately orchestrates heat effector mechanisms via the autonomic nervous system. They include increased heat production by increased muscle tone, shivering and hormones like epinephrine, and prevention of heat loss, such as vasoconstriction. Thus, fever can be differentiated from hyperthermia by the circumstances surrounding it and its response to anti- pyretic medications.

Clinical symptoms: Fever is associated with the combined effects of toxaemia and hyperthermia. There is increased body temperature, pulse rate, respiration rate, wasting, oliguria, increased thirst, anorexia, constipation, depression and muscle weakness. In farm animal practice, most common cause of fever is presence of an infectious disease such as peritonitis, pneumonia, mastitis, encephalitis, septicaemia etc. The fever may be classified as follows depending on nature of occurrence.

Simple fever: Temperature rises and remains high with variation of less than 1 to 2oF for several days.

Transient fever: Fever subsides within 24 to 48 hr after its onset as seen in bovine ephemeral fever.

Continuous fever: The temperature remains high for a longer period as in. Tick borne fever

Intermittent fever: It is a type or pattern of rise in body temperature in which there is an interval where temperature drops to normal.

Recurrent fever: It takes the form of relatively prolonged attack with non- febrile periods of about similar duration.

Atypical fever: The fever runs an irregular course

Biphasic fever: The temperature shoots to a new height and drops down to normal and again shoots to the new height. It is seen in diseases like Canine distemper, Louping ill, strangles, swine erysipelas.

Classification of Fever on the Basis of Etiology

Septic fever it is caused due to bacteria, virus, protozoa or fungi, localized infection, abscess, cellulitis, emphysema.

- Intermittently systemic as in bacteraemia
- Consistently systemic as in septicaemia

Aseptic Fever

- Chemical fever is caused by injection of foreign protein, intake of dinitrophenols.
- Surgical fever occurs due to breakdown of tissues and blood
- Fever from tissue necrosis - breakdown of muscle after injection of necrotizing material
- Severe haemolytic crisis
- Extensive infarction
- Extensive necrosis in rapidly growing neoplasm
- Immune reaction-anaphylaxis, angioneurotic edema
- Drug induced-prolonged use of antimicrobials may cause cryptic fever

Clinical Classification of Fever

1. Acute fever refers to sudden sharp elevation of body temperature as seen in acute viral or bacterial diseases, such as acute pneumonitis, acute peritonitis
2. Subacute fever - Temperature does not rise too high and remain below of 103oF. It is seen in mild inflammatory conditions like nephritis, cystitis, metritis etc.
3. Chronic fever- Temperature persists for a prolonged period such seen in pulmonary tuberculosis, chronic pyometritis, surra in horse etc.

Types of Fever

1. Simple fever: It is the type of fever where the temperature does not exceed 2°F in 24hrs. This type of fever is seen in lobar pneumonia, enteric fever etc.
2. Remittent fever: Temperature does not touch the normal limit in 24hrs and variation of temperature is more than 2oF in 24hrs. This type of fever is observed when diurnal variation is exaggerated as observed in septicaemia, broncho pneumonia, urinary tract infections.
3. Intermittent fever: It is a type of fever where the rise of temperature persists for a few hours of the day, as seen in human malaria, chronic surra & canine distemper. It is further classified as;
   a. Quotidian-daily rise and fall of temperature - *E. coli* infection, pyelitis
   b. Tertian - temperature appears and disappears every alternate day -malaria in human beings
   c. Quartan - temperature comes and goes every 4th day
   d. Hectic - sudden rise of temperature daily with chills or rigor, profuse sweating when temperature falls

4. Irregular fever (Atypical fever): Daily fluctuation of temperature does not follow a regular pattern - Louping ill, canine distemper, swine erysipelas, strangle in horses
5. Transient fever: Fever subsides within 24 hrs without any diurnal variation - Ephemeral fever
6. Continuous fever: Plateau of temperature remains for a longer period -Tick borne fever
7. Periodic fever: Periodic attack of fever alternating with a febrile and non-febrile period
8. Sthenic fever: High rise of temperature is accompanied with a strong bounding pulse as observed in acute inflammatory conditions, such as acute laminitis in horses
9. Asthenic fever: Moderate temperature accompanied with weak but quick pulse, fever in pulmonary tuberculosis
10. Pele stein fever: Special kind of fever associated with Hodgkin's lymphoma, being highest for 1 week and low for next week
11. Neutropenic fever is defined as a single oral temperature greater than or equal to 101°F (38.3°C) or a temperature greater than or equal to 100.4°F (38°C) for at least an hour, with an absolute neutrophilic count (ANC) of less than 1500 cells/microliter. The infectious etiology cannot be determined in most cases, and gets marked as a fever of unknown origin (FUO). In severe neutropenia, the absolute neutrophil count (ANC) is less than 500 per microliter. Febrile neutropenia is the most common and serious complication of cancer therapy. Febrile neutropenia occurs in the absence of normal neutrophils, commonly seen in immune-suppressing chemotherapy.

Stages of Fever

First stage (Initial stage, stadium increment)

- Body temperature rises as the result of spasms of blood vessels in the skin
- The phase of rapidly rising temperature is accompanied by chills due to stimulation of cutaneous nerve endings following fall of temperature caused by spasm of superficial blood vessels

Second stage (Hot stage, stadium fastigii)

- This is the period of maximum temperature which remains more or less constant
- There is cessation of shivering

Third stage (Stadium decrementii, stage of defervescence)

- This is the stage of decline of temperature.
- Heat loss is ensued as a result of profuse sweating due to dilatation of peripheral blood vessels
- Body regains its normal temperature.

Clinical signs of fever

First stage

- Dullness, drowsiness, disinclination to move
- Inappetance, errection of hair, arched backs rapid respiration
- Shivering
- Deep coloured urine

Second stage

- Profound dullness, drooping of ears
- Lowering of head, frequent respiration, rapid pulse
- Congested mucus membrane, hot breath
- Dry muzzle, anorexia
- Drop in milk production
- Increased thirst
- Constipation, scanty urine

Third stage

- Respiration less frequent
- Regain of appetite
- Muzzle is moist and urine turns normal

Clinicopathology

- There is no clinico pathologic changes which are specific to fever
- Hemogram reflects the changes associated with the cause of fever
- Inflammation is characterized by marked changes in total leukocyte count (TLC) in blood picture

Diagnosis

- Microbiological testing of blood samples
- Analysis of serous fluid from body cavity
- CSF analysis

- Milk sample analysis
- Reproductive tract secretion analysis
- Joint fluid analysis
- Biopsy
- Explorative laparotomy

Treatment

Antimicrobial: Antimicrobials are indicated to treat bacterial infections. The selection of antibiotics depends on cause, severity and accessibility of the drug to the lesion. In cases of fever of unknown origin broad spectrum antibiotics are used.

Antipyretics: If the fever is high enough to cause discomfort, the use of Non-Steroidal Anti-Inflammatory Drugs (NSAIDs) may be considered. Most NSAIDS like Phenylbutazone, flunixin meglumine, Acetyl Salicylic Acid, Sodium Salicylate are inhibitors of prostaglandin synthesis that lower the thermoregulatory set point. If required, fluid and electrolyte therapy should be given.

Toxaemia

It is a generalized clinical state characterized by circulation in the blood stream of the toxin produced by bacteria, parasites and/or body cells. However, it does not include the diseases caused by plants or insects or ingested organic or inorganic poisons. Theoretically, the diagnosis of toxaemia can be made only if the toxin is present in the blood stream, but in many cases, there is contributory evidence of antigenic toxin, or a probable source of toxin which in many cases, are virtually impossible to isolate or identify.

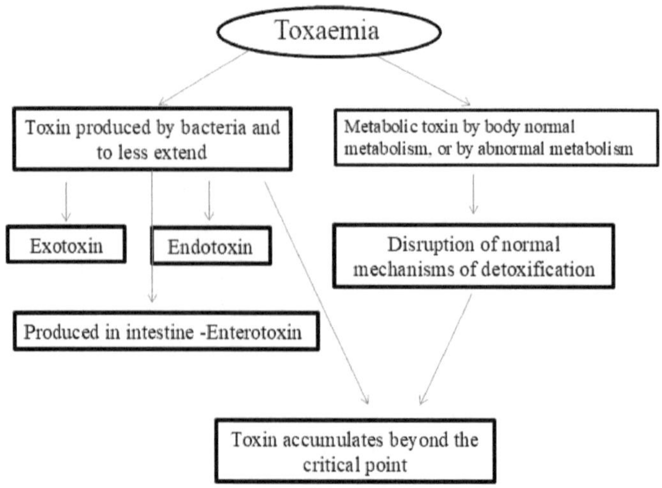

Exotoxin is protein in nature and is produced by living Gm +ve organisms. The toxin diffuses to the surrounding medium and has specific site of action. Example: tetanus toxin acts at motor end plate. The exotoxin produced in the intestine is called enterotoxin. Example: Toxin produced by *Clostridium perfringens* in the gastrointestinal tract. Endotoxins are lipopolysaccharide in nature, found within the cell as a part of cell wall of Gm -ve organism, and are liberated into the media or surrounding upon destruction of the cell wall. *E. coli, Salmonella* sp *and Pasteurella species* are the important pathogenic organisms in livestock that produce endotoxins. Endotoxemia is also one of the important causes of death in horses affected with gastrointestinal disorders in which there is physical obstruction causing strangulation and ischaemia. Endotoxins are not ordinarily absorbed through the intestinal tract unless it is injured as in case of enteritis or more commonly acute intestinal obstruction. Generally, a small amount of endotoxin that is absorbed from the intestinal tract into the circulation is detoxified in the liver. A state of endotoxemia is produced only when the endotoxin are produced in larger quantity and are being absorbed or the liver fails to detoxify it as in case of hepatic insufficiency. Endotoxins may also be produced in large amount in case of server mastitis, peritoneal abscess or from other septic foci or large areas of injured traumatised infracted tissue. Best known endotoxins are those of *E. coli* which are used extensively as a model for experimental endotoxemia.

In case of obstruction of lower alimentary tract, there may be increased absorption of toxic substance such as phenols, cresols, and amines which are normally excreted in the faeces, resulting in development of syndrome called autointoxication. Production of toxins by abnormal metabolism enhances production of histamine and histamine-like substances in damaged tissue. Ketonemia, due to an improper fat metabolism and lactic acidemia due to acute ruminal impaction are common examples of toxins produced by abnormal metabolism.

Pathogenesis: Toxaemia affects metabolism of carbohydrate, protein and minerals, and also damages vital organs including liver and kidney parenchyma. Fall in blood sugar level, sever drop in liver glycogen, increased level of blood pyruvate and lactate are some of the outcome of toxin effects on carbohydrate metabolism. Toxins also produce effects on protein metabolism, and include increased tissue break down, rise in blood non-protein nitrogen level, increased total serum protein, and enhanced serum protein as a result of increase in antibody production and alternation in relative proportion of animo acids in blood. Hypo ferremia, hypozincaemia, hypocupraemia, due to increased blood ceruloplasmin, have been recorded in patients. Weakness of myocardium with reduced response to cardiac stimuli, damage to renal tubules and glomeruli leading to albuminuria, loss of muscular tonicity, depression of haematopoiesis with increased leukocytes count, hypersensitivity reactions are seen in toxaemia.

Clinical Findings: The clinical symptoms vary with the species and severity of toxic process. In acute toxaemia, depression, anorexia, and muscular weakness are common. Lethargy, separation from group, inappetance, failure to grow or produce, weak, rapid but irregular pulse are the characteristic findings. Heart rate is increased, but the sounds are reduced. There may be albuminuria. Fever may be present if toxaemia is associated with infections. The rise in temperature may normal in early stage, but later the temperature may become subnormal. Terminally, there is muscular weakness, convulsion, coma, and death occurs in coma or during convulsion.

Clinical Pathology: It may be possible to isolate and identify the specific toxin and determine the source. Low blood glucose level, increased level of urea, total serum protein, decreased A/G ratio, albuminuria, and aplastic anaemia are the common clinic- pathological findings. Leukocyte count shows leukopenia, neutropenia and lymphopenia in severe cases of endotoxemia. However, leucocytosis along with neutrophilia is recorded in mild cases of endotoxemia.

Necropsy finding: Microscopically, liver and kidney parenchyma shows degenerative changes. Glomeruli and tubules of the kidney, myocardium and adrenal gland may show degenerative necrosis.

Treatment: Identification of toxin and administration of specific antitoxin reduces the severity of the disease. But in many cases, it is impossible to pinpoint the toxin. Intensive fluid and electrolyte therapy should be given through intravenous route to improve the excretion of the toxin and to prevent shock due to fall in peripheral circulatory volume. Parental nutrition is indicated in anorexic patients till the animal eats and drinks normally. Glucose, amino acids and fat emulsion may be added to fluid therapy for parental nutrition. Glucocorticoids are given at the dose rate of 1 mg/kg body weight intravenously at every 24 hours. Non-steroidal anti-inflammatory drugs such as acetylsalicylic acid, phenyl butazone (15 mg/kg body weight initially and 10 mg/kg body wt at 6-12 hrs internal) are administered to reduce inflammation and pain. In acute infection, the administration of antimicrobial drugs will usually control the infection and the animal will recover.

Septicaemia

Septicaemia is a general systemic state of the body where toxaemia is compounded with hyperthermia and presence of large numbers of infectious microorganisms including viruses, bacteria or protozoa in the blood stream. The presence of bacteria in the blood stream for only a transitory period unaccompanied with clinical signs is called bacteraemia. In septicaemia the causative agent is present throughout the course of the disease and is directly responsible for the appeared clinical signs.

Causes of Septicaemia in all Species

Anthrax, Pasteurellosis, Salmonellosis, Pseudomonas, Rift valley fever, Leptosporidia, Sarcosporidia.

Cattle, Sheep, Pig

Pasteurella multocida, Pasteurella haemolytica, Pasteurella (yersinia) *pseudotuberculosis*

Sheep

Histophilus ovis, Haemophilus agni

Horse

Pasteurella haemolytica

Pig

Hog cholera, African swine fever, *Streptococcus zeoepidemicus*, *Erysipelothrix insidiosa*

The exotoxins or endotoxins produced by infectious agents produce a profound toxaemia and high fever: This is due to the rapidity by which they multiply and the rapid spread to all body tissues. The microorganism also localizes in many organs and may cause direct endothelial damage and haemorrhages into tissues. In viremia, the same general principles also present the except that toxins are not produced by virus. The general clinical signs are caused by the products of tissue cells killed by the multiplying virus. Septicaemia may also result from loss of leukocyte production because of injury to bone marrow as seen in radiation injury. Immuno-incompetence due to congenital defects in the immune system, and immunosuppression resulting from treatment with corticosteroid in adult animals leads to septicaemia due to poor body resistance. Disseminated intravascular coagulation occurs in fatal septicaemic diseases. Circulation of foreign materials such as bacterial cell walls, antigen-antibody complexes and endotoxins bring about vascular injury with partial disruption of the endothelial lining, subsequent platelets adherence, and formation of platelet thrombi. The symptoms produced in septicaemia are those of toxaemia and hyperthermia. Submucosal and subepidermal petechial or ecchymotic haemorrhages are seen under the conjunctiva, mucosa of mouth and vulva.

Treatment: The animal should be treated with antibacterial drugs, sera and antitoxin as soon as possible. The body temperature should be reduced with the use of antipyretic drugs. If the animal is anorectic, parenteral nutrition reduces starvation. Hygienic precautions and prophylactic measures are to be taken to prevent the spread of the disease.

Shock

Shock is a condition that results from a depressed state of many vital body functions caused by a lack of effective circulation, decreased arterial blood pressure, inadequate supply of oxygen to the tissues, or an inability of the tissues to properly use oxygen. The state is manifested by clinical sings such as acute depression, coldness, a weak rapid pulse, fast breathing. Regardless of cause, shock is life-threatening and requires immediate attention and treatment. Decreased blood volume as occurs after internal or external haemorrhage, dehydration, burn and severe vomition and diarrhoea, reduced activity of heart, peripheral vasodilatation of veins – reduction in effective circulatory blood volume, so there is insufficient blood to fill heart during distole

Classification of Shock

Hypovolemic Shock: Reduced blood volume due to dehydration or haemorrhage, and fall in blood pressure leads to hypovolemic shock. It occurs when the volume of circulatory fluid is significantly reduced, where there is severe haemorrhages with acute loss of 35% or more of total blood volume, and in dehydration from severe diarrhoea, intestinal obstruction, especially if fluid loss is severe and over a short period.

Cardiac output is the amount of blood that the heart can pump out in one minute. It is calculated as the stroke volume (how much blood each heart beat can push out) multiplied by the heart rate (how fast the heart beats each minute). If there is less blood in the system to be pumped, the heart speeds up to keep its output steady, resulting tachycardia. So, increased heart rate is a compensatory mechanism to prevent hypovolemic shock.

Vomiting and diarrhoea due to gastroenteritis and excessive sweating due to heat exhaustion and heat stroke, and infectious disease condition, diabetic ketoacidosis lead to significant loss of water, and create a situation of hypovolemic shock. In hypovolemic shock, the animal fails to replenish the amount of lost fluid by taking in enough water, and the body is unable to maintain blood pressure and cardiac output. In all shock states, when cells start to malfunction waste products build up, a downward spiral of cell death begins, increased acidosis occurs, and a worsening body environment leads to further cell death - and ultimately organ failure.

Hemorrhagic Shock: It is a subset of hypovolemic shock and occurs due to significant bleeding that occurs relatively quickly. So, the total blood volume is reduced. Trauma is the most common example of bleeding. Blood loss has two effects such as loss of volume within blood vessels to be pumped and a reduced oxygen carrying capacity because of the loss of red blood cells.

Maldistributive shock: During severe trauma or burn injury following major surgery, after prolapse of uterus, rapid withdrawal of ascitic fluid and severe colic pain in horses, the distribution of the fluid is disorganised leading to shock.

Cardiogenic Shock: This occurs as a result of an acute reduction of cardiac output due to myocardial damage or myocardial asthenia with loss of muscular power which causes failure of tissue perfusion.

The loss of ability of heart to pump blood to the rest of the body results in fall in blood pressure, and cardiogenic shock is produced although there is enough red blood cells and oxygen and adequate blood volume. The myocardium itself requires blood supply for its functioning. The resultant decrease in stroke volume leads to fall in cardiac output.

Neurogenic shock: The involuntary muscles within blood vessel walls maintain the squeeze so that the volume within the vessels stays constant even if the body position changes. The squeeze is maintained by signals from nerves in the sympathetic trunk, a long bundle of fibres running from the skull to the tail alongside the vertebral column. The sympathetic trunk stops working in brain and spinal cord injury and blood vessels get dilated resulting in blood pooling away from the heart. Since there isn't enough blood returning to the heart, the heart has a hard time pumping blood into the body. This situation looks like hypovolemic shock, but there is enough blood but effective blood volume is reduced.

Hypoglycaemic Shock and Hyperglycaemia: The brain is very susceptible to low blood sugars. High or low blood sugars are almost associated with diabetes in pet animals. When blood sugar levels spiral high out of control, there is risk of significant dehydration due to glycosuria. If there is not enough insulin in the blood stream, cells cannot use the glucose and instead turn to an alternative anaerobic metabolism to generate energy. Since glucose can't enter cells to be used, hyperglycaemia occurs as the glucose level build in the blood stream. The kidneys try to excrete excess sugar, but because of chemical concentration gradients between blood and urine, significant amounts of water are lost as well. The body quickly becomes dehydrated and blood pressure drops, decreasing blood flow to cells. Cells which are now lacking glucose inside them are now starved of oxygen and turn to anaerobic metabolism, causing acid waste product to build up. Excess acid in the body changes the metabolism for all organs, making it more difficult for oxygen to be used. This downward spiral will continue until insulin and significant fluids are given.

Anaphylactic shock: Shock may be caused by presence of bacteria or their toxin in the blood stream. Severe allergic reaction (anaphylactic shock), overdoses with drug such as certain anaesthetic or acute stress may lead to shock. In some cases, a combination of above causes may be involved. Anaphylactic shock is an extreme generalized allergic reaction in which widespread histamine release causes bronchial constriction leading to respiratory distress, dilation of veins, circulatory collapse and possibly death.

Toxic/Septic shock: It can occur when endotoxin and microorganisms get access to the circulation and act on vascular system either directly or via mediation released by their interaction with host cells. Endotoxin is obviously important in the development of shock in G-ve bacterial sepsis.

Clinical signs of Shock

Early Signs of Shock

The animal may be either excited or subdued. Rapid heart rate

Pulse not difficult to find Gums may be normal or pale

Late Signs of Shock

Gums extremely pale or show a bluish discoloration.

Heart rate is probably elevated and irregular, but may be normal or below normal as heart muscle begins to fail.

The pulse will be weak and either difficult or impossible to locate. Coldness to touch and rectal temperature is below normal. Respiration may be slow or rapid, shallow or deep.

The eyes may take on a glazed appearance and appear not to focus normally. Mental condition deteriorates from depression to stupor to coma.

Diagnosis

1. Clinical signs
2. Estimation of arterial blood pressure is one of the best methods of measuring the severity of shock. In severe shock, the systolic blood pressure falls from normal of approx .120 mm Hg to bellow 60 mm Hg which indicates a fatal outcome.
3. Shock should be anticipated in septicaemic diseases especially in neonates in acute localized infection with intestinal diseases, severe trauma, severe fluid loss of any reason and in major surgery.

Treatment: If signs of shock are recognized, or a serious injury has taken place, supportive care, such as intravenous fluids, oxygen and other measures can help reverse shock and prevent permanent organ damage. The key to successful treatment is prompt professional care. The treatment of hemorrhagic shock depends on finding and controlling the source of bleeding. Intravenous fluids are used to help with resuscitation to increase the fluid volume within the blood vessel space, but blood transfusion is not always mandatory. If the bleeding is controlled and the patient becomes more stable, the bone marrow replenishes the lost red blood cells. In cardiogenic shock, treatment is given to restore blood supply and to support blood pressure. Intravenous fluids and medications are given in neurogenic shock to increase the tone in the blood vessel walls. Effective treatment and management of septic and toxic shock is less successful unless the sepsis can be controlled and the source of toxaemia eliminated. Appropriate antibiotic therapy should be given parenterally along with immunotherapy and corticosteroids.

Water Intoxication

When animals are very thirsty particularly after exercise, the ingestion of excessive quantity of water may result in water intoxication under field condition. Calves are most commonly affected, but the disease is also recorded in sheep. The of main clinical signs are 1. haemoglobinuria 2. hypothermia 3. salivation. There is cerebral oedema in post mortem findings. Some nervous signs may also be observed like muscles weakness, tremor, restlessness, ataxia, tonic and clonic conclusion and coma. Water intoxication does not occur commonly and can be avoided by allowing thirsty animal to have only limited access to water. Syndrome will appear to be related to salt poisoning and haemolytic anaemia that can occur in calves drinking cold water.

Treatment: Sedatives and diuretics are used. In severe cases intravenous injection of hypertonic solution is indicated to increase water loss.

Edema: It is the excessive accumulation of fluid in tissue spaces, caused by a disturbance in the mechanism of fluid interchange between capillaries, tissue spaces and lymphatic vessel.

Etiology: It results mainly from an increase in hydrostatic pressure in the capillaries, a fall in osmotic pressure of the blood, obstruction to lymphatic drainage or from damage to capillary wall. Increased hydrostatic pressure is mainly noticed in congestive heart failure and locally in portal hypotension due to hepatic fibrosis causing.

Clinical Findings: Anasarca (swelling of superficial tissue) or ascites (swelling of peritoneal cavity) or hydropericardium (swelling of pericardial sac) are the common clinical signs. Anasarca in large animals usually confined to the ventral wall of abdomen and thorax, brisket region edema. In the grazing animal, it is usually noticed in intermandibular space. Intermandibular edema may be less evident in animals which do not have to lower their head for grazing. Oedematous swelling are soft, painless and pits on pressure. In ascites, there is distension of abdomen and the fluid can be detected by a fluid thrill on tactile percussion.

Diagnosis

- Clinical signs
- Severe hypoalbuminemia.

- Marked proteinuria
- Liver function test may be indicated, when liver disease is the suspected cause of edema.

Treatment

- Congestive heart failure may be treated with digoxin
- Pericarditis –by drainage of sac and feeding of high-quality protein
- Salt restriction in the diet
- Use of diuretics and aspiration of fluid. Fluid aspiration must be carried out very slowly to avoid peripheral circulatory failure.

Pain

Pain is a state of consciousness, a feeling sensation and a mental event, and as such, it can never be observed. Behaviour like writhing, screaming, or drawing the affected part away from pain inducing agent are observed during pain. Thus, pain in other individual or animal cannot be directly experienced by anyone else including the clinician. Nearly all the external signs that lead to infer pain in other humans are seen in animal and birds. The behavioural signs include writhing, facial contortions, moaning, yelping or other forms of calling, attempts to avoid the source of the pain, appearance of fear at the prospect of its repetition, and so on. The animals have nervous systems similar to human beings and respond physiologically like human beings do when the animal is in circumstances in which we would feel pain. The impulses, emotions, and feelings nuclei are located in the diencephalon, which is well developed in many other species of animals, especially mammals and birds.

Pain is one of the more difficult manifestations of the disease to conceptualize because it is necessary to describe, quantify and characterize outward physical responses that have a sensitive, as well as physiologic basis. It is generally stated that free nerve endings of small-diameter serve the function of perception and production of pain. These nerve endings have been identified or found in the superficial layers of the skin, joint surfaces, periosteum, muscle, tooth pulp, and viscera. Stimulation of these receptors results in transmission of pain impulses to the spinal cord, hypothalamus, brainstem and cerebral cortex. The signals are processed in the manner that resembles the processing of information from other types of receptors. When the stimulus is of sufficient intensity to exceed the pain detection threshold, these impulses is centrally processed which permits characterization of the intensity and location of the stimulus. The end result of pain perception is elicitation of multiple motor responses generally indicative of arousal and avoidance.

Classification of Pain

Superficial pain refers to painful responses that are initiated by stimuli applied to the skin or subcutaneous tissues.

Deep pain originates from muscles, joints, bones and connective tissues.

Visceral pain originates from stimulation of nerve endings in the viscera due to localized damage to the visceral organ.

Some common causes of abdominal pain include traumatic reticulo peritonitis (TRP), tympany, abomasal ulcer, gastric ulcers, urolithiasis, spasms, abomasal displacement, peritonitis, abomasal impaction, cystitis, ruminitis, cholelithiasis in ruminants and intestinal volvulus, intestinal torsion, obstruction, intussusceptions, indigestion, colitis, enteritis, ileal impaction, pyelonephritis, meconium impaction, peritonitis. Pleuropneumonia, lung abscess, pleurisy, traumatic reticulo pericarditis pneumonia, choke, fractured ribs, pleuritis, ruptured oesophagus are associated with chest pain.

Management of Pain: Proper management of pain and mitigation of suffering speaks about the animal welfare status. Animal in pain should be given immediate attention and should be administered with pain killers. Non-steroidal anti-inflammatory drugs (NSAIDs) primarily are anti-inflammatory due to inhibition of prostaglandin production. So, while prostaglandin production rapidly diminishes, any previously present prostaglandin must be removed to subside inflammation. NSAIDs such as phenylbutazone, flunixin, meloxicam, and carprofen have delayed peak concentrations at the site of inflammation and persist in inflammatory exudates for long periods after plasma concentrations are negligible. This explains the delayed onset and prolonged duration of anti- inflammatory action that does not correlate with plasma pharmacokinetics. In addition, cyclooxygenase inhibition does not explain all of the anti- inflammatory activities of NSAIDs. Some anti-inflammatory action appears to be related to their ability to enter into the lipid bilayer of cell and disrupt normal signals and protein-protein interactions in cell membranes. NSAIDs are more lipophilic at a low pH. In the cell membrane of neutrophils, NSAIDs inhibit neutrophil aggregation, decrease enzyme release and superoxide generation, and inhibit lipoxygenase.

NSAIDs act as analgesics by inhibiting COX and preventing the production of prostaglandins that sensitize the afferent nociceptors at peripheral sites of inflammation. However, there is increasing evidence that some NSAIDs have a central mechanism of action for analgesia and act synergistically with opioids. NSAIDs are likely to be more effective as analgesics when inflammation is a part of the pain process and are more effective as analgesics when given prior to the onset of the inflammatory processes. The time of onset and duration of analgesic properties of NSAIDs does not correlate well with their anti- inflammatory properties—the analgesic effect is likely to have a more rapid onset and shorter duration of action. Therefore, dosage regimens for effective analgesia may need to be different than for anti-inflammatory properties.

Ketoprofen: Ketoprofen is a propionic acid derivative available as an injectable formulation and tablets. Typically, the plasma elimination t½ is short, approximately one hour in horses, 1.6 hr in cats, and five hr in dogs. Cats and dogs can be given a dose of 2 mg/kg by injection (IV, IM or SC) the first treatment, followed by 1 mg/kg PO (tablets) q24h. The label dose is for five days of therapy, but 0.25 mg/kg has been recommended for chronic therapy. Adverse effects of ketoprofen are rare. There are reports of haemostasis problems and acute renal failure in animals following anaesthesia and surgery when ketoprofen was administered in the perioperative period.

Carprofen: Carprofen, like ketoprofen, is a racemic mixture of "S" and "R" enantiomers. Most of the COX inhibition anti-inflammatory action is attributed to the "S" enantiomer and analgesia is attributed to the "R" enantiomer. Carprofen has greater activity against COX-2 than COX-1, but its overall cyclooxygenase inhibition is weak, so the anti-inflammatory and analgesic activity may be due to central the effects. Carprofen may cause an idiosyncratic hepatopathy in dogs.

Meloxicam: Meloxicam is available as an injectable solution and oral syrup for chronic therapy of osteoarthritis in livestock and pets. Meloxicam has high activity against COX-2. It has no effect on platelet aggregation or renal prostaglandin synthesis, showing a sparing of COX-1. The elimination t½ in the dog is 12–36 hr. Meloxicam is highly efficacious and well tolerated, with few reports of adverse effects. The oral dose in dogs is a one-time loading dose of 0.2 mg/kg followed by 0.1 mg/kg PO q24h. Once a therapeutic effect has been attained, the dose can be titrated to the patient for the lowest possible daily dose. In recent studies injectable, meloxicam given pre-operatively was more efficacious than butorphanol for pain control and did not cause adverse renal or haematological effects.

8

Allergies, Anaphylaxis, Oedema

Allergy is a condition wherein the immune system reacts abnormally to a foreign substance in the environment that is generally harmless to others. These substances are known as allergens that are found in dust mites, pets, pollen, insects, ticks, moulds, foods, and drugs (medications). Atopy is the genetic tendency to develop allergic diseases.

The common symptoms of allergy are a runny nose or sneezing, pain or tenderness around your cheeks, eyes or forehead, coughing, wheezing or breathlessness, itchy skin or a raised rash termed hives, diarrhoea, feeling or being sick, swollen eyes, lips, mouth or throat and fever. In many cases, the allergic reaction is mild. Severe cases can lead to shock and may be life-threatening.

A drug allergy is an adverse reaction to a drug or medication that occurs when the body's immune system produces a substance against the drug. Symptoms can include red and warm skin, blotchy rash, hives, itching, swollen mucous membranes, fluid retention. In case of Symptoms appear within an hour of taking the medication, but can also occur hours, days, or even weeks later. Immediate allergies occur within one to six hours of taking the medication. Non-immediate allergies take place more than 24 hours after taking the medication.

Etiology

Anaphylaxis is triggered by a severe allergic reaction, when the immune system reads a food or substance as something harmful by mistake. In response, the immune system releases chemicals to fight against it. These chemicals are responsible for symptoms of an allergic reaction. Allergy symptoms usually aren't life-threatening, but a severe reaction can lead to anaphylaxis. The most common triggers of anaphylaxis are food allergens like peanuts, milk, fish, and shellfish. In adults, stings from insects, latex, and some medications can cause anaphylaxis. Certain medications, including antibiotics, aspirin and other pain relievers and the intravenous contrast used in some imaging tests, stings from bees, yellow jackets, wasps, hornets and fire ants and latex can act as allergens. However, the cause of anaphylaxis is not identified in many cases, and those are termed as idiopathic anaphylaxis.

The immune system produces antibodies that defend against foreign substances. This is favourable as long as the foreign substance is harmful, such as certain bacteria or viruses. But some people's immune systems overreact to substances that don't normally cause an allergic reaction.

The risk of anaphylaxis is more if the animal had this reaction before or it has allergies or respiratory distress. Conditions like heart disease or a build-up of white blood cells can also increase the risk. There are not many known risk factors for anaphylaxis, but some reasons that might increase the risk include:

- Previous anaphylaxis-The risk of having a serious reaction increases with past experience. Future reactions might be more severe than the first reaction.
- Allergies - Animals who have either condition are at increased risk of having anaphylaxis.
- Certain other conditions. These include heart disease and an irregular accumulation of a certain type of white blood cell (Masto cytosis).

Clinical Signs

Anaphylaxis symptoms usually occur within minutes of exposure to an allergen. However, anaphylaxis may occur a half-hour or longer after exposure. In rare cases, anaphylaxis may be delayed for hours. In anaphylaxis, the immune system releases a flood of chemicals that can cause the body to go into shock. Blood pressure drops suddenly, and the airways get narrowed, blocking free breathing. The pulse becomes fast and weak, and there may be skin rash, nauseous and vomiting. Anaphylaxis must be treated right away with an injection of epinephrine to preclude deadly outcomes. Common triggers include certain foods, some medications, insect venom and latex. Symptoms of anaphylaxis also include hives and itchy, pale, or flushed skin. Symptoms usually happen minutes after the exposure to allergen, but they might not appear for a half hour or longer. Symptoms may recur, even without more exposure to the allergen. This second reaction is called biphasic anaphylaxis.

Treatment

The diagnosis and long-term management of anaphylaxis are complicated. Referral treatment is mostly required from the specialists. Anaphylaxis is the most severe, life-threatening allergic reaction that requires immediate treatment with adrenaline or epinephrine. It may happen seconds or minutes after the exposure to the allergen like peanuts or bee stings. Treatment for drug allergies depends on the type of symptoms and can include use of antihistamines for mild symptoms like rash, hives, and itching, bronchodilators for asthma-like symptoms like wheezing or cough, and corticosteroids for skin, mouth, or intravenous use, and epinephrine injection for anaphylaxis.

Prevention: The best way to prevent anaphylaxis is to stay away from substances that cause this severe reaction.

Edema, also spelled oedema, is otherwise termed as dropsy, hydropsy and fluid retention. It causes a swelling due to the build-up of fluid in the tissue. The legs or arms are most commonly affected. Symptoms may include tightness of the skin, heaviness of the affected area, and joint stiffness. Oedema is caused by medicines, pregnancy and in diseases, such as congestive heart failure, kidney disease, venous insufficiency or cirrhosis of the liver.

The blood travels throughout the body to distribute vital nutrients, adequate oxygen, and fluid, and to maintain tissues levels of nutrients. The hydrostatic pressure and osmotic pressure of the blood itself control fluid movement across blood vessels. Osmosis is the diffusion of water across a semipermeable membrane to balance different concentrations of solutes on two sides of the membrane. The osmotic pressure prevents water from moving across the capillary membrane, whereas hydrostatic pressure pushes water across the membrane. Proteins in blood plasma draw water from interstitial spaces due to build up osmotic pressure. Capillaries bring in extravascular fluid to maintain interstitial fluid levels during proper functioning. The fluid builds up in the interstitial tissues due to pressure imbalance as capillaries fail to drain these tissues, resulting in oedema. The obstruction of lymphatic drainage also leads to the same result, but such fluid accumulation in tissues is more localized. Another important variable in fluid movement is capillary permeability. The increased blood flow increases hydrostatic pressure that raises capillary permeability, causing passage of fluid into the interstitial space resulting in edema. More fluid is drawn to the tissue space and retained in the interstitial space with enhanced blood protein in the tissues. The surrounding tissues become inflamed with the fluid accumulation, which obstructs blood and lymph vessels, disrupting fluid movement into and out of the vessels.

Udder Edema

It is a periparturient disease, characterized by excessive accumulation of fluid in the intracellular space of mammary gland, swelling of udder, teats and sometimes of lower abdomen.

The etiology of udder edema is unclear, and may be associated with reduced mammary blood flow and increased intravenous pressure. Excess prepartum intake of energy, sodium or potassium also increases the likelihood of edema. Excessive intake of potassium aggravates the edema via the renin-angiotensin system by increasing potassium ion excretion accompanied by retention of sodium ion and water.

Physiological udder edema is a non-infectious metabolic disorder noticed in a high percentage of dairy cows. It results from accumulation of lymphatic fluid in the interstitial space of the mammary gland and surrounding tissues. It has been recorded in 48% of Holstein dairy cows both prepartum and postpartum, whereas 30% of the dairy cows had never experienced udder edema. Parity is negatively associated with prevalence of edema. The increasing foetal pressure in the pelvic area during pregnancy causes impaired blood and lymphatic fluid circulation, resulting in fluid buildup in the udder tissues. This fluid accumulation leads to deterioration of the support structures in the udder, quarters and teats, increased risk of udder cleft dermatitis and mastitis. Tenderness of the udder and malformed teats offset the milk let down, and negatively affect production performance throughout an animal's lifetime.

The physiologic udder edema has been associated with excessive salt in the diet, greater age at first calving, above-ideal body condition score (BCS), genetic traits, physiological changes during maturation and udder development, oxidative stress, increased stress from pen movements when taking out or bringing in new animals to the group, overcrowding, and heat stress. Calving in winter increases the chances of udder edema by 3.68 times more than in summer. The sites of fluid accumulation are brisket, udder, navel, and, in extreme cases, the legs and vulva. Edema becomes less severe and the prevalence decreases as the parity increases. Dairy cows with swollen udders from mastitis exhibit behaviours such as spending less total time lying, less time lying on the affected quarter or quarters, increased standing time, and increased stepping behaviour. Cows with udder edema exhibits greater kicking behaviour during milking. Physiological changes such as tenderness of the teats and swelling of udder tissues usually present as painful. Pain has been associated with mastitic udders.

Dairy cows with udder edema may have tender teats. Milking becomes an unpleasant experience for both the milkers and cow, further resulting in potential injury, infection, or early culling, greatly affecting the profitability and wellbeing of that animal. The following factors contribute to the development of udder edema.

Physiological Changes and Mammary Development

The transition of pregnant heifer to the milking string brings about substantial physiological changes. Already in place mammary system develops significantly during the last stages of gestation. The internal anatomy begins to change. The physiological demand of blood for mammary gland development is the precursor to udder edema in primiparous heifers. Dairy cows already on milk have prominent subcutaneous abdominal vein, also known as the milk vein, whereas, in heifers, this vein is not present. Young cattle have a cranial epigastric vein, which flows anterior to the mammary gland, and a caudal epigastric vein, which flows posterior of the mammary gland. In the latter stages of gestation, a heifer's cranial epigastric and caudal epigastric veins merge through anastomosis, creating the milk vein, clearly demonstrating immense vasculature of the mammary gland. The creation of the subcutaneous abdominal vein allows for an increase in the amount of blood flow from the mammary gland. Two types of oedemas occur; generalized and local. Generalized edema can be seen throughout the body. It is most prominent in the ventral areas, such as the barrel, limbs, and udder. Localized edema occurs because of venous stasis, defined as decreased blood flow in the veins due to an increase in capillary pressure, or because of impaired lymphatic drainage. The blood flows in reverse direction due to the increase in amount of blood needed for mammary gland development. It is precisely the opposite of the blood flow in young heifers. The veins respond by increasing in size with the increase in blood flow. Histamines, released by cells to respond to inflammatory reactions, results in its higher concentration in colostrum and plasma in dairy cows with udder edema. Higher-than-normal histamine concentration in the udder tissue can lead to self-destructive behaviour of animals licking the udder tissue and teats until raw, creating significant lesions that can lead to dermatosis. The periparturient udder edema has been associated with increased risk of clinical mastitis.

The physiological changes such as increased blood flow, decreased fluid movement between the tissues and capillaries, and increased histamine levels present a challenge for dairy farmers. The best approach is prevention before udder edema becomes severe. Due to the natural physiological, metabolic, and vascular functions throughout the last stage of gestation, It is required to manage nutrition, genetic, and calving-in factors such as body condition and age at calving to provide the most optimal scenario to minimize udder edema. Massaging the udder during milking and the use of diuretics reduces edema.

The high-yielding cows have increased genetic potential to develop udder edema, due to the substantial genetic correlation found in their study. Cows with high production pedigrees had an 11% higher incidence of edema compared with those of low production pedigrees. The high production pedigree animals show greater edema above the rear udder. Calf birthweight is inversely proportional to incidence of udder edema. However, seasonal fluctuations in ambient temperature had no relation to edema. However, there is a direct association between age at calving and increased occurrence of udder edema.

Congestion of udders was phenotypically related to deep, low-hanging udders that lacked cleavage. Cows with deep udders were more susceptible to mastitis. Both edema and calving ease had significant and positive estimates of genetic correlation with stature and strength, which is how well the body shape supports production and longevity.

The endocrine system of cows prepare the body for parturition and lactogenesis during the transition period, when dairy cows are provided with periparturient nutrition. The growth hormone increases and plasma insulin decreases as a dairy cow shifts from late gestation to early lactation. These alterations in the endocrine system coupled with decreased dry matter intake (DMI) affect the metabolism leading to mobilization of fat from adipose tissue and glycogen from the liver. Low DMI during the last week of gestation causes impairment of the liver, resulting in decreased lipid and lipoprotein concentrations, resulting in occurrence of udder edema, as the serum concentrations of total proteins, triglycerides, cholesterol, and lipoproteins were lower in cows with udder edema.

The severity of an udder edema can be reduced through salt restriction. Different forms of anionic salts include sodium chloride, magnesium sulfate, calcium sulfate, ammonium sulfate, calcium chloride, ammonium chloride, and magnesium chloride. Supplementing with anionic salts into the diet is the only way to achieve a negative dietary cation and anion difference (DCAD). Potassium chloride, which is used as a replacement for sodium chloride, results in the same severity of udder edema as animals supplemented with sodium chloride. Animals supplemented with these salts also developed edema sooner, and the edema took more time to clear from the udders. Feeding diets containing $CaCl_2$ to cows during early lactation may extend the beneficial effects of $CaCl_2$ in moderating edema into the early-lactation interval, due to feeding $CaCl_2$ decreasing severity of edema prepartum.

Oxidative Stress

Oxidative stress occurs when oxygen is not reduced to water during metabolic processes. Partially reduced oxygen becomes a free-radical superoxide (O^2). Increased metabolic rate, rapid growth, high milk production, or extreme aflatoxin exposure could elevate superoxide creation. Superoxide is reduced to hydrogen peroxide (H_2O_2). Both of these occur naturally during the metabolic process and are not harmful when broken down properly. Damage from these is suspected to be the result of converting into more free radicals and the catalytic transition elements, such as iron, which have harmful effects. Release of catalytic Fe becomes more likely under conditions of dietary imbalance, trauma, or stress, often accompanying calving. The body has a normal antioxidant capacity, responsible for controlling oxygen radicals throughout the body. Enzymes such as catalase, peroxidase, and superoxide dismutase scavenge the free-radical. Nonenzymatic antioxidants (tocopherols, ascorbic acid, and glutathione) also occur within the body. Exogenous antioxidant molecules, which enter the body through the diet, include vitamin E, vitamin C, carotenoids, and flavonoids. When in oxidative stress, reactive oxygen species surpass the defence capability of the antioxidants. Free radicals without the control by antioxidants and enzymes, inflict damage on DNA, proteins, and lipids involved in fundamental metabolic processes These damaged cellular components result in cytotoxicity, genotoxicity, and carcinogenesis when damaged cells proliferate. This is a major concern for the stability and functionality of the entire body. Oxidative stress suppresses production of androgens and oestrogens, impairing reproduction mechanisms, increasing the occurrence and severity of milk fever, increasing the amount of placental retention, and increasing sodium and water retention, leading to udder edema. Eliminating oxidative stress reduces the risk of periparturient disorders, reproductive failure, mastitis, and udder edema. This approach is beneficial for longevity of the udder and the animal as a whole.

Udder Edema and Udder Cleft Dermatitis

Udder edema is associated with mastitis and udder cleft dermatitis. Both diseases can cause detrimental effects to the animal's welfare, milk production ability, and longevity on the dairy operation. Udder cleft dermatitis (UCD), otherwise known as udder rot or necrotic dermatitis, is a skin lesion that appears in areas of tightly adjacent skin. Such areas include the udder cleft along with the medial (inner) aspect of the thigh and the lateral (outer) aspect of the udder. Friction between these 2 spaces due to a swollen udder or poor confirmation can lead to chafing, dermatitis, and, if persistent enough, necrosis of the tissues. Good udder confirmation with "a strong anterior udder attachment was a protective factor." These udder sores present with pus, thickened and inflamed skin, crust, and wounds that easily bleed. A putrid odour has also been noted in advanced cases. Along with udder edema, other risk factors associated with udder cleft dermatitis on a herd level include high yearly milk production, a mean SCC > 200,000 cells/mL, and an existing prevalence of digital dermatitis. On an individual cow basis, increasing parity is linked to greater risk of UCD, along with breed, milk production, and previous clinical mastitis cases. Animals with UCD had a 3.3 times greater risk for clinical mastitis. It is also suggested that skin lesions, hock lesions, and udder lesions can be a source for *Staphylococcus aureus,* the most isolated pathogen in intramammary infections in dairy heifers. A reduction in udder edema is linked to a decrease in SCC. Cows with teat and udder edema have an increased risk of clinical mastitis. The udder edema at calving was positively correlated with teat edema, milk leakage one-week prepartum, milk leakage at calving, blood in milk at calving, and skin lesions between udder and thigh. Mastitis diagnoses increased the odds of concurrent or subsequent diagnosis of udder edema, disorder of the abomasum, ketosis, and nonparturient paresis. The milk yield at the first DHIA test day was 3.6 kg lower in cows with udder edema. Longevity of dairy cows is an economically important trait for farmers and has gained in importance as a global indicator for animal welfare. Reduction of udder edema, UCD, and mastitis will not only affect milk production, quality of product, and revenue, they will also increase animals' welfare and longevity potential on the dairy operation.

The treatment includes administration of diuretic, antihistaminic, anti-inflammatory, Vitamin A, D3, E, broad-spectrum antibiotic along with massage and hot fomentation to alleviate post-parturient udder edema.

Brisket edema: It is a localized edema, observed in high-altitudes, where fluid collects in a specific anatomic location instead of accumulating in the whole body.

Edema in Dogs

Edema in dogs is more complicated than commonly thought of as a swelling. It refers to the accumulation of abnormally large amounts of fluids in the interstitial space, and swelling is the visual result of excess fluid accumulation. This swelling occurs either because too much fluid moves from the blood vessels into the tissues or because not enough fluid moves from the tissues back into the blood vessels. Both scenarios result in a fluid imbalance. Edema also commonly occurs as a side effect of significant disease such as heart failure, renal insufficiency, liver disease, or malnutrition.

Symptoms

The symptom can help identify the underlying cause of edema. Inflammation is usually the underlying cause, when the swelling is restricted to just one small part of the body, such as part of a leg or one side of the face. Sometimes, the dog looks puffy all over, possibly due to inadequate protein, too much sodium, or a severe bacterial infection. Another common symptom is swollen abdomen, sometimes looking as though the dog has swallowed a large ball. The fluid accumulation is more commonly associated with organ swelling or the accumulation of blood in the abdomen. The heart disease is one of the most common causes of swollen abdomen with body fluids. Most commonly, the swelling isn't visible, but signs of the edema show up in different forms, such as difficulty breathing, caused by pulmonary edema or effusion, aka fluid in the chest or neurologic symptoms/ seizures due to cerebral edema, or fluid in the brain cavity. These are the most challenging cases, as the early signs may be more subtle.

Etiology

Edema caused by inflammation is usually less serious than other forms of edema. Resolving the cause of inflammation after diagnosis and identification of the cause of the swelling usually eliminates the edema. Other forms of edema are much more serious and significant. Excess fluid accumulation in the abdomen is most commonly caused by heart disease, liver disease, kidney failure, or cancer. Accumulation of excess fluid in the lungs occurs frequently with heart disease. Fluid in the nervous system accumulates as a result of trauma, toxins, or severe metabolic disorders such as diabetes and electrolyte abnormalities.

A thorough history-taking and a complete physical examination frequently gives enough information to know where to start the diagnostic testing. Most dogs need a complete blood count, biochemistry profile, urinalysis and a thyroid test. Abdominocentesis and thoracocentesis is performed to aspirate the accumulated fluid for its examination. Sometimes, X-rays and ultrasound examination are also useful. An echocardiogram is required to get a full diagnosis of heart disease. A CT scan or MRI is conducted in case of cerebral edema to look for the underlying cause, and to remove some fluid for examination.

Treatment for Edema in Dogs

The line of treatment for edema depends on the underlying cause. If the was caused by inflammation, Removing the cause of the inflammation, responsible for the swelling, and treating with anti-inflammatories is the strategy to reduce edema. Edema secondary to organ failure such as heart, liver, or kidney diseases is more complicated. The proper diagnosis and treating the underlying disease facilitate to get rid of the fluid. In case of heart diseases, the treatment requires a combination of drugs to help the heart, as well as additional medications to reduce excess fluid accumulation. The treatment of cerebral edema involves a combination of antibiotics, anti-inflammatories, and anti-seizure medications, as well as drugs tailored to remove the fluid. Many cases of edema require hospital stays. Edema caused by inflammation is often the quickest and simplest to resolve, and can often be done in pets at home. Edema caused by organ failure or that affects the nervous system, however, can require multiple-day hospital stays. Almost all cases of edema require care and medications at home after the visit to animal healthcare institutions, while edema due to heart failure requires lifelong treatment. Edema in dogs other than that affecting a small area of the body, such as single leg or ear, is usually considered serious, requiring thorough investigation and aggressive care.

Wearing compression garments and reducing salt in the diet often relieves edema. The disease responsible for edema needs treatment as well. Wrapping the affected area with a compression bandage can help reduce the swelling right away. Water therapy, or hydrotherapy, is great for dogs and their owners, if done together.

Lymphedema in Dogs

Lymphedema is a condition characterised by collection of lymph fluid in tissues of a dog because of obstructions in the lymphatic system such as lymph nodes, vessels, and organs. Generally, the lymph vessels absorb fluids that leak out into interstitial space from the capillaries, and deliver the lymph fluids back into the blood stream, where they are able to be used in the immune system functions. The causes of lymphedema may be from a chronic illness or an acute condition such as an injury. It may also occur as a secondary condition stemming from another illness or, it can itself be the primary illness. The symptoms in a primary lymph disorder are usually noticed in puppies less than two months of age. The most common and obvious sign of lymphedema is the swelling of one or all extremities or the abdomen. The swelling anywhere in a dog can be a scary thing to see. It may not be painful, but it gives a very uncomfortable and strange feeling. Most of the time, the lymphedema in a dog is caused by another illness or injury. For example, swelling of the abdomen may be from cancer or heart disease, while swelling of the face, neck, or tongue could possibly be an allergic reaction.

The higher occurrence lymphedema in dogs is seen in certain breeds such as Borzoi, Rottweiler, Bulldog, Poodle, German Shepherd, Tervuren, Labrador Retriever, Old English Sheepdog, German Shorthaired Pointer, Great Dane. It is caused by injury or burn, tumours, infections, cardiac disorders, or hepatic insufficiency. There are several types of lymphedemas. The most common are primary, caused by a defective or missing tissue or organ such as lymph vessels, and secondary that occurs due to a disease such as cancer, trauma or an infection.

Symptoms: Lymphedema results in a fluid build-up somewhere in a dog. It can also cause other symptoms, such as swelling of any leg or anywhere on the body, including the face and abdomen, manifested by pain, weakness, change in skin colour, refusing to eat or walk and any abnormal behaviour.

Diagnosis: An extensive history taking and physical examination including weight, reflexes, body temperature, heart rate, blood pressure, respirations, breath sounds, skin and coat condition, and pupil reaction time are required to make a diagnosis. Some diagnostic tests, such as a complete blood count, blood urea nitrogen (BUN), urinalysis, packed cell volume (PCV), liver panel, heartworm test, and an analysis of the fluid from the affected area are performed for a diagnosis. Lymphangiography may be done by injecting dye into the lymph nodes before x-rays are performed. This is one of the most accurate tests in determining the reason for the lymphedema. Additionally, an ultrasound, CT scans, or MRI may be required for a more detailed view.

Treatment: The treatment for lymphedema is based on the cause or the lack of a cause. The best choice in cases of mild swelling with no obvious cause, may be to wait and see if it goes away on its own or if it gets worse. The following treatments may be instituted.

Pressure wraps: Wrapping the affected area with a compression bandage can help reduce the swelling right away.

Warm water massage: Water therapy, or hydrotherapy, is great for dogs and their owners if done together. The heat of the water and weightlessness can instantly make your dog feel better. Continued therapy may be able to lessen the lymphedema or get rid of it completely, but that depends on the cause.

Treatment: If infection is imminent, antibiotics are prescribed. Corticosteroids can help get rid of the swelling right away, or a type of benzopyrone, which can also reduce swelling. Surgery along with radiation or chemotherapy may be the best treatment. if cancer is the underlying condition. In the case of heart disease, medication such as beta blockers or an ACE inhibitor are prescribed.

Edema disease in pigs: It is a common cause of illness and death loss in pigs during the first 2 weeks after weaning. The disease is an enterotoxaemia, caused by E. coli that colonizes the small intestine and produce Stx2e. Bacterial colonization is mediated by F18ab fimbriae. The disease susceptibility is determined by the presence of receptors for these fimbriae on small intestinal epithelial cells that is inherited as a dominant trait. Clinical signs and lesions are largely the result of Stx2e, which causes necrosis of endothelial and smooth muscle cells in small arteries and arterioles. Vascular damage in the brain stem with resultant infarction and encephalomalacia is the main cause of death in affected pigs. Identification of the virulence factors has provided a basis for effective preventative measures, including development of vaccines.

9

Dehydration, Acid Base Imbalance and Fluid Therapy

The two different compartments of the body fluids are extracellular fluid (ECF) and intracellular fluid (ICF) with variation in solute concentrations. The extracellular fluid is the fluid in the cardiovascular system or in between the cells, called interstitial fluid. The fluid present in the gastrointestinal tract or urinary bladder are not included in the either of the fluid compartments. The inter-compartmental distribution of water is maintained by hydrostatic, oncotic, and osmotic forces and the loss of fluid in either of the compartments affects other compartment fluid volume and its composition. The distribution of water in the body and different fluid compartments are given below in a tabular form.

	% of the body weight
Total body water	60
Intracellular water	40
Extracellular water (plasma water + interstitial water)	20
Interstitial water	15
Plasma water	5

Fluid loss: The initial response to negative water balance is withdrawal of fluid from the tissues and the maintenance of normal blood volume. The fluid is drained primarily from the extravascular compartment and the interstitial fluid space. The osmotic pressure of extracellular fluid (ECF) increases with water losses exceeding water intake that activates hypothalamic osmoreceptors and an antidiuretic hormone (ADH) is released from the posterior pituitary gland. Both the increased ECF osmotic pressure and ADH elicit the feeling of thirst. The receptors that elicit thirst have an osmotic threshold higher than the osmoreceptors involved in ADH release. Thus, ADH can act on the kidneys to increase water reabsorption before thirst is elicited. The loss of fluid from these spaces is manifested by loss of skin elasticity, dryness of skin and mucous membranes and reduction and retraction of eyeball (enophthalmia) due to reduction in the volume of postorbital fat deposits. The second outcome of fluid loss is reduced circulating blood volume and an increased blood concentration, called haemoconcentration, increased blood osmolarity, sodium, potassium, protein and RBC count, leading to peripheral circulatory failure. Increased blood level of potassium induces irregularities in heart beat which may lead to death due to heart block.

Dehydration

It is defined as disturbance of body water and electrolyte balance, where more fluid is lost from the body than is absorbed, resulting in reduction in circulating blood volume and dehydration of the tissues. Dehydration is also defined as the loss of water and salts that are essential for normal body function.

Electrolyte imbalances occur commonly as a result of loss of electrolytes, shifts of certain electrolytes or relative changes in concentrations due to loss of water. Common electrolyte imbalances include hyponatremia (reduced blood sodium), hypokalemia (reduced blood potassium), hypocalcaemia (reduced blood calcium) and hypochloraemia (reduced blood chloride).

Acid-base imbalance, either acidosis or alkalosis, occurs as a result of the addition of acid and depletion of alkali reserve (acidosis), or the loss of acid with a relative increase in bicarbonate (alkalosis).

The reasons of dehydration can be grouped under two broad categories.

1. **Failure of water intake**- Deprivation of water, lack of thirst due to toxaemia, and the inability to drink water as in oesophageal obstruction etc.
2. **Excessive loss of water**: It is the most common reason of dehydration, as seen in diarrhoea, vomition, polyurea, loss of fluid from extensive skin wound etc. Severe dehydration also occurs in acute carbohydrate engorgement, acute intestinal obstruction and diffuse peritonitis and in dilatation and volvulus of the abomasum.

Pathogenesis

The initial response to negative water balance is the withdrawal of fluid from the tissues in order to maintain the normal blood volume. The fluid is drained primarily from the intracellular and interstitial fluid spaces. Vital organs including the central nervous system, heart and skeleton contribute a little and the major loss occurs from connective tissue, muscle and skin. The loss of fluid from the interstitial and intracellular spaces results in loss of skin elasticity, dryness of the skin and mucosa, and a reduction and retraction of the eyeball (anophthalmia) due to reduction in the volume of the postorbital fat deposits. The secondary response to a continued negative water balance is the reduction in the fluid content of the blood causing a reduction in circulating blood volume (volume depletion) and an increase in the concentration of the blood (haemoconcentration). Haemoconcentration increases the viscosity of the blood that impedes the blood flow, and may exacerbate peripheral circulatory failure. The loss in circulating blood volume also contributes to the mental depression of dehydrated animals, which is also due to varying degrees of acidosis and toxaemia depending on the cause of the dehydration. During dehydration, kidneys compensate effectively by decreasing urine output and increasing urine concentration. Water is also preserved by reduced faecal output and increased water absorption from gastrointestinal tract which results in dehydration of the contents of the rumen and large intestine and the faeces become dry and scanty. Dehydration exerts important effects on tissue metabolism. There is an increase in breakdown of fat, then carbohydrate and finally protein to produce water of metabolism. The increased endogenous metabolism under relatively anaerobic conditions results in the formation of acid metabolites and thus, metabolic acidosis. Urine formation decreases because of the restriction of renal blood flow and together with the increased endogenous metabolism, causes a moderate increase in blood levels of non-protein nitrogen.

Dehydration may also cause death, especially in acute intestinal obstruction, vomiting and diarrhoea, but it is mainly a contributory cause of death when combined with other systemic states, such as acidosis, electrolyte imbalances, toxaemia and septicaemia.

Assessment of dehydration

Body weight loss (%)	Sunken eye & face	Skin fold test persist for (sec)	PCV (%)	Total serum solid (g/l)	Fluid required to replace volume deficit (ml/kg b.wt)
4-6	Barely detectable	-	40-50	70-80	20-25
6-8	++	2-4	50	80-90	30-50
8-10	+++	6-10	55	90-100	50-80
10-12	++++	20-45	60	120	80-120

Clinical Findings

The important clinical findings in dehydration include

- Dryness and wrinkling of the skin
- Shrunken appearance of the eyes: The eyes recede into the sockets
- Skin subsides slowly after being picked up into a fold (skin fold test).
- Rapid loss of body weight
- Muscular weakness
- Inappetance or anorexia
- Decrease in milk yield
- In primary water deprivation, dehydrated animals are very thirsty.
- In dehydration secondary to enteritis, associated with severe inflammation, acidemia and electrolyte imbalance, there may be no desire to drink.

- The best indicator of hydration status in the farm animal is elasticity of the skin of the neck and lateral thorax. This is assessed by pinching the skin between the fingers, rotating the skin fold by 90° and noting the time required after release of the skin fold to disappear (normally < 2 seconds).

Clinical Pathology

Increase in the packed cell volume (PCV)

Increase in total serum protein, which may be modified by the presence of severe enteritis, peritonitis, or proteinuria.

Types of dehydration

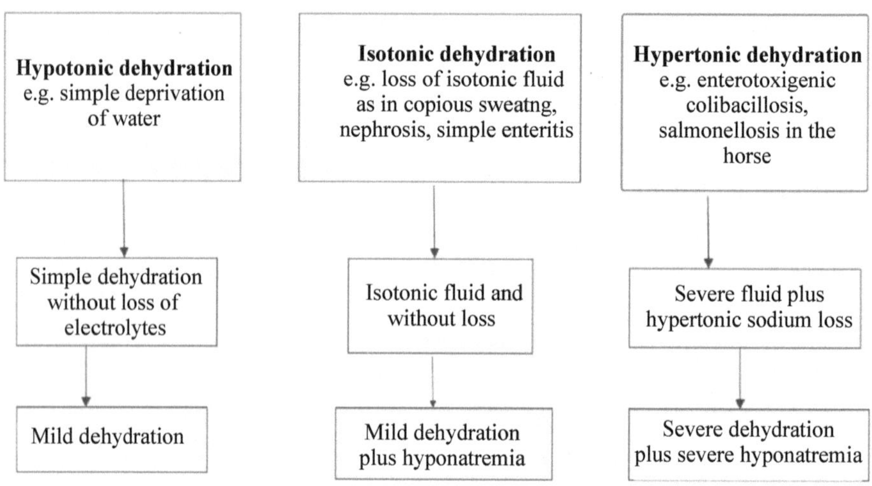

Electrolyte Imbalance

Hyponatremia refers to abnormally low level of sodium in the blood associated with dehydration. Sodium is the most abundant ion in the extracellular fluid (E.C.F) and is responsible for the maintenance of osmotic pressure of the extracellular fluid. The most common cause of hyponatremia is increased loss of sodium through the intestinal tract in enteritis, particularly marked in acute diarrhoea. In calves with acute diarrhoea due to enterotoxigenic *E. coli*, the sodium concentration of the intestinal fluid secreted in response to the enterotoxin is similar to that of plasma, and hyponatremia usually occurs (hypotonic dehydration). Animals affected with diarrhoea for several days continue to lose large quantities of sodium, resulting in severe hyponatremia. Hyponatremia can become severe when sodium-free water or 5% dextrose is used as the only fluid therapy in already hyponatremic animals. Hyponatremia can also occur in animals with proximal tubular dysfunction.

Etiology and Pathogenesis of Hyponatremia

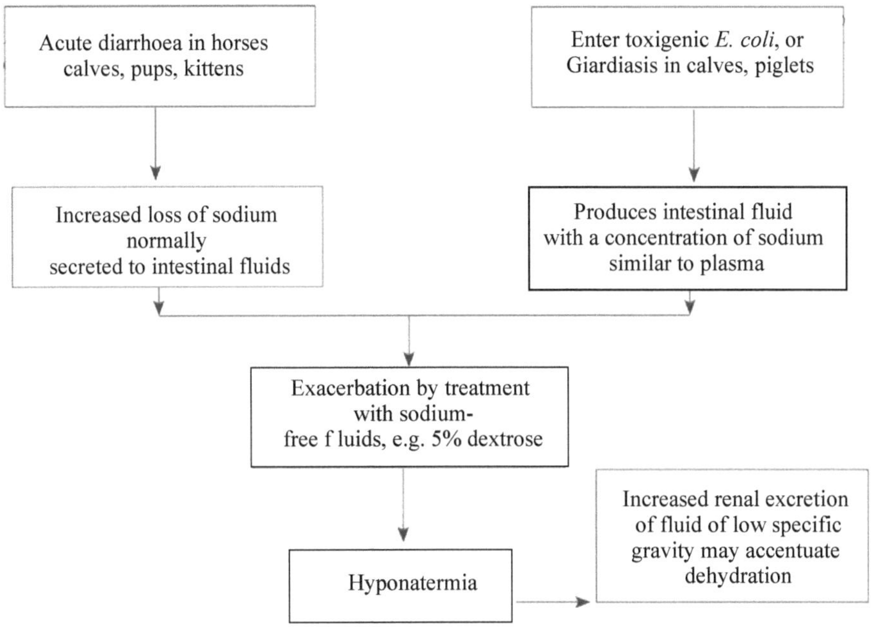

Clinical findings

The clinical findings include marked dehydration, hypothermia and muscular weakness. Hyponatremia causes an increase in the renal excretion of water in an attempt to maintain normal osmotic pressure, which results in a decrease in the extra-cellular fluid, leading to decreased circulating blood volume, hypotension, peripheral circulatory failure and ultimately, renal failure. Hypernatremia may occur in the initial stages of vomiting, diarrhoea or renal disease when the water loss exceeds the corresponding loss of electrolytes. Food and water deprivation in normal animals is also associated with reduced urine and faecal output; but continued cutaneous and respiratory water loss may result in hypernatremia due to only water loss. Hypernatremia also results from administration of hypertonic saline or sodium bicarbonate when water intake is restricted. Hypernatremia can also result in salt poisoning in cattle.

Hypochloraemia

Hypochloraemia occurs as a result of an increase in the net loss of the chloride electrolyte in the intestinal tract as seem in acute intestinal obstruction, dilatation, enteritis and impaction and volvulus of the abomasum. Normally a large amount of chloride is secreted in the abomasum by the mucosal cells in exchange for bicarbonate, which moves into the plasma.

The hydrogen, chloride and potassium ions secreted in gastric juice are normally absorbed by the small intestine. The failure of abomasal emptying and obstruction of the proximal part of the small intestine will result in the sequestration of large quantities of chloride, hydrogen and potassium ions which leads to a hypochloremic, hypokalemic metabolic alkalosis. A severe hypochloraemia can be experimentally produced in calves by feeding them a low chloride diet and daily removal of abomasal contents.

Etiology and Pathogenesis of Hypochloremia

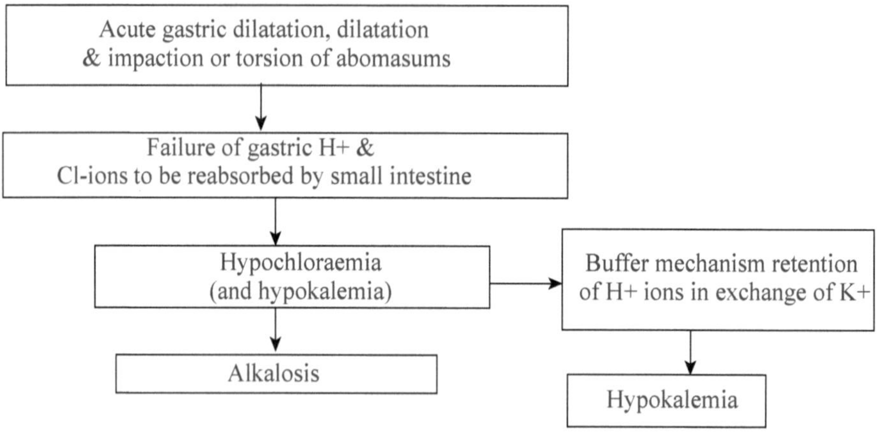

Clinical Findings

Clinical findings include anorexia, weight loss, Lethargy, Mild polydipsia, Polyuria. A marked metabolic alkalosis with hypokalemia, hyponatremia, azotemia and death

Etiology & Pathogenesis of hypokalemia

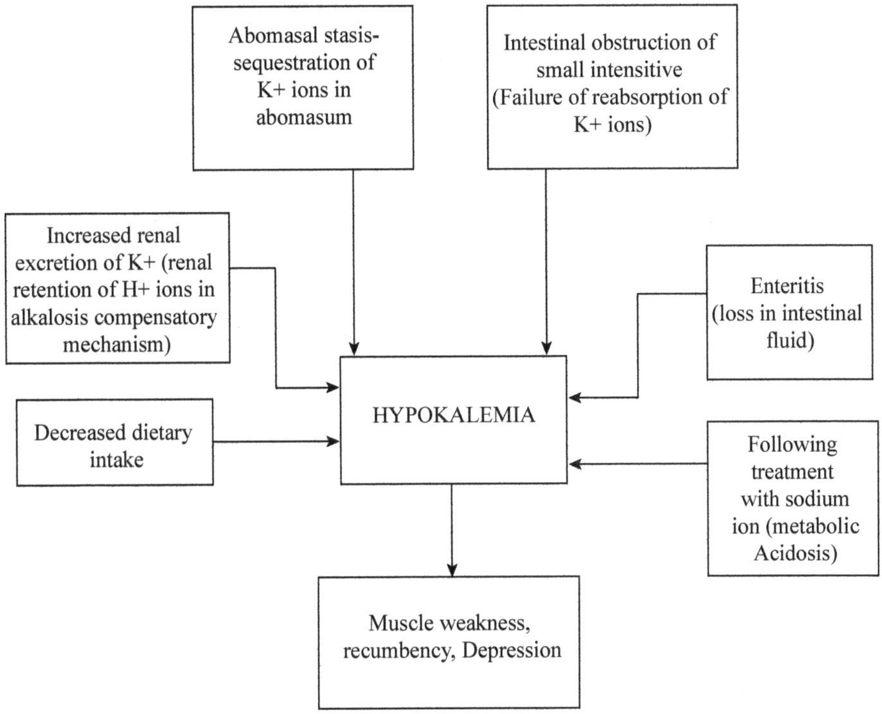

Hyperkalemia, increased serum potassium concentration, is a more serious systemic state than hypokalemia. Short-term exercise of horse at high intensity results in marked, but transient hyperkalemia (9-10 mEq/L). It is often associated with a metabolic acidosis, particularly when the acidosis results from volume depletion, and is complicated by renal shutdown. It is usually associated with muscular necrosis,

Hypocalcaemia: This may occur in recently calved cows which have been anorexic for few days. Also, this can be due to reduction in feed intake because of any illness which may have affected appetite, decreased absorption in diarrhoea it may be the earliest stage of hypocalcemic parturient paresis.

Clinical findings: Anorexia, tachycardia with a reduction in the intensity of the heart sound and occasional arrhythmia, a decrease in the frequency and amplitude of rumen contractions or complete ruminal stasis

Hypophosphatemia- It also occurs in cattle under conditions similar to those of hypocalcemic. A decrease in feed intake or alimentary tract stasis will result in serum inorganic phosphate concentration. The condition becomes normal when the feed intake and alimentary tract function are restored to normal.

Acid-Base Imbalance

The pH of the blood is maintained within the normal range of 7.35 – 7.45 by its buffer systems, of which the bicarbonate system is the most important. Any change in the pH may cause alkalosis (increase) or acidosis (decrease).

Alkalosis

It can be caused by an increased absorption of alkali, excessive loss of acid or deficit of carbon dioxide.

Etiology and Pathogenesis of Alkalosis

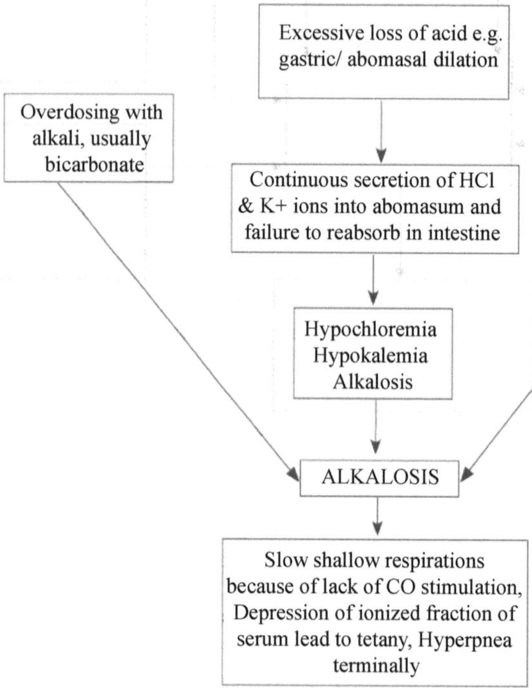

Acidosis

The general causes of acidosis can be classified into 3 categories on the basis of their pathogenesis.

- Excessive loss of base (bicarbonate)
- Accumulation of endogenous or exogenous acid
- A combination of both processes

Some common specific causes include acute diarrhoea in new born animals, acute enteritis in adult cattle and horses, carbohydrate engorgement in ruminants and horses. Acidosis also occurs where there is retention of carbon dioxide in the blood due to interference with normal respiratory air exchange.

Etiology and Pathogenesis of Acidosis

The factors leading to acidosis is described as follows

Shock & peripheral circulatory failure leading to anaerobic oxidation Overtreatment with acidifying solution in alkalosis, Retention of CO in the blood due to interference with normal respiratory exchange as in pneumonia, sever pulmonary emphysema, prolonged parturition causing asphyxia in new born, Excessive loss of HCO - ion in acute enteritis, Acute Intestinal obstruction, Retention of acid metabolites due to renal disease, Absorption of large quantity of fixed acids as in carbohydrate engorgement in ruminants, grain engorgement in horses, ketosis in ruminants

Acidosis leads to

Uncompansated e.g. when respiratory process is depressed by hypovolemic shock or enterotoxigenic collibacillosis

Fluid Therapy

Cardiac function, intravascular volume, and vascular tone, integrity, and patency are critical to normal circulation. An abnormality in one or more of these components of circulation leads to compensatory changes to maintain perfusion. The hemodynamic and cellular changes that develop as a result of these abnormalities are called shock. As shock progresses, oxygen and substrate delivery to the tissues becomes insufficient to meet energy requirements for cellular maintenance and repair. If shock progresses and cellular energy demands cannot be met, the ensuing organ failure leads to death. Early recognition of the type and stage of shock is vital to establishing a successful fluid therapy plan.

Shock is typically classified into 3 categories: hypovolemic, cardiogenic, and distributive. Hypovolemic shock develops when there is a blood volume deficit of $\geq 15\%$. Cardiogenic shock results when the heart fails as a pump. Common aetiologies include pulmonary emboli, cardiac tamponade, valvular insufficiency, cardiomyopathy, and cardiac arrhythmias. Distributive shock is caused by maldistribution of blood flow away from the central circulation as a result of peripheral vasodilation. The different types of shock may have different hemodynamic profiles during the early and middle stages. Frequently, more than one type of shock is present

Changes in fluid and electrolyte balance in response to disease processes may vary widely

- Hypovolemia is common and electrolyte changes are variable. Goal is to correct fluid and electrolyte imbalances before anaesthesia, if possible. Many anaesthetic agents produce vasodilation and hypotension relative hypovolemia that alters sympathetic nervous system activity and the endocrine system.
- Redistribution of blood flow with changes in vascular resistance
- Reduction in urinary flow rate, renal blood flow, and glomerular filtration rate seen with withholding water (fasting), anaesthetic drug effects, and increased ADH levels. These effects can be eliminated or reduced by "filling the tank" with crystalloids.

General Principles of Fluid administration

Maintenance vs. Replacement Therapy

- Maintenance fluid therapy is designed to meet the patient's ongoing sensible and insensible fluid losses with normal fluid volume over 1 – 2 days; in the normal animal. This is primarily water loss, with a lesser degree of electrolyte loss.
- Replacement fluid therapy is designed to replenish existing fluid deficits; this usually requires replacement of both water and electrolytes
- The optimal fluid type for each of the above therapy depends upon serum electrolytes, acid-base status, and concurrent administration of drugs and blood products.

Precautions for rapid fluid administration

- In dogs, maximum fluid administration rate is 90 ml/kg/hr (1 blood volume/hour)
- In cats, maximum fluid administration rate is 60 ml/kg/hr (1 blood volume/hour)
- Cardiopulmonary status should also be monitored carefully during fluid administration.
- Packed cell volume and total plasma protein levels should be monitored to maintain PCV > 20% and TPP > 4 g/dl

Normal fluid administration rates during anesthesia

- Dogs, cats: 10 - 20 ml/kg/hr
- Horses, cattle: 5 - 10 ml/kg/hr

Fluid therapy is used to replenish intravascular volume or interstitial fluid volume, or to correct electrolyte abnormalities. More accuracy is required in formulating the fluid therapy plan for the critically ill than in the usual hospital patients, because the kidneys and cardiovascular system, which can correct most fluid therapy miscalculations, may be in a compromised state. It is important to understand the dynamics between the fluid compartments and the specific components of each fluid to select the best fluid for the individual situation. The volume and rate of administration are guided by pathologic conditions such as cardiac disease, systemic inflammatory response syndrome diseases, fluid loss into the third body fluid space, blood pressure, and pulmonary and cerebral edema.

Fluid Compartment Deficits

Fluid deficit in the intravascular space results in poor perfusion and inadequate tissue oxygenation. This volume deficit results in a lower vessel wall tension and stimulation of the baroreceptors. Physical parameters that reflect perfusion status include heart rate, pulse intensity, capillary refill time, mucous membrane colour, and rectal temperature. Fluids administered must remain in the intravascular space, increase the vessel wall tension, and obliterate the need for the baroreceptor compensatory response. Most animals with an intravascular deficit (poor perfusion) also have concurrent extravascular deficits. This warrants the simultaneous administration of crystalloid solutions when colloids are used for correction of the intravascular deficit.

Fluid deficit in the extravascular space (interstitial and intracellular) causes tissue dehydration. This results in tenting of the skin, dry mucous membranes, sunken eyes, and dullness of the cornea. Perfusion parameters should be evaluated simultaneously to determine whether poor perfusion is compounding the dehydration. Crystalloid fluids with the same tonicity as plasma are used to replenish the extravascular spaces. Crystalloids are water-based solutions with low molecular weight molecules that are freely permeable across the capillary membrane. The concentration of these molecules in relation to normal plasma indicates the tonicity of the fluid. The solution is called isotonic if the concentration is similar to normal plasma.

Physical and laboratory parameters are used to quantify the degree of dehydration. Semi-dry oral mucous membranes, normal skin turgor, and eyes maintaining normal moisture indicate 4-5% dehydration. Dry oral mucous membranes, mild loss of skin turgor, and eyes still moist indicate 6-7% dehydration. Dry mucous membranes, considerable loss of skin turgor, and eyes retracted indicate 8-10%. Very dry oral mucous membranes, complete loss of skin turgor, severe retraction of the eyes, dull eyes, and possible alteration of consciousness indicate 12% dehydration.

The physical guidelines for estimating dehydration are misleading in two common clinical situations. Chronically emaciated animals may have metabolized the fat from around the eyes and the collagen in the skin, resulting in poor skin turgor and sunken eyes despite normal hydration.

Animals with rapid fluid loss into a third body fluid space (a space within the body cavity where fluid from the local interstitial and intravascular spaces leak) have rapid fluid shifts from the intravascular compartments into these spaces before clinical evidence of interstitial fluid loss is seen. Both situations require evaluation of mucous membrane and eye moisture, PCV, and total plasma solids before dehydration can be estimated.

The intravascular space can be replenished using crystalloid solutions alone. However, large quantities of crystalloids may be required to provide volumes sufficient to blunt the baroreceptor response after 1 hr of infusion. The extra interstitial volume for a short period does not cause a problem in most animals, and if renal function is normal, the additional volume is eventually excreted. However, the brain and heart do not have "forgiving" interstitial spaces in disease, and additional interstitial fluid can cause organ failure.

Many conditions can increase capillary permeability and cause systemic inflammation (e.g. Parvo viral diarrhoea, pancreatitis, septic shock, massive trauma, heat stroke, cold exposure, burns, snake bite, and systemic neoplasia). Increased vessel permeability is an additional factor that favours movement of crystalloid fluid from the intravascular space into the interstitial space and complicates maintenance of intravascular volume. A combination of a large molecular weight colloid (to hold the water in the vessels) and crystalloids (to replenish the interstitial spaces) is recommended. In addition, many of these animals have a third body fluid space, most likely due to significant regional inflammation. The end result is massive fluid requirements, which make maintaining fluid balance (especially intravascularly) difficult. It is almost impossible to predict how much fluid will be required to maintain perfusion and hydration in these critical patients. Fluid selection becomes critical, with the rate and volume titrated until the goals of resuscitation have been met (ie, normal heart rate, blood pressure, perfusion parameters).

Three Principles of Fluid Therapy

(1) Replacing lost fluid: In practice, the problem arises to decide whether to give saline or dextrose. Most body fluids contain salt but less than plasma. Indicators of water (5% dextrose) requirements

- Thirst, Low urine output with high ratio of urine: plasma osmolarity, Dry maxilla and groin, raised plasma sodium Indicators of saline requirement
- Fluid loss (excess diuretic, vomit, burns, diarrhoea, sweat), Decrease tissue turgor, postural hypotension, tachycardia, low jugular vein pulse, low urine output, ratio of urine and plasma osmolarity, raised haematocrit Indicators for another fluid

- Blood should be transfused in acute blood loss. If patient is shocked, a Plasma expander' or 'colloid' is normally given in case of shock.

(2) Allowance for future abnormal loss: Moderate continuous sweating requires an extra 500 ml of 0.9% saline and 500 ml 5% dextrose per 24 hour as in cases of fever with body temperature above 39.5^{o}C.

(3) Maintain normal daily requirements. About 2500 ml fluid containing 100 mmol sodium and 70 mmol potassium per 24 h are required. A good regimen is 2500 ml dextrose-saline with 30 mmol potassium per litre.

Postoperative patients usually need more fluid and more saline.

Patients with heart failure should not be size normal saline. If intravenous fluids has to be given 5% dextrose should be administered cautiously.

1. Plasma Osmolality

Plasma osmolality is a function of the ratio of plasma solute to water. It is regulated by changes in water balance. Water intake is derived primarily from three sources: Ingested water, water contained in food, and water produced from oxidation of carbohydrates, proteins, and fats. Water losses occur in the urine and stool, as well as evaporation from the skin and respiratory tract. Alterations in plasma osmolality of as little as 1% - 2% are sensed by osmoreceptors in the hypothalamus. These receptors initiate mechanisms that affect water intake (via thirst) and water excretion (via antidiuretic hormone, ADH) to return plasma osmolality to normal.

2. Effective Circulating Volume

Effective circulating volume is defined as that part of the extracellular fluid (ECF) that is in the vascular space and effectively perfusing tissues. It varies directly with ECF volume and also with total body sodium, since sodium salts are the primary ECF solutes holding water in the extracellular space. Therefore, regulation of sodium balance by changes in renal sodium ion, and the maintenance of effective circulating volume, are closely related.

3. Major Effectors of Effective Circulating Volume

Three major effectors altering effective circulating volume: 1) The sympathetic nervous system, 2) angiotensin II, and 3) renal sodium excretion. Volume depletion, sensed by arterial baroreceptors as hypotension, causes an increase in peripheral sympathetic tone. Increased sympathetic tone returns volume to normal by initiating specific compensatory changes. These compensatory changes include the followings:

- Venous constriction: Increased venous return
- Increased myocardial contractibility and heart rate: Increased cardiac output

- Arterial vasoconstriction: Increases systemic vascular resistance and blood pressure
- Increased renin secretion: Increases levels of angiotensin II which is a potent vasoconstrictor
- Increased renal tubular sodium resorption (due to increased levels of angiotensin II and aldosterone)

Sympathetic tone induced changes in effective circulating volume are transient and compensatory; appropriate changes in renal sodium excretion are required to restore normal volume.

4. Body Defence Against Hyperosmolality

The major corrective measures against hyperosmolality (accumulation of excess solute in of body water) is increased thirst. Although the kidney can minimize water losses via the action of ADH, water deficits can be corrected only by increased dietary intake.

5. Hypoosmolality

Hypoosmolality can result from excessive body water retention with subsequent dilution of body solutes or from solute loss in excess of water loss (e.g. diarrhoea). The kidney excretes large volumes of water daily, persistent water retention resulting in hypoosmolality occurs only in the presence of decreased renal water excretion. In patients with normal renal function, hypoosmolality must therefore be due to solute loss in excess of body water loss.

6. Hypovolemia (i.e., dehydration)

Hypovolemia causes an increase in renin secretion. The subsequent increase in angiotensin II causes an increase in blood pressure (as a result of arterial vasoconstriction), as well as renal sodium retention (this is both a direct effect and also the result of increased aldosterone secretion). Water is also retained with sodium retention.

7. Degree of Dehydration in an Animal

Clinical assessment of dehydration is best accomplished by serial body weight monitoring. The physical examination findings often underestimate the degree of dehydration. During the acute phase of volume depletion, these classical physical examination findings are all that are available. The chart below details a general guideline to estimate dehydration percentage.

Table 1: Estimating the percentage dehydration by physical examination findings

Estimated % of Dehydration	Physical Examination Findings
<5	History of fluid loss but no findings on physical examination
5	Dry oral mucous membranes but no panting or pathological tachycardia
7	Mild to moderate decreased in skin turgor, dry oral mucous membranes, slight tachycardia, and normal pulse pressure
10	Moderate to marked degree of decreased skin turgor, dry oral mucous membranes, tachycardia, and decreased pulse pressure
12	Marked loss of skin turgor, dry oral mucous membranes, and significant signs of shock

8. Fluid administration Through Subcutaneous Route

In mild dehydration, subcutaneous fluids administration is useful. Isotonic fluids should be used and no more than 5 to 10 ml/lb should be given at each injection site. The rate of subcutaneous fluid flow usually is governed by patient comfort. These fluids are aseptically administered and multiple sites are required to provide adequate fluid volume. Generally, all subcutaneous fluids are resorbed within 6 to 8 hours. The use of intravenous fluids to reestablish peripheral perfusion should be considered if fluids are still noted in subcutaneous after this time.

9. Intra-peritoneal Route of Fluid Administration

The intraperitoneal route of fluid administration is quick and easy. The fluids are generally resorbed thus increasing the circulating volume. However, there is potential threat of bacterial peritonitis, perforating viscera, and decreasing ventilation from impeding diaphragmatic excursion. Experience with peritoneal dialysis in dogs has shown that peritoneal fluids often traverse the diaphragm, entering the thoracic space, and further affecting ventilation.

10. Intravenous Fluids Administration

In general, intravenous fluid administration is indicated in dogs and cats with a 7% or greater degrees of dehydration. There are numerous potential routes for intravenous fluid administration such on peripheral veins, jugular veins and intra-osseous veins.

11. Estimation of Fluid Volume

The amount of fluid needed for replacement depends on the patient's hydration status and blood volume, restoration of total body water and electrolytes.

12. Phases of Fluid Therapy

- Emergency phase
- Replacement phase
- Maintenance phase

13. Replacement Therapy

The volume of fluid administered during the dehydration phase is based on an assessment of fluid needs for the following:

Returning the patient's status to normal (**deficit volume**) Replenishing normal ongoing losses (**maintenance volume**) Replacing continuing abnormal losses (**continuing volume loss**)

14. Deficit Volume Calculation

The deficit volume is an estimate based on findings from the physical examination (Table I) or from known changes in body weight. The estimated dehydration is multiplied by the body weight. It must be remembered that it is difficult to replace all deficits in a 24-hour period. An attempt to do so may result in urinary losses furthering dehydration. Thus, it is recommended that only 75% to 80% of the deficit volume be replaced during the first 24 hours. Also, "daily maintenance volumes" has to be added to your calculated deficit volume if the animal is not eating nor drinking.

Example 1

A 22-lb (10 kg) dog is assessed to have 7% dehydration. What volume of fluid should be given during the first 24 hours?

Total Deficit Replacement Volume = Deficit Volume plus Maintenance Volume

Deficit replacement volume (ml) = % dehydration x body weight (lb) x 454[a]

x 0.80

Deficit replacement fluid volume (ml) = 0.07 x 22 lb x 454a x 0.80 = 560 ml

a = 454 ml = 1 pound or

Deficit replacement volume (ml) = % dehydration x body weight (kg) x 1000[b]

x 0.80

Deficit replacement fluid volume (ml) = 0.07 x 10 x 1000 x 0.80 = 560ml

b = 1000 ml = 1 kilogram

15. Maintenance volume

Maintenance volume is the amount of fluid required to meet the normal ongoing losses. Ongoing losses are divided into **sensible** and **insensible** losses. Sensible losses can be measured, refers to are water losses in the urine and faeces. Insensible losses are normal but are not easily quantitated. These water losses occur during panting or sweating. One-third of the maintenance volume is made up of the insensible volumes and two-thirds, sensible volumes.

16. Fluid for Maintenance Volume if the Animal is Not Eating or Drinking

There is a paucity of data regarding water needs for the dog and cat. Many authors have recommended dramatically different fluid and energy requirements. Water and energy requirements are numerically the same (1 Kcal of energy = 1 ml of water). Data from recent research has been used to document that energy expenditure or consumption is less than published formulas and recommendations. Estimation of water needs include 50ml/kg/day, 132 kcal X $Kg^{0.75}$, 156 X $Kg^{0.667}$, (30 X kg) + 70, 70 X $Kg^{0.75}$. Indirect calorimetry is being used to estimate energy (and thus water) needs for the dog and cat. These studies reinforce that previously recommended formulae overestimate the energy (water) requirements of the cat and dog.

Example 2

A 22-lb (10 kg) dog is assessed to be 7% dehydrated and has been vomiting.

How much fluid should be given during the next 24 hours?

Volume of fluid required = Deficit volume + Maintenance volume

= [0.07 x 22 lb x 454 x 0.80] + [(10 x 30) + 70]

= [560] + [370] = **930 ml**

or = [0.07 x 10 kg x 1000 x 0.80] + [(10 x 30) + 70]

= [560] + [370] = **930 ml**

17. Water Needs in a Sick Animal

The question regarding energy (water) requirements in sick animals continues to elicit controversy. Illness, injuries, and surgery require more energy (water). However, there is mounting evidence in dogs that increased energy requirements are not common in sick or injured dogs. In fact, there are publications documenting the lower energy (water) requirements for both the normal and sick or traumatized dog. The reserves in injured or sick animals are already minimal and it makes little sense to increase metabolic requirements in order to survive. The available energy is conserved to reduce metabolic requirements. The presence of significant hypothyroid function and reduced metabolic requirements have been documented in dogs in critical care unit.

18. Continuing Losses

A crude but effective guideline for replacing continuing abnormal losses is to estimate the volume of fluid loss and then double this estimate. The result will be surprisingly close to the actual volume of vomitus, diarrhoea, or urine.

19. Inadequate Fluid Volume

Any acute change in body weight results from losses or gains in water. If the animal is losing body weight while being given crystalloid fluids, the animal is likely receiving inadequate volumes of fluid. If renal function is adequate, a dehydrated animal will have urine specific gravity above 1.025.

20. Overhydration

Signs of over hydration are:

- Restlessness
- Shivering
- Tachycardia
- Tachypnoea
- Nasal discharge
- Vomition
- Diarrhoea

Clinically, pulmonary oedema is the terminal event of overhydration. An increased serous nasal discharge, followed by chemosis, and finally pulmonary congestion on auscultation can be noted before pulmonary oedema.

The common parental crystalloid fluids, their electrolyte composition (m Eq/l), pH, and osmolarity

Solution	Na+	K+	Cl-	Ca++	Mg++	Buffer (q/L)	Calories Kcal/L	Osmolality mOsm/L
Dextrose 5% in Water	-	-	-	-	-	-	170	278
Dextrose 2.5% in 0.45% Saline	77	-	77	-	-	-	85	280
Ringer's Lactate Solution	130	4	109	3	-	Lactate 28	9	272
Ringer's Solution	147	4	156	4.5	-	-	-	309
Normosol-R (Multisol-R)	140	5	109	-	3	Acetate 27 Gluconate 23	15	294
Dextrose 5% in Ringer's Lactate	130	4	109	3	-	Lactate 28	179	525
Normal Saline (0.9%)	154	-	154	-	-	-	-	308
Dextrose 50%	-	-	-	-	-	-	1700	2525
Dextrose 5% in Saline (0.9%)	154	-	154	-	-	-	170	-
Potassium Chloride solution	-	2	2	-	-	-	-	-

ii. Colloids

1. **General**: The term colloids is used to describe fluids that exert oncotic pressure because they contain large particles that cannot leave vascular space. Particle size becomes important if capillaries are leaky as in inflammation or septicemia. If particles leak into interstitium, they take fluid with them that worsen hypovolemia plus cause edema in lungs and peripheral tissues. Synthetic colloids may inhibit coagulation. Hence, it should be used with caution in congestive heart failure and oliguria, where volume expansion may be detrimental.
2. **Plasma**: Fresh frozen plasma contains all the clotting factors, antithrombin III, a few platelets and alpha-2 macroglobulin as well as albumin. Albumin is the source of oncotic pressure. It is relatively small, 69K Daltons, and can leak from inflamed capillaries.

Plasma transfusion is a viable choice in inflammation or pancreatitis. It takes 22.5ml/kg of plasma to raise recipient serum albumin by 5g/L.

3. **Hetastarch**: Large molecular weight synthetic colloid in 0.9% NaCl

Dose: Cat 5 - 10 ml/kg/day

Dog 20 - 40 ml/kg/day

4. **Pentastarch**

Dose: Cat 5 – 10 ml/kg/day

Dog 10 – 25 ml/kg/day

Less tissue retention, shorter half-life than HET. More potential for coagulopathy.

5. **Dextran:** Dextran 40 MW - 40K Daltons Dextran 70 MWw - 70K Daltons Dose: Cat 5 – 10 ml/kg/day

Dog 10 – 20 ml/kg/day

21. Hemoglobin Based Solutions

Oxyglobin R: – Long shelf life, typing and cross matching not required. It may improve tissue oxygenation and perfusion.

22. Selection of the Parenteral Fluid

In selecting a fluid, it is important to know which electrolytes are lost and the knowledge on the pathophysiology of the disease is essential to institute replacement therapy. The table given below provides an overview of electrolyte changes and replacement recommendations.

Serum

Condition Diarrhoea	**Na^+D**	**Cl^-D**	**K^+ D**	**HC D**	**O^{-3}**	**Volume D**	**Fluid of Choice NormosolR-R + KCl or**
							Lactated Ringer's + KCl
Pyloric obstruction	D	D	D	I		D	0.9% NaCl + KCl
Dehydration	I	I	N	N/D		D	NORMOSOL^R-R + KCl, Lactated Ringer's + KCl, 0.9% NaCl + KCl, 5% dextrose
Congestive heart failure	N/D	N/D	N	N		I	0.45% NaCl + 2.5% dextrose + KCl, 5% dextrose
End-stage liver disease	N/I	N/I	D	D		I	0.45% NaCl + 2.5% dextrose + KCl
Acute renal failure	I	I	I	D		I	0.9% NaCl,
-Oliguria -Polyuria	D	D	N/D	D		D	NORMOSOL^R-R + KCl, Lactated Ringer's + KCl,
Chronic renal failure	N/D	N/D	N	D		N/D	NORMOSOL^R-R, Lactated Ringer's solution, 0.9% NaCl
Adrenocortical insufficiency	D	D	I	N/D		D	0.9% NaCl
Diabetic ketoacidosis D-Deficiency / Loss	D	D	N/D	D		D	0.9% NaCl (± KCl)

Intravenous Therapy

Intravenous therapy refers to the administration of liquid substances directly into vein. The word intravenous simply means "within a vein". Therapies administered intravenously are often called specialty pharmaceuticals. It is commonly referred to as a drip because many systems of administration employ a drip chamber, which prevents air entering the blood stream. The intravenous route is the fastest way to deliver fluids and medications throughout the body as compared to other routes of administration. Some medications, as well as blood transfusions and lethal injections, can only be given intravenously.

Complications

If the cannula is not sited correctly, or the vein is particularly fragile and ruptures, blood may leak into the surrounding tissues. This situation is known as "tissuing" or "blown vein". Using cannula to administer medications cause may extravasation of the drug which can lead to edema, causing pain and tissue damage, and even necrosis depending on the properties medication.

The person attempting to obtain the access must find a new access site proximal to the "blown" area to prevent extravasations of medications through the damaged vein. For this reason it is advisable to place the frost cannula at the most distal at appropriate vein. If a patient needs frequent venous access, the veins may scar and narrow, making any future access extremely difficult or impossible. A peripheral intravenous cannula cannot be left in the vein indefinitely, because of the risk of insertion-site infection leading to phlebitis, cellulites and sepsis. The US Centres for Disease Control and Prevention updated their guidelines and now advise the cannula need to be replaced every 96 hours. This was based on studies organised to identify causes of Methicillin- resistant *Staphylococcus aureus* (MRSA) infection in hospitals. In the United Kingdom, the UK Department of health published their finding about risk factors associated with increased MRSA infection; now include intravenous cannula, central venous catheters and urinary catheters as the main factors increasing the risk of spreading antibiotic resistant strain bacteria.

Intravenous Drip

An intravenous drip is the continuous infusion of fluids, with or without medications, through an intravenous access device. This may be to correct dehydration or an electrolyte imbalance, to deliver medications, or for blood transfusion.

Intravenous Fluids

There are two types of fluids that are used for intravenous drips; crystalloids and colloids. Crystalloids are aqueous solutions of mineral salts or other water-soluble molecules. Colloids contain larger insoluble molecules, such as gelatin. Blood itself is a colloid. Colloids preserve a high colloid osmotic pressure in the blood. Crystalloids are generally much cheaper than colloids. The most commonly used crystalloid fluid is normal saline, a solution of sodium chloride at 0.9% concentration, which is close to the concentration in the blood

(isotonic). *Ringer's lactate* or *Ringer's acetate* is another isotonic solution often used for large-volume fluid replacement. A solution of 5% dextrose in water, sometimes called D5W, is often used, if the patient is at risk for having low blood sugar or high sodium. The choice of fluids may also depend on the chemical properties of the medications being given. Intravenous f luids must always be sterile

Composition of common crystalloid solutions

Solution	Other Name	[Na^+] (mmol/L)	[Cl^-] (mmol/L)	[Glucose] (mmol/L)	[Glucose] (mg/dl)
D5W	5% Dextrose	0	0	278	5000
2/3D & 1/3S	3.3% Dextrose / 0.3% saline	51	51	185	3333
Half-normal saline	0.45% NaCl	77	77	0	0
Normal saline	0.9% NaCl	154	154	0	0
Ringer's lactate	Lactated Ringer	130	109	0	0
D5NS	5% Dextrose, Normal Saline	154	154	278	5000

28 mmol/L lactate, 4 mmol/L K^+ and 3 mmol/L Ca^{2+}.

28 mmol/L acetate, 4 mmol/L K^+ and 3 mmol/L Ca^{2+}.

Infusion equipment

A standard intravenous infusion set consists of a pre-filled, sterile container (glass bottle, plastic bottle or plastic bag) of fluids with an attachment that allows the fluid to flow one drop at a time, making it easy to see the flow rate (and also to reduce air bubbles); a long sterile tube with a clamp to regulate or stop the flow; a connector to attach to the access device; and connectors to allow "piggy backing" of another infusion set onto the same line, e.g., adding a dose of antibiotics to a continuous fluid drip.

Oral Fluid Therapy

Whenever possible, the oral route is used to deliver the maintenance requirements. The oral route is preferred for maintenance therapy, provided there are no abnormalities of the digestive tract which interferes with the oral administration or the absorption of fluids. In ruminants such as adult cattle, rumen function must be within normal range for significant absorption of fluids and electrolytes. Oral electrolyte solutions and water should be made available at all times to animals affected with diarrhoea and other diseases in which there are continuous loss of fluid and electrolytes. The provision of water and salt together with glucose and glycine facilitates the absorption of sodium and water in calves with diarrhoea. A wide variety of oral and parental electrolyte replacement solutions are available commercially. Most preparations are made available as powders for mixing with water. They contain sodium, chloride, potassium, glucose and glycine for the correction of dehydration. The glucose and glycine facilitates the absorption of sodium across the mucosa. Oral fluids also contain alkalinizing agents for the correction of acidosis. Bicarbonate is effective immediately but there is speculation that it may interfere with milk clotting in the abomasum which requires an acidic environment for the action of rennin. Some oral electrolyte solutions for diarrheic calves contain citrate, gluconate, acetate or citric acid, which are metabolized to carbon dioxide and water after absorption into the body, and hydrogen ions are consumed in the process. A bicarbonate-rich solution induces the best alkalizing effect and is highly recommended. The use of oral electrolyte solutions without sodium bicarbonate for the treatment of acidosis in diarrhoeic calves may explain the syndrome of acidosis without dehydration. The close relationship between net fluid retention and urine output indicates that urinary excretion may eliminate most of the administered fluid and electrolytes in certain circumstances.

10

Diseases of Digestive Systems

The primary functions of the digestive system include prehension of food material, its digestion, and absorption, excretion of undigested material and maintenance of internal environment. All these activities are accomplished through the following four major functions of the gastrointestinal system.

Motor functions: This function is required for movement of the food material and ingesta and for contraction of the musculature of gastrointestinal tract.

Secretory function: It is responsible for secretion of gastric juice and pancreatic juice

Digestive function: Motor and secretory functions are required to perform the digestive function of alimentary tract.

Absorptive function: Failure of the absorptive function leads to non- absorption of nutrients from the gastrointestinal tract.

Diagnosis of GI Tract Disorders

The history of abdominal distension, difficulty in prehension, diarrhoea or constipation, reduced intake of food, pain during mastication, cud drooping etc. give an idea on involvement of gastrointestinal tract disorders. Examination of body parts, particularly mouth, abdomen, per-rectal examination, physical examination such as palpation, percussion and auscultation are then carried out to assess the abnormality. Examination of body fluid such as pH of rumen fluid, biochemical and physical examination of rumen fluid including sediment activity test and cellulose digestion test, medical imaging such as radiography, ultrasonography, and exploratory laparotomy may be conducted to pin-point the disorder.

11

Manifestations of Digestive Tract Dysfunction

Abnormality in Prehension, Mastication and Swallowing

Difficulty in prehension is seen in jaw disorder, painful condition of tongue, mal-apposition of teeth and diseases of buccal cavity. Mastication problem is seen in stomatitis, inflammation of and mucous membrane. Swallowing disorder is noticed due to foreign body, tumour and inflammation of oesophagus and pharynx, and paralysis of oesophagus.

Drooling of Saliva/ Excessive Salivation

These are seen in cases of salivary gland problem, foot and mouth disease, hyperthermia etc. The drooling of saliva may be local or systemic in origin.

Vomition/ Regurgitation

True vomiting is seen in monogastric animals, which is accompanied by retching and coordinated contraction of certain group of muscles. On the contrary, projectile vomiting occurs due to overloading of stomach with food and water, and ingesta are ejected out with a little effort.

Diarrhoea, Constipation and Scant Faeces

Increased frequency of defecation or fluidity of faeces is called diarrhoea. It is seen in cases of gastrointestinal tract infection, gastroenteritis, and gastrointestinal parasitism. Diarrhoea is generally associated with metabolic acidosis due to loss of bicarbonate. Decreased frequency of defecation and reduced water percentage in the stool is called constipation, and it is associated with difficulty in passage of stool. Scant faeces refer to less quantity of faeces but the consistency is normal.

Paralytic Ileus

Paralysis of gastrointestinal tract is seen in peritonitis, excessive handling during operation, dehydration, acid base imbalance and hypokalemia.

Alimentary Tract Haemorrhage/ Bleeding

Blood in vomitus is called haematemesis, and the presence of digested occult blood in stool is called melena. In melena, the stool colour becomes black. The rupture of blood vessels of GI tract leads to bleeding into the tract and passage of occult (hidden) blood in faeces

Abdominal Pain

The abdominal pain in ruminants is manifested by frequent closing of eye-lid, stamping, looking towards flank, frequent lying down and getting up, frequent bellowing, grinding of teeth. In dogs, abdominal pain is manifested by pressing on ground, frequent lying down and getting up, and animal feels uncomfortable on mild digital pressure.

Tenesmus

Tenesmus refers to persistent straining, may or may not be associated with involvement of gastrointestinal tract. Colitis, coccidiosis, and enteritis in calves are the alimentary tract dysfunctions associated with tenesmus.

Shock and Dehydration

Vomition and diarrhoea are associated with loss of body fluid and electrolytes, and imbalance in acid-base equilibrium. The rapid loss of the body fluid may be associated with clinical signs of shrinkage of eye ball, reduced urine output and elasticity of the skin. The loss of fluid to the tune of 10-20% in a short period of time, as seen in acute diarrhoea, may result in hypovolemic shock.

Abdominal distension

This is commonly noticed when there is accumulation of gas, fluid or impacted material in the gastrointestinal tract. E.g. Bloat, impaction of rumen, intestinal tympany.

Complete or partial loss of appetite

Partial (inappetance) or complete (anorexia) loss of appetite is a common sign in many affections of the digestive system. However, such symptoms are more commonly observed in fever due to both the infectious or non-infectious diseases.

12

Diseases of Buccal Cavity and Oesophagus

The disease may involve one or more mouth parts such as tongue, palates, gum, teeth, muscles and salivary gland resulting in difficulty in prehension, mastication of the food material and inadequate buffering of the food with saliva.

Stomatitis

Inflammation of the oral cavity that includes glossitis (tongue), palatitis (palates), gingivitis (gums), is characterized by masking of mouth, painful condition of the oral cavity, difficulty in mastication, and swallowing. Simple inflammation of mouth is frequently observed in horses. Ulcerative or infectious inflammation are commonly seen in young, and occasionally in old, debilitated animals.

Etiology

Non-infectious causes of stomatitis in all species of domestic animals include

- Trauma caused by physical agents such as injury by sharp objects, chewing and ingesting caustic agents, electrical cords etc.
- Malocclusion of teeth
- Renal failure and associated uraemia
- Eating/ drinking of frozen or too hot food/ water
- Chemical agents
- Licking and drinking of irritants, accidental exposure to strong acids and alkali, thallium intoxication in dog and cat
- Immune mediated stomatitis
- Pemphigus in dogs, lupus erythematosus

Species-specific infectious causes of stomatitis include

Cattle: Foot and Mouth Disease (FMD), Bovine Malignant Catarrhal (BMC), Bovine Virus Diarrhoea (BVD), Actinobacillosis, Actinomycosis, Rhinosporidiosis

Dogs and Cats: Parvo virus, Corona virus, leptospirosis, candidiosis, feline viral rhinotracheitis, feline calicivirus, secondary infection to immunosuppression (feline leukaemia virus, feline immunodeficiency virus), tooth root abscesses, severe periodontitis, osteomyelitis.

Sheep and Goat: Contagious ecthyma, pox, PPR (Pestes des petit ruminants)

Clinical signs: Stomatitis is associated with partial or complete loss of appetite, painful mastication/ ingestion, empty chewing movements, smacking lips, copious salivation and saliva with blood cells, pus or epithetical tissue, foul and foetid odour from mouth, halitosis, enlargement of local lymph node, rubbing of mouth against hard objects, frequent movement of jaw, increased desire for water and refusal of animal for examination of the mouth.

Diagnosis: The mouth parts of the animal showing above clinical signs has to be examined for presence of vesicles, ulcers, catarrhal inflammation, erosions and the sings associated with such lesions.

Treatment

Identification of the cause helps in specific treatment of the problem. Non-specific supportive treatment includes avoidance to further exposure to irritant objects, application of antiseptic lotions such as boroglycerine, chlorhexidine, washing with potassium permanganate solution. Systemic antibiotic should be administered in severe cases. Non-steroidal anti-inflammatory drugs may be administered to reduce inflammation and pain. Polaprezinc (Chelated form of zinc and L-carnosine) has been found effective in treatment of acetic acid- induced stomatitis and may help in promoting healing of stomatitis in the clinical setting. Animal should be provided with soft palatable diet till recovery.

Diseases of Teeth

Diseases of teeth are characterized by discolouration or deformity of the teeth table, inability to chew, or painful mastication. The causes of teeth disorders may be congenital or acquired as in fluorosis (hydro-fluorosis/ industrial fluorosis). The teeth colour becomes yellowish and carries, decaying of cement surface appears with chronic exposure to fluoride. Infectious agents like canine distemper in dogs with the enamel formation, and tooth colour becomes dirty yellow. In very young animals, the teeth become bluish in colour due to deficiency of calcium.

Teeth disorders are clinically manifested by missing of teeth, presence of damaged, displaced or wearing teeth. There may be malocclusion, presence of foreign body, presence of sharp object between the teeth, discolouration of teeth, and foul smelling from mouth (halitosis).

Diagnosis can be made from history, clinical examination, X-ray examination and clinical pathology. The animal should be treated with the specific drug depending on the cause of the problem. In case of fluorosis in endemic areas, animal should be given chemoprophylactic drug such as aluminium sulphate, calcium carbonate etc. and they should be removed from the affected area after exposure for a certain duration. Antibiotic can be given for 4 to 5 days to check the bacterial growth, and anti-inflammatory should be given to reduce pain.

Diseases of Muzzle

Harelip/ split lip /cleft palate is a hereditary condition, characterized by opening in the hard palate between mouth and nasal cavity, and is associated with difficulty in prehension and mastication of the food material. The disease can be diagnosed by history, clinical signs and physical examination of the patient.

Photosensitization upon exposure to ultraviolet light following ingestion of *Lantana camera* that contain photodynamic substances, is characterized by sever inflammatory changes and the affected area looks reddish purple. The lesions are seen in hairless areas like teat, ear, vulvar region including muzzle. Animal with photosensitized lesions should not be exposed to direct sunlight, and should be treated with antihistamines, liver supporting drugs such as liver tonic, glucose and calcium and antibiotic should be used if the inflamed area has been affected with secondary bacterial infection.

Affections of Gland

Deficiency of vitamin A leads to affection of the salivary gland. Involvement of salivary gland results in the interference with head movement and swallowing process. Affection of glands is treated with antibiotic and anti-inflammatory drug and in case of fistula, surgical intervention is required.

Affections of Pharynx

Involvement of pharynx may include inflammation like pharyngitis, pharyngeal obstruction, and pharyngeal paralysis.

Pharyngitis: It is the inflammation of the pharynx caused by infectious agents such as *Actinobacillus* sp. canine distemper, corona virus infection, faulty drenching, ingestion of hot or cold substances, and penetration of the sharp objects. The inflammation of the pharynx is characterized by refusal to take food, painful swallowing, and coughing induction upon external pressure on pharyngeal area, regurgitation of food and water, drooling of saliva, swelling of muscles and lymph nodes. The animal takes a posture with head in an extended position to get relief form pain.

Pharyngeal Obstruction: It may be complete or incomplete depending on the obstructing element. The obstruction may occur due to some fruit, vegetable or bones. Swelling due to actinobacillosis, abscess and chronic growth (polyp) may cause partial or complete obstruction of the pharynx. Appetite remains normal, but it fails to swallow, food comes back to mouth when attempted to swallow. Animal can drink water in case of partial obstruction. Inspiratory noise can be heard on auscultation of pharyngeal region. There may be foul breathe, purulent discharge, laceration of tongue and reduced body condition.

Pharyngeal Paralysis: It may be primary or secondary. Injury to nerve, trauma, encephalitis, central nervous system disorder as seen in rabies may lead to paralysis. The animal has desire to eat but prehension of food and water is not possible and there will be cud dropping. Regurgitation may occur through nostril and there is coughing and aspiration pneumonia and loss of body condition.

Treatment: The underlying cause of pharyngeal affections has to be identified for successful treatment. In case of pharyngitis, animal has to be treated with parenteral antibiotic along with anti-inflammatory drug. If the cause of pharyngeal paralysis is other than rabies, nervine tonic, a combination vit B1, B6 and B12, should be tried. Pharyngeal obstruction due to food material has to be removed or pushed towards rumen. However, forcible push must not be given as it may result in tearing of the tissue requiring surgical intervention.

Diseases of Oesophagus

Diseases of oesophagus in farm animal practice include oesophagitis, oesophageal diverticulum, oesophagial achalasia and choking. Oesophagitis may result from downward movement of infection from mouth or pharynx. Mechanical causes include drinking of irritant liquid or chemicals. Oesophageal diverticulum is a pocket like structure and an outpouching of mucosa through the muscular layer of the oesophagus. It can be asymptomatic or they may cause dysphagia and regurgitation. Diagnosis is made by barium meal contrast radiography. Achalasia is characterised by failure of the smooth muscles of the oesophagus to relax, thus impeding the passage of food from mouth to stomach. Choking is a common accident in cattle and horses. The object may be lodged in pharynx or oesophagus. Greedy eating habits and swallowing foods hastily without proper mixing with saliva result in choke in animals. Cattle frequently become choked on pieces of food as roots and apples that are too large to pass through the oesophagus smoothly. The most common form of choke in horses is seen due to accumulation of dry food in the oesophagus. Animals that have choked should not be given access to feed of any kind to prevent probable aspiration and pneumonia. The administration of drug that stimulates the secretion of saliva such as pilocarpine is a very successful method of relieving the choke in horses and sometimes in cattle.

13

Simple Indigestion

Simple Indigestion is also called mild dietary indigestion, mainly seen in adult animals during lactation. Stall-fed greedy animals suffer more from such indigestion than animals reared on grazing. This is a minor disturbance of gastrointestinal functions in ruminants, mostly in cattle, and rarely in sheep and goats.

Etiology: A sudden change in quantity, quality, composition of food or change in feeding time that brings about alteration in the ruminal environment, can cause simple indigestion. The disease is more common in stall fed animals because of variability in the quality and quantity of their feed. Increase in feed intake especially concentrates beyond the capacity of digestion in dairy animals, and cattle, sheep and goat forced to eat large quantities of poor- quality straw, bedding, or scrub during drought and flood are the common cause of indigestion. Simple indigestion can result from sudden change in the feed, using spoiled or frozen feeds, mouldy feed, introducing urea to a ration, turning cattle on to a lush cereal grain pasture, or introducing feedlot cattle to a high-level grain ration. The common predisposing factors include supply of less quantity of drinking water, depraved appetite resulting in intake of inedible things as occurs in phosphorus deficiency, prolonged oral antibiotic administration, ingestion of placenta, and allergy. Simple indigestion is primarily ruminal atony that usually follows a sudden change in pH of the ruminal contents due to excessive fermentation or putrefaction of ingested feed. The simple accumulation of excessive quantities of relatively indigestible feed may physically impair rumen function for 24-48 hr. The disease is characterised by cessation of primary ruminal contractions, disturbance in ruminal flora and fauna, and stagnation of food in the rumen.

Clinical Signs: The signs depend on cause of the disorder. Silage-overfeeding causes anorexia and a moderate drop in production in dairy cattle. There will be cessation of rumination and the left flank may be moderately distended. The rumen is usually full, firm, and doughy on palpation. The primary contractions are almost absent, but secondary contractions may be present. There may be moderate degree of contraction. Temperature, pulse, and respiration rates are normal. The faeces are normal to firm in consistency but reduced in amount. Recovery is usually spontaneous within 24-48 hr.

Simple indigestion due to excessive feeding of grain results in anorexia and ruminal stasis. The rumen is not necessarily full and may contain excessive fluid. The faeces are usually soft and foul smelling. The affected animal is bright and alert and usually begins to eat within 24 hr. A more severe upset of ruminal disorder is ruminal acidosis, which is discussed later.

Diagnosis: This is based largely on elimination of other possibilities and a history of a change in the nature or amount of the diet. The disease should be differentiated from following disorders.

Traumatic reticuloperitonitis: It is associated with systemic reaction and painful responses to deep palpation of the xiphoid. There will be mild temperature, and haematological findings reveal neutrophilia and shift to left. Traumatic pericarditis is manifested by brisket edema and painful walking, particularly on elevated slope.

Ketosis: The history and the absence of ketonuria help to eliminate ketosis, in which case the intake of roughage will be normal.

The possibility of displaced abomasum can be eliminated by auscultation. Vagal indigestion and abomasal torsions become more readily detectable as they progress because they have a longer course, but initial differentiation may be difficult. Grain overload is distinguishable by its greater severity and the pronounced fall in the pH of the rumen contents. In case of bloat, the abdominal swelling will be more intense and there is respiratory distress.

Phytobezoars cause partial or complete anorexia and scant faeces. Per-rectal examination reveals distended loops of intestine and the firm phytobezoar masses are palpable.

Toxaemic diseases: There will be increase in body temperature above the critical limit.

Treatment: Treatment is aimed at correcting the suspected dietary factors. Spontaneous recovery within 24-48 hour is usual. Administration of 20-40 L of warm water or saline via a stomach tube, followed by vigorous kneading of the rumen, may help restore rumen function. Magnesium hydroxide orally seems to be useful when excessive amounts of high-energy feeds have been ingested. If too much urea or protein has been ingested, acetic acid or vinegar may be administered through oral route. If the activity of the ruminal microbes is reduced, administration of 4-8 L of ruminal fluid from a healthy cow will help restoring the rumen function. Besides, other rumenotoric drugs containing probiotic, copper, cobalt, iron, yeast, magnesium oxide and magnesium hydroxide may be administered. The animal should be provided with soft palatable diet

14

Alkaline Indigestion

Indigestion resulting from change in pH of ruminal fluid exceeding 7.5 is referred to as alkaline indigestion or ruminal alkalosis. An increased production of ammonia in the rumen, because of faulty feeding of proteins or accidental ingestion of urea, associated with rise in ruminal pH above 7.5, results in alkaline indigestion,clinically manifested by signs of urea or ammonia toxicity. Thus, alkaline indigestion occurs with the generation of excessive ammonia and a dramatic elevation in ammonia concentrationin the rumen following feeding of high-protein diets, and non-protein nitrogen sources such as urea, biuret and ammonium phosphate. Alkalosis that occurs in rumen is caused by a decrease in hydrogen ion concentration, and an increase of pH in the rumen liquor. It is an important problem in cattle and buffaloes, but a less common indigestion than other types.

Causes of Ruminal Alkalosis

Alkaline indigestion occurs usually due to excess and sudden intake of protein rich feed, urea feeding, abrupt change of ration, drinking of contaminated and sewage water and excessive feeding of straw over a prolonged period of time. Most of the times, soyabean meal is commonly fed to dairy cows and growing cattle due to its high energy and nitrogen content leading to alkalosis.

Abomasal atony due to dilatation, impaction or torsion of the abomasum is one of the commonest causes of alkalemia (alkalosis) in cattle. There is continuous secretion of hydrochloric acid and potassium into the abomasum, with failure of evacuation of the abomasal contents into the duodenum for absorption. Sequestration of hydrochloric acid and potassium occurs in the abomasum, along with reflux into the rumen, all of which results in a hypochloremic, hypokalemic alkalosis.

Pathophysiology of ruminal alkalosis: Excess feeding of protein-rich feed leads to excess production of ammonia in rumen beyond the utilizing capacity of the microbes. The non-utilized ammonia for microbial synthesis is absorbed by the gastrointestinal tract. The increased level of ruminal ammonia increases its rate of absorption. The excess absorbed ammonia reaches the portal circulation, and are transported to liver and converted to urea. The urea is excreted via kidney through urine, or recycled into the rumen by way of saliva or through blood. Excessive accumulation of ruminal ammonia or nitrogen decreases in volatile fatty acid (VFA) production, resulting in alkaline pH of ruminal fluid that inhibits ruminal contractions and causes paresis of rumen musculature.

The range of ruminal pH in alkaline indigestion remains between 7.5 and 9.5. The rumen pH exceeding 7.5 decreases the population of ruminal microflora. The rumen fluid appears watery and dark brown in colour. It may become slimy with putrid odour. The high ruminal fluid pH brings about inappetence and interferes with cellulose digestion and paresis of rumen musculature. Urea decomposed paddy straw or protein decomposition releases a large amount of ammonia that makes the rumen pH highly alkaline (about 8 to 9.5). Ammonia is absorbed from the rumen, and the animal exhibits signs of ammonia toxicity.

Clinical Signs

All clinical case of alkaline indigestion had history of ingestion of large amount of protein rich and low carbohydrate diet. Ruminal alkalosis in cattle can cause a number of symptoms including slow, shallow breathing in an attempt to preserve carbon dioxide by breathing slowly and shallowly, muscular tremors, convulsions, tetany, tonic and clonic convulsions. The animal suffering from ruminal alkalosis may develop disturbances in gastroenteric, hepatic, renal, circulatory and nervous systems. The animal is dehydrated. The heart and respiratory rates are elevated with heart rate of 90–120 beats/min and respiratory rate of 50–80 breaths/min. The rumen is enlarged, static, and splashy. The ruminal fluid may be watery and brown in colour. The animal may have diarrhoea or loose faeces, often with whole grain in it. The animal may also have constipation and semi-solid or pasty faeces. The principal signs of alkaline indigestion include anorexia, moderate decrease in milk yield, depression, and nervous disturbances along with leucocytosis. Besides, the affected animals show dry muzzle, tachycardia, tachypnoea, ruminal atony, grinding of teeth, salivation, twitching of ears and facial muscles, significant increases in Hb and PCV, hypoglycemia, increased blood urea nitrogen (BUN), lowered protein, albumin and globulin, low serum Ca and P level. Reduction in milk production causes economic loss to the farmers. The animal may have a cool skin surface, subnormal temperature, depression, and weakness. Hyperpnea and dyspnoea may also occur in the terminal stages.

Diagnosis and Differential diagnosis: The diagnosis of ruminal alkalosis can be made from the clinical signs, history and by determining pH of the rumen liquor. The history may reveal an access to urea or exclusive feeding of straw. The clinical signs suggestive of alkaline indigestion include atonic rumen, complete anorexia, signs of abdominal pain, passing of pasty faeces, convulsions. The laboratory examinations reveal dark brown coloured ruminal fluid with watery consistency and putrefied, fishy or ammoniacal smell and pH ranges from 8.5 to 10. Microscopically, there is absence of rumen microflora, and methylene blue reduction test (MBRT) and sedimentation activity test (SAT) time is elevated. The increased BUN in blood and cerebrospinal fluid (CSF) to about 2 to 3 times is noted during biochemical testing.

The disease has to be differentiated from lead poisoning, which is associated with nervous signs, and history reveals a licking point. The poisoning with Chlorinated hydrocarbon (CHC), applied as ecto-parasiticide or pesticide, is manifested by more pronounced convulsive episodes are and response to calcium therapy along with history may help in differentiation.

Ruminal Acidosis: Acidosis is the counter disease of alkalosis. A wrong diagnosis in these two diseases may cause animal loss. Sub-acute ruminal acidosis (SARA) can affect the health and performance of dairy cows fed on high carbohydrate ration, even those that are high yielding and well managed. SARA can cause a number of consequences, including reduced milk yield, lameness, and gastrointestinal damage. Metabolic alkalosis can be chloride-responsive or chloride non-responsive. Chloride-responsive alkalosis is treated with an intravenous administration of a fluid that contains NaCl. Chloride-unresponsive alkalosis is typically caused by a severe deficiency in magnesium or potassium, or by mineralocorticoid excess. Acidosis and alkalosis refer to the pathophysiologic process that cause net accumulation of acid or alkali (base) in the body, whereas acidemia and alkalemia refer specifically to the pH of the blood. Alkalosis is caused by an increased absorption of alkali, excessive loss of acid or a deficit of carbon dioxide.

Treatment

The treatment is aimed at targeting the primary cause by changing the ruminal pH and stopping further ammonia production. The treatment of alkaline indigestion in cattle includes use of acidifier like 5 % Acetic acid in 250 ml water orally once a day. Commonly, vinegar bottles are available in 1 litre, and commercial lactic acid preparations are also used. Normal Saline 0.9% is used for chloride responsive metabolic alkalosis @ 50 – 100ml/hour until urinary chloride rises to > 25 mEq/L (> 25 mmol/L) and normalization of urinary pH. Underlying conditions are treated with the correction of hypovolemia and hypokalemia. The blood pH is urgently correctedby hemofiltration or hemodialysis is an option. Preferably, intra ruminal antibiotic is advised to check the growth of unwanted microbes (Sterptopenicillin @1 large dose). Liver tonics play a vital role, and B-complex preparation with liver extract is more effective in improving impaired liver function. Cud transplantation can be performed for the rejuvenation of microflora.

15

Ruminal Acidosis

Ruminal acidosis is a gastrointestinal disorder caused by ingestion of large quantity of easily fermentable carbohydrate by ruminants, not earlier habituated with such type of diet. The disease is manifested by dehydration, anorexia, diarrhoea, lethargy and neurological signs. It is otherwise termed as D-lactic acidosis, overeating, acute impaction, grain engorgement, grain overload, founder, carbohydrate intoxication and false milk fever etc. The disease is seen both in acute and subacute forms. The subacute form is more common in cows with high level of production as they are to be maintained with high plane of nutrition particularly carbohydrates to meet the energy requirement.

Aetiology and pathogenesis: Ruminal acidosis most commonly occurs following engorgement of large amounts of starch or other rapidly fermentable carbohydrate. Excessive intake of readily fermentable starch often occurs when animals are first being adapted to a high-concentrate diet, or after an accidental ingestion of large quantity of highly fermentable carbohydrate following a period of starvation. The population of Gm +ve bacteria, *Streptococcus bovis* increases following ingestion of large quantity of easily fermentable carbohydrate. This leads to production of more volatile fatty acids (VFA) and lactic acids, thereby reducing the ruminal pH and destruction of cellulolytic bacteria. Finely ground carbohydrate feed exposes large surface area for rapid microbial digestion and acid production leading to fall in rumen pH. Both the acute and chronic forms of the disease are seen depending on kind of food ingested, earlier experience of feeding on that feed, and ruminal environment.

Anaerobic microbes in the rumen and caecum ferment carbohydrates to VFA and lactate. The daily ruminal production of VFA may go above 55 mol. Herbivores absorb these organic acids from the rumen for metabolism by tissues. Abrupt increase in carbohydrate supplies following grain engorgement or during adaptation to high-concentrate diets increases the total acid including lactic acid in the rumen. Normally, lactate is present in the digestive tract at only low concentrations, but when carbohydrate supply is increased abruptly, lactate can accumulate, resulting in ruminal lactic acidosis.

Sub-acute ruminal acidosis is a condition characterised by a rumen pH between 5.0 and 5.5, where the total concentration of short chain fatty acids (SCFA) has been increased, and the ratio between acetic acid, propionic acid and butyric acid has been shifted towards propionic acid and butyric acid, and the accumulated concentrations of lactic acid in the rumen fluid does not exceed 5-10 mmol per litre. The rumen microbial flora is characterised by a dominance of Gram-negative bacteria though the number of Gram-positive bacteria is higher. In contrast, the acute clinical ruminal acidosis is characterised by pH below 5.0, and the accumulation of lactic acid up to 300 mmol per litre, and absence of protozoa (defaunation) in the rumen fluid and the rumen microbial flora is dominated by Gram-positive bacteria.

Rumen fluid osmolality normally ranges 240-265mOsm/L, with roughage diets, and 280-300 mOsm/L with concentrate diets. Minerals, VFA, lactate, and glucose are the primary solutes in ruminal fluid those contribute substantially to the osmotic pressure. The blood osmotic pressure of 285 to 310 mOsm is maintained by dissolved protein, glucose, BUN and other solutes. Ruminal osmolality may go as high as 515mOsm in lactic acidosis resulting in water being drawn from the blood vascular system, leading to dehydration. Rapid influx of water to neutralize osmotic pressure swells the ruminal papillae and can pull patches of the ruminal epithelium into the rumen by stripping the internal surface layers of the rumen wall from the underlying layers. The damage to the wall of the rumen or small intestine leads to sepsis and liver abscess. Repair of digestive tract leads to thickening of ruminal and reticular wall (hyperkeratosis or parakeratosis) that inhibits rate of VFA absorption for months or years after the damage has occurred. Passage of VFA post ruminally for absorption is possible, but abomasal presence of VFA hinders acidification, and protein and mineral digestion. Thus, a single bout of non-fatal ruminal acidosis may have prolonged effects. In reduced pH range, fungi such as Mucor, Absidia, Rhizopus etc. multiply and causes mycotic or fungal rumenitis. Acid reaches liver following absorption along with Corynebacterium species, Fusobacterium species, and cause hepatic abscess.

As the ruminal pH declines, the amplitude and frequency of rumen contraction decreases, and there is ruminal atony at about pH 5. The entry of lactic acid to duodenum exerts a reflex inhibitory action on rumen causing ruminal stasis. Vasoactive substances such as histamine and endotoxins are released during a low ruminal pH and alter hemodynamic of peripheral vasculature. These substances cause vasoconstriction and dilation, which injures microvasculature. The resultant ischemia causes reduction in oxygen and nutrient supply to the extremities, and physical degradation of junctures between tissues that are structurally critical for locomotion, and animal shows clinical signs of laminitis.

Usually, absorbed lactic acid is buffered by plasma bicarbonate buffering systems. The L form of the absorbed lactic acid is being used preferentially, and D lactic acid accumulates in the body fluid. So, the disease is also called D lactic acidosis. The pH of body fluids may or may not be depressed with decreased ruminal pH during acidosis as body fluid pH is buffered by bicarbonate buffering system. Central nervous system function can be disturbed by low bicarbonate concentrations even if blood pH is not depressed. The other clinical signs include anorexia, variable feed intake, diarrhoea, and lethargy.

Type of Acidosis

Depending on the intake of digestible carbohydrate, acidosis is classified in to three different types; Acute rumen acidosis (ARA), Sub acute rumen acidosis (SARA), and Chronic latent rumen acidosis (CLRA)

Clinical Signs

The affected animals shows signs of anorexia, dullness, ruminal stasis, diarrhoea, nasal discharge, increased heart and respiratory rate, restlessness, lameness, staggering gait, hyperpnoea, abdominal respiration, falling down, grinding of teeth, flank watching, tremor and severe tetanic movement of hind and fore limbs. However, severity of clinical signs starts receding in two days in surviving animals. Subacute ruminal acidosis is manifested by reducing milk production and poor feed digestibility.

Clinical Pathology

Biochemical changes in rumen liquor include decreased pH and increased concentration of lactic acid and TVFA. Changes are most severe between 12 and 72 hours after feeding. The fall in rumen pH is associated with increased production of lactic acid in the rumen due to faster and complete fermentation of starch by amylolytic bacteria in the ruminoreticular compartment, leading to fall in the ruminal pH between 5.6 and 5.2, often being used as benchmarks for chronic or acute acidosis, respectively. Acute ruminal acidosis may be associated with decreased blood pH below 7.35. There may be increased serum level of liver specific enzymes likes AST, ALT, GGTP and sorbitol dehydrogenase.

Necropsy finding: The visceral organs show congestion. There may be sloughing of epithelial layer from the rumen. Rumenitis and reticulitis are commonly seen and chronic cases of ruminal acidosis are manifested by parakeratosis, liver abscess.

Differential Diagnosis

The ruminal acidosis has to be differentiated from traumatic reticulo peritonitis (TRP) that is associated with a high rise in temperature, leucocytosis, neutrophilia and abducted elbow. Milk fever is characterized by hypocalcaemia that mostly occurs within 72 hours after parturition, manifested by sternal or lateral recumbency. The nervous form of ketosis is associated with ketone bodies in urine, sweet smell from the breath and mucous coated faeces.

Treatment

The treatment is given to correct both the ruminal and systemic acidosis. Ruminal acidosis is corrected by administration of sodium bicarbonate @ 1-2 gm/kg body weight orally. The systemic acidosis is rectified by intravenous administration of 200-500 ml of 7.5% sodium bicarbonate solution in an adult animal. The administration of oxytetracycline with equal quantity of distilled water in to rumen suppresses further production of lactic acid by inhibiting lactic acid producing microbes in the rumen. Antihistaminic drug, 10-15 ml for large animal and 3-5ml for small animal, is given intramuscularly to suppress elevated histamine level. Corticosteroid like Betamethasone or Dexamethasone @ 3-5ml per large animal should be given to prevent shock. Vitamin B-complex injection should be given to enhance ruminal motility. Rumenotoric preparations are given to regain normal fauna and flora. Gastric lavage is performed in acute ruminal acidosis, and fresh ruminal cud collected from slaughtered animal may be transplanted to restore ruminal micro flora and fauna.

Control: Appropriate feeding management can reduce the incidence of acute ruminal acidosis in a dairy farm. The ruminants should not be subjected to have abrupt change to an easily digestible concentrate rich ration to which they do not have prior acquaintance. The goal of producing higher milk yields from dairy animals requires feeding of high-energy diets to dairy cattle. The modern dairy cows are able to produce more milk, in turn, requiring an excessively high intake of energy rich dry matter. Therefore, feeding of rations high in concentrate is necessary, thus predisposing animals to develop some kind of pathological conditions associated with overfeeding. The prevention of SARA can be described by two principles: first, to allow a proper adaptation of the ruminal mucosa in the post-partum period and the micro flora in order to absorb and metabolize the high levels of small chain fatty acids (SCFA) and secondly, to keep the ruminal pH in physiological ranges in spite of high-energy intake. In other words, the mechanisms applied consist of either prevention of a fermentation disorder or regulation of the fermentation process by direct intervention. The first part can be achieved mainly by management of husbandry practices and feeding, the second pathway is achieved by maintaining fibre-content of the ration within defined borders or adding alkalizing and buffering substances to the ration.

16

Subacute Ruminal Acidosis

Subacute Ruminal Acidosis (SARA) is a common but serious problem in high yielding dairy cows, where the management system has a major focus on higher productivity, for which the ruminants are kept on high plane of nutrition and fed with carbohydrate-rich diet to exploit the production potential of the animals. This a chronic disease and differs from overt cases of ruminal acidosis that occurs following ingestion of highly fermentable and finely ground carbohydrate ration to which the animal is not earlier accustomed. The later is characterized by signs of anorexia, dullness, nasal discharge, distended abdomen, ruminal stasis, increased respiratory and heart rate, incoordination, blindness, tremor and lameness. Sudden change of feeding to high grain diets in ruminants that are adapted to digest predominantly forage diets, accidental feeding on grain diet in large quantity or overeating following periods of feed deprivation leads to excessive production of lactic acid in rumen, drop in ruminal pH, preferential absorption of D-lactic acidosis and systemic acidosis. Lactic acidosis is widespread in ruminant production systems, and it has severe consequences in ruminants. Some of the problems associated with lactic acidosis include lameness, polioencephalomalacia, ruminitis, poor immune functions, reduced feed intake, reduced nutrient absorption, liver abscess, grain bloat, clostridial infections, transient diarrhoea, milk fat depression and poor milk production, moderate rumen distension, haemorrhages in lung, and sudden death. In India, 18% cases of anorexia in sheep and goats have been accounted for ruminal acidosis.

Subacute ruminal acidosis (SARA) is a well-recognized digestive disorder in dairy herds with high yielding animals. It is a ruminal fermentation disorder characterized by extended periods of depressed ruminal pH (≤ 5.5). The problem is insidious in nature and more difficult to diagnose. The demand for increased milk production with higher grain and lower fibre diets to maximize energy intake during early lactation leads to drop in ruminal pH. The pH should lie between 6.5 and 6.8 for optimum ruminal fermentation and fibre digestion, and it fluctuates below this level for short periods during the day. The drop in ruminal pH accompanies breakdown of dietary carbohydrates, particularly from cereal grains such as corn and barley. Grains are high in readily fermentable carbohydrates that are rapidly broken down by ruminal bacteria, leading to the production of volatile fatty acids (VFA) and lactic acid. Under normal feeding conditions, VFA are readily absorbed by papillae on the rumen wall. Once absorbed, VFA enter in to the bloodstream and can be used for synthesis of milk constituent. Dairy herds experiencing SARA will have a decreased efficiency of milk production, impaired health and high rates of involuntary culling.

Ruminal pH

The production of acids and inherent buffering capacity of the rumen determines the ruminal pH. In spite of number of complex mechanisms to keep the ruminal pH above the biological threshold of about 5.5 and affected animals' own correcting measure of eating less aftermath of onset of acidosis, the disease continues to be a problem for high yielder on high grain diets. High intake of energy (grain) rich diet in dairy cattle is associated with lowering of ruminal pH. The total daily intake of fermentable carbohydrates depends on total dry matter intake and the density of non-fibre carbohydrates in the diet. This suggests that ruminal acidosis could become an increasing problem with genetic progression aimed at enhancing milk productivity. The better feeding management allows cows to eat more to sustain high level of production.

Ruminal pH varies considerably during the course of a day, and is particularly driven by time and the amount of fermentable carbohydrate meal in each day. As pH drops, lactate-synthesizing bacteria such as *Streptococcus bovis* begin to ferment glucose to lactate instead of VFA. This is a dangerous situation, since lactate has a much lower pK (3.9) than VFA (4.8) and lactate is 5.2 times less dissociated than VFA at pH 5.0. As a result, lactate stays in the rumen longer and contributes to the downward fall in ruminal pH. But, the problem of drop in ruminal pH cannot be solved by frequent feeding of the animals as repeated feedings enhance the total quantity of dry matter intake.

Ruminants possess highly developed mechanisms to maintain ruminal pH within a physiological range from 5.5 to 7. The low ruminal pH that follows intake of large amount of fermentable carbohydrate may be associated with increased osmolarity of the ruminal contents inhibiting the feed intake, the first response after drop in ruminal pH. Ruminants possess a complex system for buffering the organic acids in the rumen, though the buffering is relatively too small. Saliva produces large amount of buffer that contains sodium, potassium, bicarbonates and phosphates and is secreted during the chewing activity. Thus, fibre content of the feed increases saliva secretion as it enhances the chewing activity. However, saliva secretion is not influenced by the lowered ruminal pH. The ability of the rumen to rapidly absorb the organic acids from the rumen contributes greatly to the stability of the ruminal pH. The passive absorption of VFA across the ruminal wall is enhanced by finger-like papillae that provide huge surface area. Ruminal papillae increase in length when cattle are fed high grain diets. This presumably increases the surface area and absorptive capacity that protect the animals from acid accumulation in the rumen. Thus, large dietary changes may not affect the mean ruminal pH but lowest pH level is greatly influenced.

Genesis of SARA

SARA develops when the ruminal pH drops below the physiological threshold of about 5.5 for a protracted period. Lactic acid in the rumen is rarely the cause of depressed pH in the rumen in SARA unlike acute acidosis. It results from excessive VFA production that exceeds the ability of the ruminal papillae to absorb them. Abrupt change from high fibre diets to high concentrate does not allow ruminal bacteria and ruminal papillae adequate time to adjust, thereby leading to the rapid production and accumulation of VFA. Too small particle size of the feed is another cause of SARA as this suppresses rumination and the production of saliva that helps in buffering the rumen pH.

Severe grain-induced SARA is dominated by *Streptococcus bovis* and *Escherichia coli*, whereas mild grain-induced SARA was dominated by *Megasphaera elsdenii,* and alfalfa pellet-induced SARA is dominated by *Pasteurella albensis*. The severity of SARA and degree of inflammation are highly correlated with the abundance of *E. coli* and not with lipopolysaccharide in the rumen. Periodic abundance of available starch allows amylolytic bacteria (e.g., *Ruminobacter amylophilus*, *Streptococcus bovis*, *Lactobacillus* spp.) to proliferate and produce excessive quantities of fermentation acids. Unfortunately, gains in VFA absorption at ruminal pH below 5.5 can be offset by lactate production. As pH drops, *Streptococcus bovis* begins to ferment glucose to lactate instead of VFA. At ruminal pH of 5.0, lactate is 5.2 times less dissociated than VFA, as a result lactate stays in the rumen and contributes to further fall in ruminal pH, leading to acute ruminal acidosis. Ruminal acidosis is known to be related to diseases such as rumenitis, rumen parakeratosis, liver abscesses (LAs), laminitis and bloat. Laminitis in cattle has been linked to nutrition, specifically in acute acidosis and SARA. Although the exact relationship between SARA and laminitis is not known, one of the theories relates SARA-induced damage to the ruminal epithelium, allowing absorption of histamine and endotoxins. These chemicals and possibly, other compounds disrupt normal circulation and cause inflammation within the hoof, leading to the condition commonly referred to as laminitis. A mild episode of SARA does not affect ruminal, epithelial barrier function immediately after the episode but a rapid and more severe acidification (pH 5.2) *in vitro* increases epithelial permeability following the insult.

The periparturient cows are at risk of subacute ruminal acidosis because of the time required for the rumen microflora and papillae to adapt to increased intake of concentrates immediately before parturition and during early lactation when feed intake increases rapidly to meet the energy needs of high-producing dairy cows. The adaptation of the ruminal microflora and papillae from a system appropriate for forage to a system capable of utilizing high-energy lactation rations requires a gradual change over a period of 4-6 wk. Therefore, any increase in energy intake before parturition should take in to account the period of time required for adaptation of the rumen epithelium to dietary changes. Further, cows become more prone to acidosis over time even though intake of grain is decreased to avoid acidosis. The severity of each subsequent bout of acidosis increases, especially for cows fed diets low in physically effective fibre and at high risk of acidosis. Therefore, a bout of acidosis that occurs due to improper feed delivery or poor diet formulation can have long-term consequences on cow health and productivity.

Sings and Diagnosis of SARA

Ruminants suffering from SARA do not exhibit overt clinical signs. However, the consequence of SARA is complex and diverse. The major sign is reduced feed intake. Decreased feed consumption or reduced digestibility of grain also confuse for diagnosis of SARA. On the other hand, interruption in normal consumption pattern also precipitates acidosis. Other signs include lethargy, diarrhoea, appearance of undigested grain in the faeces, foamy faeces containing gas bubbles, excessive salivation, panting, kicking at the belly and general signs of discomfort and stress. The secondary signs in chronic cases of SARA include episodes of laminitis, weight loss, poor body conditions despite adequate energy intake and unexplained abscess. The inflammation and abscess throughout the ruminant's body may occur due to the negative impact of organic acids on the ruminal wall leading to parakeratosis enabling translocation of pathogens in to the blood stream. Thus, SARA should be investigated, if secondary signs are occurring for no apparent reasons. In order to achieve proper diagnosis, SARA has to be considered as herd management problem. Screening of the herd for SARA by rumenocentesis, performed on a sample group, preferably on 12 individuals, may reveal the presence of SARA. The herd screening should include the suspected risk group. Rumenocentesis for diagnosis of SARA may be ethically questionable but the technique remains the most reliable means of diagnosing SARA.

Effects on Production

Milk fat depression is a major sign of SARA and is caused by incomplete saturation of fatty acids within the rumen. Incompletely saturated fatty acids (particularly certain *trans* forms) are potent inhibitors of milk fat synthesis in mammary gland. Feeding high levels of unsaturated fatty acids increases the *trans* fatty acid uptake from the rumen. Ruminal acidosis may also facilitate *trans* fatty acid uptake from the rumen, perhaps by inhibiting certain bacteria, which are normally responsible for biohydrogenation in the rumen.

Consequences of SARA include depression in feed intake, reduced fiber digestion, milk fat depression, diarrhoea, laminitis, liver abscesses, increased production of bacterial endotoxin and inflammation, characterized by increases in acute phase proteins. Presumably, heightened volatile fatty acid production stimulates satiety receptors in cattle, which in turn results in the commonly observed "off-feed" or low intake syndrome. Rumen digestion of neutral detergent fibre (NDF) from grass hay, legume hay and corn silage decrease in animals suffering from SARA. In addendum, dietary neutral detergent fibre (NDF) content higher than 33.8% of DM had lower ($P < 0.01$) fecal endotoxin than those with a dietary NDF content lower than 33.8% of DM. Poor feed digestibility and absorption also lead to reduced production. The associated secondary problems are also responsible for poor production performance of the animals.

Public Health Concern

Ruminal acidosis can have direct public health concern. Low ruminal and intestinal pH due to high grain diet have been reported to increase the risk of shedding enterohemorrhagic *E. coli*. It has been established that switching cattle to a high forage diet just prior to slaughter decreases such shedding.

Management of SARA

Monensin either as controlled release capsule or premix was tried for attenuation of SARA in Holstein cows. Neither of them had effect on ruminal VFA concentration but reduces the acetate and propionate ratio. Monensin- premix treated cows produce more milk as a result of higher dry-matter intake but with reduced milk fat content. Similarly, supplementing the diet with direct-fed microbials that enhance lactate utilizers in the rumen may reduce the risk of SARA. Yeasts, Propionobacteria, Lactobacilli and Enterococci have been used for enhancing lactate utilization.

The risk of developing SARA can be reduced by adopting a feeding regime, which balances ruminal buffering with the production of volatile fatty acids from fermentation of carbohydrates. This can be accomplished by feedstuff selection and/or by the addition of dietary buffers such as sodium bicarbonate or potassium carbonate. Cows do not prefer voluntary intake sodium bicarbonate to attenuate SARA. Therefore, adjustments of suboptimal feeding and management routines are believed to be essential prophylactic steps. It is recommended to avoid the following events to minimize the risk of SARA,

1) a daily mean ruminal pH lower than 6.16 and 2) a time period in which ruminal pH is <5.8 for more than 5.24 hr/day. Thus, the key aspect of SARA prevention is careful control of the intake of rapidly fermentable carbohydrates. This is accomplished by providing adequate chemical fibre, processing grains properly, including high fibre concentrates as needed and offering feed to cows in a manner that allows them to consume small, frequent meals on a regular basis. Feed additives and supplements can provide an additional margin of safety for preventing SARA.

Cows receiving the MFC (medium concentrate: forage) treatment have lower daily mean ruminal pH (below 6·0) for a longer duration compared with the HFC (high forage: concentrate) diet. Additionally, cows receiving the MFC diet have a greater reduction in milk fat concentration (g/100 g FA) of FA C16 (17 vs. 9%), trans-10 18:1 (159 vs. 21%) and trans-9, cis-11 conjugated linoleic acid compared with cows receiving the HFC diet.

Increasing dietary cation and anion difference (DCAD) increases urine pH and modifies the urinary excretion of minerals. With low DCAD, 70% of Cl and only 16% of Na are excreted in urine whereas with high DCAD, 33% of Cl and 53% of Na are excreted. Increasing DCAD allows cows to increase dry matter intake because of the ability of positive DCAD to maintain blood acid-base status. A localized rumen buffering effect could not be excluded and could be linked with a higher amount of HCO_3 recycled into the rumen. The main mechanism involved in regulating blood pH might be renal excretion of protons and strong ions and renal HCO reabsorption. Dietary supplementation with acarbose in dairy cattle fed high-production rations may be effective in reducing the time for which rumen pH is suboptimal, with no negative effects on ruminal fermentation and blood metabolites. Supplementing feed of the dairy cows with active dry yeast increases the mean, minimum and maximum ruminal pH; decreases time spent in subacute rumen acidosis, and decreases total VFA concentration in the rumen.

Conclusion

SARA is a major production limiting health problem in dairy cattle, expected to produce more with feeding of larger quantity of highly fermentable carbohydrates. Milk production may increase temporarily by over-feeding of grain. However, the long-term health effects due to development of SARA and its economic consequences may become quiet frustrating. The prevention of SARA includes the establishment of feeding and management guidelines designed to minimize rumen acid load. Limiting the intake of rapidly fermentable carbohydrates, providing adequate ruminal buffering and allowing the ruminal adoption to high grain diet are the basic principles of preventing SARA. Regular monitoring may facilitate early recognition of the condition and limit economic losses. Some degree of SARA may be inevitable, and presents a challenge to the dairy industry as consumers become increasingly concerned about the welfare of production animals.

17

Tympany

It is otherwise known as bloat or hoven. It is classified as primary or secondary, based on its etiology. Continual production of large amount of gas in the rumen through the process of microbial fermentation is a physiological process and they are eructed out. Anything that interferes with eructation leads to over- distension of rumen and reticulum due to accumulation of gas. Such problem is termed as ruminal tympany or bloat. The disorder is perhaps most commonly seen in cattle, but certainly is not uncommon in sheep and goats.

Frothy bloat, otherwise termed as primary tympany, results from failure of eructation mechanism due to trapping of gases of fermentation as stable, persistent foam. As quantities of this foam build up, the rumen becomes progressively distended and bloat occurs. The legumes, such as alfalfa or clover (pasture bloat) are rapidly digested in the rumen, which seems to results in a high concentration of fine particles that trap gas bubbles. Additionally, some of the soluble proteins from such plants may serve as foaming agents that help in trapping of gases inside the bubbles. Also, feeding of a diet with high levels of grain, especially finely grounded (feedlot bloat), undergoes rapid digestion. The abundance of small particles appears to trap gas in bubbles. Some species of rumen bacteria produce an insoluble slime that promotes formation of stable foam.

Free gas bloat, also called secondary tympany, occurs due to failure in eructation process of the free gas produced in the rumen. The obstruction of the oesophagus by foreign bodies, abscesses, tumours, or interfere once with ruminoreticular motility as seen in reticular adhesions, damage to motor functioning of the rumen clearly can lead to failure in expulsion of gases. The posture of the ruminant also affects eructation. A ruminant cannot eructate when lying on its back. A cow in a ditch and unable to make itself up right bloats rapidly. Ruminant that has to undergo surgery in dorsal recumbency should be starved for 12 to 24 hours prior to surgery. Otherwise by the time the surgeon is ready to make the incision, the abdomen is already distended.

Regardless of frothy or free gas bloat, distension of the rumen compresses thoracic and abdominal organs. Blood flow in to the abdominal organs is compromised, and pressure on the diaphragm interferes with lung function. The cause of death is usually hypoxia due to respiratory failure.

Clinical Signs: Bloat is an acute disease with a short course in many instances, animals are "found dead". Although it is primarily an acute disorder, chronic, recurrent forms are also seen. The common clinical signs include abdominal distension, and the distension is typically more prominent on left side because of anatomical location of the forestomach. The entire abdomen may become distended with increased in accumulation of gas in the rumen. Animals show reluctance to move and there will be cessation of feeding. Signs of general and respiratory distress such as anxiety, vocalization, rapid breathing, neck extended with protruding tongue are noticed. Respiratory distress and poor oxygenation to tissue may result in staggering. Once an animal with bloat is recumbent, death occurs rapidly due to respiratory failure.

Diagnosis of bloat, irrespective of primary or secondary, is typically straightforward and the clinical picture largely reflects how long the condition has existed. The secondary bloat can be differentiated from the frothy bloat by puncturing a large bore needle or trocar and canula in the paralumbar fossa on the left side or passing a stomach tube into the rumen through the oral cavity in an attempt to expel the gases from the rumen. Gases come out readily in secondary bloat, whereas lot of froth comes out in primary bloat and expulsion of gases is being hindered by the foams.

Necropsy: The post-mortem examination reveals characteristic lesions, including congestion and haemorrhages in the cranial thorax, neck and head, and compression of the lungs. The pressure from the distended rumen causes congestion and haemorrhage of the oesophagus in the neck region, while the oesophagus in the thorax remains pale. This demarcation between congestion and pallor seen in the region of the thoracic inlet is called the "bloat line".

Usually, the liver is also pale because of displaced blood and interruption of blood supply. Obvious distension of the rumen is certainly observed in animals that die of bloat. But bloat also occurs rapidly after death from almost any cause in ruminants, and is not a useful diagnostic lesion.

Treatment: Bloat is an emergency problem, and must be relieved with haste. In acute condition, rumen gas should be released immediately by emergency rumenotomy. Insertion of a rumen trocar through the left flank into rumen is sometimes advocated, but usually not very effective unless it has a large bore (i.e. 1 inch), and is often followed by complications such as peritonitis. In less severe cases, a large bore stomach tube should be passed down the oesophagus into the rumen. But, passing of stomach tube must not be attempted if rumen is highly inflated and the gas in rumen is seriously compromising the respiration. In such case, an incision should be given in the left flank to relieve pressure immediately. Use of large probang or stomach tube releases the free gas readily, although it may need to be repositioned repeatedly to effectively relieve the pressure. In the case of frothy bloat, antifoaming medications can be delivered directly into the rumen through the tube. The animal should then be closely observed to ensure that the treatment is effective and the animal begins to belch gas, otherwise a rumenotomy may be indicated.

A variety of antifoaming agents have been used to relieve frothy bloat. These include vegetable oils (corn, peanut) or mineral oil (100-300 ml). A number of effective commercial products are available that include surfactant like poloxalene or a detergent like alcohol ethoxylate.

Control

Control of bloat relies on management coupled sometimes with medications, but despite best efforts, it is rarely totally effective. Some of the techniques advocated may be applicable to small herds, but are too labour intensive to use in large herds. Many of the techniques used are based on reducing the rate of fermentation in the rumen. Examples of control strategies include maintaining pastures that have grasses mixed with legumes such as alfalfa, feeding of hay before turning out on bloat-inducing pastures, feeding roughage such as straw or grass hay in addition to concentrate, feeding of cracked or rolled grain rather than finely grounded.

Ruminants allowed grazing in the cities and semi-urban areas ingest polythene bags along with food that chokes the cardiac end of the rumen resulting in chronic secondary tympany. Disposal of polyethene and plastic bags should be proper to avoid ingestion of non-biodegradable materials by the animals. Although not well defined, a genetic component to susceptibility to bloat has been identified, which might be exploited to some extent in reducing herd prevalence of this condition.

18

Vagus Indigestion

The lesion on the vagal nerve, possibly due to traumatic reticuloperitonitis, leads to distension of rumen, reticulum and abomasums, and is characterized by diminished appetite, decreased ruminal contraction, gradual reduction in milk production, and scanty and sticky faeces, and papple shaped abdomen. However, vagal nerve lesions are not noticed in all cases in post-mortem examination. Affection of dorsal part of the vagal nerve hinders passage of food from rumen and reticulum; whereas affection of ventral part does not allow food to pass beyond abomasum.

Etiology and Pathogenesis

Injury, inflammation, orpressure on the vagal nerve or mechanical obstruction of the cardia or reticulo-omasal orifice due to papilloma or ingested placenta, results in ruminoreticular distention syndrome leading to distension of the rumen, reticulum and abomasum. There are 4 types of vagal indigestion based on the site of the functional obstruction. Type I vagal indigestion, or failure of eructation, results in free-gas bloat, and occurs most commonly due to inflammatory lesions in the vicinity of the vagus nerve, such as localized peritonitis, adhesions (a sequelae of traumatic reticuloperitonitis), or chronic pneumonia. Less common causes include pharyngeal trauma, which affects a more proximal part of the vagus nerve, and oesophageal compression by abscesses or neoplasia, such as lymphosarcoma. Type II vagal indigestion, or failure of omasal transport, develops as a result of any condition that prevents ingesta from passing through the omasal canal into the abomasum. Adhesions and abscesses (reticular or single liver abscesses) are the most common causes.

They are usually seen on the right or medial wall of the reticulum near the route of the vagus nerve. Reticular abscesses and adhesions are almost invariably due to traumatic reticuloperitonitis. Mechanical obstruction of the omasal canal by ingested material like placenta, or masses due to lymphosarcoma, squamous cell carcinoma, granulomas, or papilloma, can also cause chronic ruminoreticular distention due to failure of omasal transport. Type III vagal indigestion is abomasal impaction that develops due to feeding of dry, coarse roughage, such as straw, in a chopped or ground form with restricted access to water, usually during extremely cold environmental temperature. Vagal indigestion can develop in cattle after abomasal volvulus without abomasal impaction. Type IV vagal indigestion, or partial forestomach obstruction, is poorly defined, and develops in cattle during gestation. It may be related to the enlarging uterus shifting the abomasum to a more cranial position, which inhibits normal motility.

Clinical signs

There is a gradual development of ruminoreticular and abdominal distension over a period of days to weeks. Distention of the dorsal and ventral sacs of the rumen results in an "L-shaped" rumen on per rectal examination. Left dorsal and left and right ventral distention of the abdomen causes a "papple" (pear plus apple) shape of the abdomen. There will be reduced appetite, gradual loss of milk production, faecal output is reduced, faeces become scant and sticky, and often contains long hay particles.

The strength of rumen contractions is decreased, but the motility may be increased. Ruminal contraction sounds are not audible as contents become frothy due to the prolonged contractions and failure of the rumen to empty. Temperature and respiration rate are usually normal. In some cases, there will be decreased heart rate. Tachycardia develops as the disease progresses. In chronic cases, the animal develops a rough hair coat, loses condition, and becomes weak, dehydrated and recumbent.

Per rectal examination reveals distended rumen filled with gas or froth.

Clinical Pathology

Hematologic examination may reveal increased PCV suggesting dehydration, or decreased PCV because of bone marrow depression. Lymphocytosis can be seen in vagal indigestion due to lymphosarcoma. Leukopenia may be present with diffuse peritonitis. Increased serum globulin and total protein can be seen with abscesses.

Vagal ingestion causes hypochloric, hypokalemic metabolic alkalosis. Low chloride indicates reflux of chloride from the abomasum into the rumen and obstruction at the level of the abomasum. The chloride levels of the rumen fluid may be increased. Metabolic alkalosis is typically present if serum chloride is decreased. The chloride is usually normal if the lesion is cranial to the abomasum. Potassium is usually low due to decreased potassium intake in the feed.

Diagnosis

The subacute to chronic ruminoreticular and abdominal distension creates suspicion for vagal indigestion. This is associated with change in abdominal silhouette and L-shaped rumen on per rectal examination. Vagal indigestion is a subacute to chronic disease and must not be confused with acute ruminal tympany. Other causes of abdominal distension, such as ascites and uterine enlargement can almost invariably be ruled out by rectal palpation due to the absence of ruminoreticular distension. Occasional cases of longstanding obstruction of the cecum or small intestine can cause severe ruminoreticular and abdominal distension and can be differentiated by per-rectal examination of palpable enlarged caecum or small intestine. Diagnosing the specific cause of vagal indigestion is more difficult but is important because of differences in treatment and prognosis. Physical examination, rectal examination, blood cell count, blood acid-base determination, and serum chemistry values are often useful. Peritoneal fluid analysis can support the diagnosis of peritonitis if total protein or nucleated cells are increased. Radiographs of the reticulum should be taken to identify a radiopaque linear foreign body (eg, wire) or reticular abscess. Definitive diagnosis often requires exploratory surgery. The prognosis of the disease is unfavourable or unpredictable.

Treatment

Medical management alone is almost ineffective. A left para lumbar fossa laparotomy and rumenotomy provides the opportunity for definitive treatment in some cases. Emptying the rumen at the time of surgery may help restore normal rumen motility. Stimulation of low-threshold tension receptors in the reticulum occurs under normal circumstances and causes reflex reticulo- ruminal contractions. However, severe distension causes stimulation of high- threshold receptors that have the opposite effect and inhibit contractions. Ruminal lavage may help in removal of the rumen content, thus reducing abdominal distension, thus reducing abdominal distension.

Supportive or symptomatic therapy should be provided in all cases, which typically involves correcting dehydration as well as calcium and electrolyte deficits, commonly with oral fluids and electrolytes. The affected ruminant may be treated with 20L of Dextrose saline for three days. Laxative may be given for 5 days in cases of impaction. Prokinetic drugs such as metoclopramide injection may be administered.

Control

The most common cause of vagal indigestion is traumatic reticuloperitonitis, which causes adhesions and abscesses that interfere with vagal nerve function. Therefore, prevention of traumatic reticuloperitonitis is important. Good management practices will prevent some cases of vagal indigestion resulting from chronic pneumonia. Early diagnosis of right-sided abomasal displacements and abomasal volvulus, and prompt surgical correction may prevent some cases. Prompt removal of the placenta from the cow's enclosure after parturition will prevent the ingestion of placenta by the animal and thereby preventing the obstruction of cardia, reticulo-omasal orifice, or pylorus.

19

Traumatic Reticuloperitonitis

This is an inflammatory condition of the reticulum and peritoneum due to piercing of the sharp foreign body into the reticular wall. Such clinical condition may result without complete perforation of the reticular wall. The common sequelae of traumatic reticuloperitonitis (TRP) include acute local peritonitis, acute diffuse peritonitis, or acute pericarditis. Either of these results following perforation of the reticular wall by the sharp foreign objects such as needle, nail etc, allowing the reticular contents to come out and infect the surrounding. Chronic local peritonitis may result in vagus indigestion or diaphragmatic hernia. Acute pericarditis may lead to chronic pericarditis or death due to congestive heart failure. Uncommon sequelae include rupture of left gastroepiploic artery or rupture of coronary artery causing internal bleeding, cardiac temponade and death.

Clinical Signs

Acute local peritonitis is manifested by grunting while walking down hill, rigidity of back muscles called tucked up, recumbency and inability to rise, and acute abdominal pain. In case of chronic local peritonitis, the pain is not evident although gait is slow and careful. Chronic or moderate bloat may be present. Acute diffuse peritonitis is associated with cessation of alimentary tract movement, and increased pulse rate.

Clinical Pathology

Traumatic reticuloperitonitis is associated with leucocytosis and neutrophilia with shift to left. The neutrophil count may go as high as 95%. The chronic diffuse peritonitis is associated with leukopenia due to associated toxaemia.

Diagnosis: The sudden onset of grunting pain, ruminal stasis, complete anorexia, severe pain on palpation of ventral abdomen, moderate degree of fever, leukocytosis and fall in milk yield arouse suspicion for traumatic reticuloperitonitis. Bamboo test, sharp lifting of bamboo on either side just behind xyphoid cartilage elicits sharp pain, is used to diagnose of traumatic reticuloperitonitis.

Treatment: The animal should be allowed to stand on an elevated floor so that the sharp foreign body does not pierce further and perforate the reticular wall. Anti-inflammatory and antibiotic should be administered for healing of the inflamed area. Magnet feeding should be practised in order to restrict the sheep foreign object within the reticulum & preventing from penetration into the reticular wall, this reducing the incidences of TRP.

Traumatic Pericarditis

Perforation of pericardial sac by sharp material coming from the reticulum through the reticular wall causes inflammatory condition of the pericardium. The problem is mostly seen during advance pregnancy during last trimester of pregnancy or at parturition, due to pressure on pericardium.

Clinical Signs

Engorgement of jugular vein, swollen ventral abdomen, easily palpable jugular pulse, pericardial friction rubs that increases or decreases with respect to respiratory sound are the commonly seen clinical signs in affected animal. Heart sound is muffled i.e. not clearly audible, and there will be dyspnoea, depression, diarrhoea, and anorexia. In some cases, acute condition may turn to chronic.

Diagnosis can be made from the clinical signs and the history of traumatic reticuloperitonitis. Metal detector test can diagnose the problem if the foreign body is metallic. Differential count shows increased neutrophil with shift to left with increased total leukocyte count. Bamboo test/ Fist test elicit pain reflexes. The prognosis is generally unfavourable.

Treatment

Antibiotic should be administered for 5 to 7 days. Anti-inflammatory preferably non-steroidal drug should be administered in non-pregnant animals. Diuretics should be given to reduce oedema. Vitamin B complex should be administered to improve general health condition of the animal. Dextrose (40ml/Kg body weight), around 12L in adult animal may be administered. Avoiding ingestion of metals can reduce the incidence of the problem

20

Omasal Impaction and Caecal Dilatation

Omasal impaction is a common disease in cattle where the omasum, one of the four stomachs in the compound stomach of ruminants becomes impacted with feed. Caecal dilatation is a common and economically important abdominal disorder that affects mainly dairy animals. In cattle with caecal dilatation, there is distension of the caecum, which may be accompanied by displacement, torsion or retroflexion of the organ and additional distension of the spiral colon. With distension alone, the apex of the caecum is displaced toward or into the pelvic inlet.

Etiology

The feeding of tough or fibrous feed, such as alfalfa stalks, or provision of insufficient water may cause stacking of feed materials in the spherical stomach between reticulum and abomasum resulting in omasal impaction. However, the common causes of secondary omasal impaction are traumatic reticuloperitonitis/pericarditis, intestinal obstruction, peritonitis and theileriosis. Similarly, the common causes of abomasal impaction are traumatic reticuloperitonitis/pericarditis and abomasal ulceration. Ruminal impaction may also occur due to the ingestion of foreign substances or the accumulation of ingested materials such as hair or plastic, causing the rumen to expand with passage of negligible or no dung.

Impaction may also occur due to indiscriminate feeding habits, swallowing indigestible plastic foreign bodies that accumulate in the rumen leading to ruminal impaction, indigestion, recurrent tympany, and many other adverse health effects. The disease is primarily noticed in stray animals residing in urban areas of developing countries. Ingested plastic materials in the rumen slowly release the chemicals in rumen fluid, which intern enter the food chain through milk and meat products. These chemicals have detrimental effects on human health.

Animals ingest these plastic materials due to erratic feeding behaviour, confusing for food and when trying to eat leftover feed materials in plastic wrappings. Ingestion of these indigestible plastic materials over a period of time leads to the development of ruminal impaction.

Rapid urbanization has reduced the resources and grazing lands of animals. Besides, extensive construction works have been carried out in urban areas, without proper care for the disposal of plastic waste. The waste is thrown anywhere on roads and near fences, making it easy for animals to graze on them and develop impaction due to plastic materials. Therefore, the ruminal or omasal impaction is more prevalent in urban and peri-urban areas.

Ruminal impaction due to plastic materials is observed in all domestic and wild ruminants. Among domestic ruminants, cattle are more susceptible to the development of ruminal impaction due to plastic materials followed by buffalo, sheep, and goat, based on the prehensile nature of these animals. Bovine species lack highly sensitive prehensile organs such as lips and tongues that discriminate sense of taste and making them indiscriminate feeders. Bovines also graze close to the ground making them more vulnerable to ingestion of plastic foreign bodies. Caprine are less susceptible to the ingestion of foreign bodies because of their well-developed sensitive prehensile organs and browsing and selective feeding behaviour.

Epidemiology

The omasum is a spherical to ovoid organ in bovine situated to the right of the midline in the middle third of the abdomen. It is one of the four compartments of the compound stomach and it is located between the reticulum and the abomasum. Its main function is the absorption of water, dissolved minerals and short-chain fatty acids. Approximately, 40 to 60 per cent of the water in the ingesta entering the omasum is absorbed, whereas 8 to 15% of the water is absorbed in sheep and goats only. Besides, the omasum acted to reduce the size of feed particles, by a mechanical function of minor importance. The primary diseases of the omasum, such as omasal impaction and omasitis, are rare although the omasum is involved in many disorders of the gastrointestinal tract, clinical signs due to omasal dysfunction are seldom observed. The omasum is not easily accessible for clinical examination by palpation, percussion and auscultation because of its position under the costal part of the abdominal wall, and it cannot be palpated through the rectum. The omasal impaction is the most common among diseases of the omasum. In most cases the omasum is only secondarily involved in any other diseases of the digestive tract. Omasal impaction is difficult to define as a clinical entity, and is usually diagnosed at necropsy when the omasum is found distended and unreasonably hard. It seems unlikely that it could cause death and is frequently observed in animals, dying of other disease. It is supposed to occur when feed is tough and fibrous, particularly alfalfa stalks.

Fodder scarcity, environmental contamination with plastic wastes, and poor standards of animal rearing compel goats to ingest and accumulate plastic foreign bodies in their rumen. Cases of plastic foreign body impaction have also been observed in wild ruminants due to the contamination of their habitat by plastic waste.

Breed

Crossbreed cattle are more susceptible than local cattle to the development of ruminal impaction due to plastic materials, as they require more feed and fodder, predisposing them for the ingestion of foreign bodies.

Sex

Sex of the animal strongly influences the ingestion of foreign bodies in animals. Female animals are more susceptible to the development of ruminal impaction due to plastic materials than male animals. Various physiological factors contribute to the ingestion of foreign bodies in female counterparts. Greater nutritional demand, negative energy balance, and mineral deficiency during pregnancy and lactation stages increase the appetite of female animals leading to consumption of foreign bodies. Farmers keep female animals longer than male animals because of their long reproductive period. Therefore, female animals are at a higher risk of exposure. Abandoning of unproductive female animals and ban of slaughter of female animals like cow in countries like India have compelled them to graze on roadsides and waste disposal areas leading to the ingestion of foreign bodies.

Age

Old animals are more susceptible to the development of impaction due to plastic materials than young animals. Cattle above 10 years of age and sheep and goat above 4 years of age were more frequently affected with indigestible materials than the other age group. This might be associated with an increase in exposure through life and animals will ingest and accumulate foreign bodies gradually over a period of time. Aged animals are culled due to low productivity and poor economic returns. These animals end up surviving on garbage leading to the ingestion of foreign bodies.

Body condition score: Animals with poor body condition are the most affected groups compared to that of good body condition. Poor nutritional status and negative energy balance are responsible for capricious appetite in animals leading to the ingestion of indigestible food materials and development of ruminal impaction due to plastic materials.

Season: The incidence of impaction due to plastic materials is noticed during dry period of the year during March to July. The fodder scarcity during this period, is the major problem in the areas of dryland and rainfed agriculture. Ruminants roam from one place to another in dry season in search of feed and fodder, thus making them prone to the ingestion due to foreign bodies. During lean seasons of agriculture, livestock owners along with their animals migrate to urban areas in search of job. In urban areas, they leave their livestock freely to find their food. Due to non-availability of feed and unselective disposal of polythene materials with some remnant food in it, livestock will graze on this waste leading to ingestion of plastic bags. Thus, most cases of impaction due to plastic materials occur sporadically in and around urban areas.

Economic Loss: Ruminal or omasal impaction due to plastic materials causes economic loss to farmers in terms of low milk yield, poor weight gain, reduced draft ability, other comorbid disease conditions, and mortality.

Pathogenesis

Ruminants feed on plastic waste materials such as polybags and plastic covers in confusion for food. The indigestible plastic materials are lodged in the rumen and then move to reticulum and omasum. The omasum absorbs water, dissolved minerals, and short-chain fatty acids. Omasal impaction is a controversial disease because the size and consistency of a normal bovine omasum can vary considerably. Various pathological conditions are encountered in animals depending on the type and amount of plastic waste ingested, type of material in plastic waste, duration of plastic waste accumulated in forestomach, and location of this plastic foreign body in gastrointestinal tract. Indigestion, impaction, tympany, polybezoars, traumatic reticulopericardtis, chemical leaching and immunosuppression are the pathological conditions encountered in animals with impaction due to plastic materials. Besides, heavy metal toxicities, endocrine disruption, carcinogenicity, teratogenicity, and urolithiasis may occur due to ruminal impaction with plastic materials in ruminants.

The pathogenesis of the caecal dilatation remains poorly understood. The cecum may become abnormally dilated with gas or distended with ingesta and it is associated with partial or complete cessation of the passage of intestinal contents which further leads to complete absence of defecation.

Clinical Signs

Omasal impaction in cattle is characterized by an enlarged and firm omasum, associated with clinical signs of complete anorexia, absence of defaecation, abdominal pain, constipation, dullness, ruminal distension, dehydration, congested mucous membranes, absence of rumen motility and initial abdominal pain. The general condition and demeanour are abnormal in most of cases of omasal impaction, around one-third cases show signs of colic. Ruminal motility is reduced or absent in two-third of cases in cattle. In 82.6% of cases, swinging and/or percussion auscultation shows positive results on the right side, and 82.4% of cases have little or no faeces in the rectum. Animals suffering from caecal dilatation and caecal impaction shows normal rectal temperature (100.58±0.50°F), heart rate (81.63±5.55 beats/min), and respiratory rate (24.09±1.32/min). Haematological findings reveal normal haemoglobin (11.26±0.32 g%) and packed cell volume (31.60±1.1%) within the normal reference range. The total leukocytic count may be in the high normal range (11.72±1.683×10^3/μL) along with marked neutrophilia (71.54±1.71%) and lymphopenia (28.45±1.71%). The serum biochemistry shows hypoproteinaemia (5.89±0.24 g/dL), hypoalbuminemia (2.37±0.27 g/dL), hypokalemia (3.24±0.16 mmol/L), hypochloraemia (90.81±4.9 mmol/L), and hypocalcaemia (7.53±0.23 mg/dL). Sodium and phosphorous levels are within the reference range for healthy animals (138.71±5.37 mmol/L and 3.24±0.16 mg/dL), while lactate is slightly increased (2.73±0.23 mmol/L). Peritoneal fluid examination reveals could be collected pre- or intraoperatively in five animals only. an increase in total protein (3.15±0.04 g/dL), a decreased albumin (1.26±0.06 g/dL), and normal total leukocytic count (48.348±4.804×10^3/μL). Hematological & Serum Biochemical Alterations Marked neutrophilia. Lymphopenia, Hypoproteinemia, Hypoalbuminemia, Hypokalemia, Hypochloremia, Hypocalcemia. In addition, there may be an increase in total protein and total leukocyte count in the peritoneal fluid

Diagnosis

Omasum in bovines is a spherical to ovoid organ situated, right of midline in middle third of abdomen. Its position under the costal part of abdominal wall does not allow its easy accessibility for clinical examination by palpation, percussion or auscultation methods. It can also not be palpated through the rectum. Majority of omasal impaction cases are found excessively enlarged and hard at postmortem examination. This makes the clinical diagnosis of omasal impaction difficult under field conditions. Omasal impaction is often only discovered during a necropsy. Ultrasonography can be used to evaluate the omasum, but some reports suggest it is not reliable. Omasal impaction has a poor prognosis, particularly if it is not diagnosed early and treated successfully.

Dilatation of the caecum is a condition of mild severity and may be asymptomatic, with the diagnosis being made during routine examination of the cow for other purposes. The caecum becomes distended with gas and expands into the pelvic inlet with the apex directed caudally. Per rectal examination is employed as an important diagnostic tool to palpate dilatation or displacement of the caecum. Caecal dilatation may be associated with foreign body syndrome. Radiographic evaluation of the caecum in bovine is not possible due to the massive viscera and body girth.

Caecal dilatation may be diagnosed via rectal palpation in cattle. Ultrasonographic examination of the dilated cecum may be done from the right flank region using Ultrasound Machine, with 5 MHz convex transducer. Caecal dilatation is confirmed when a gas-filled caecum is palpated during per rectal examination and gas is seen in more than half of the caecum on ultrasonogram. Caecal impaction is confirmed when a small gas cap along with doughy caecal body is palpated during per rectal examination, and only a gas cap involving the dorsal part of cecum could be seen on ultrasonogram.

Treatment

Treatment for omasal impaction includes kneading of the omasum to make the contents pliable, administration of four litres of mineral oil intra-ruminally for 3–5 days postoperatively and omasotomy, involving exteriorizing the omasum through an incision in the abdomen, opening it, and flushing it with water. Removal of indigestible foreign materials from the omasum gives positive effect on animal health. Besides, treatment for omasal impaction also include oral fluids, fluid transfusion and laxatives, injecting fluid, mineral oil, and laxatives into the omasum through an omasotomy. Control measures include good animal husbandry practices and proper disposal of plastic waste materials.

Early diagnosis and surgical intervention are required for achieving a successful outcome. Conservative treatment is not rewarding in the bovine with caecal affection and surgery should not be delayed in these patients. At time, animals with caecal dilation and torsion are required to be euthanised because of grave prognosis. The conservative therapy in cases of omasal impaction includes intravenous fluids, neostigmine, calcium borogluconate administration.

Surgical treatment is performed in cattle that had a poor general condition and demeanour, colic, absence of faeces, caecal torsion or retroflexion of the caecum. Operations were done on standing cattle using a proximal paravertebral local anaesthesia. A right flank laparotomy allowed exteriorization of the dilated caecum. An incision approximately 4 to 5 cm long was made in the apex of the caecum and the contents were emptied into a bucket. The remainder of the caecum, the proximal loop of the ascending colon and, if possible, the obstipated parts of the spiral colon were also emptied via gentle massage. The caecal incision was closed using monofilament absorbable suture material in a double Cushing suture pattern. The incisional area was thoroughly washed with lukewarm physiological saline solution and diluted iodine solution. After evaluating the vitality of the caecum by administering drops of neostigmine solution onto the organ, it was replaced into the abdomen. Before closing the laparotomy incision, the position and degree of filling of the caecum is assessed.

21

Diseases of Abomasum, Ulceration, Impaction, and Displacement

The glandular compartment of compound stomach in ruminants, called abomasum, may be affected by a number of diseases, including displacement or volvulus, impaction, and abomasal ulcers. The abomasal displacement or volvulus are common disorders in high-producing dairy cattle, but can also occur in calves and bulls, and is characterised by clinical signs of decreased milk production and anorexia, necessitating a prompt surgery in cases of right abomasal displacement. Impaction can be caused by a number of factors, including ingestion of non-food items, decreased water intake, and poor-quality roughage. The primary impaction is more common in beef cattle, whereas secondary impaction is more common in dairy cows. Other uncommon diseases of the abomasum include abomasal neoplasia, incarceration of the abomasum in umbilical hernias and abomasitis geosedimentosa, caused by ingested sand or gravel. Abomasal ulcers are seen in dairy and beef cattle, in calves, and in South America camelids. However, they are rarely diagnosed in small ruminants. Abomasitis is a less common disorder include caused by ingested sand or gravel (abomasitis geosedimentosa), Clostridial species and other bacteria.

Abomasal Displacement & Volvulus

Epidemiology: Abomasal displacement and volvulus are common disorders of high-producing dairy cattle, characterised clinically by anorexia and decreased milk production. Except for abomasal volvulus, abomasal displacement is rare in beef cattle and essentially remains undiagnosed in small ruminants. Impactions can be primary, which is the most frequent in beef cattle, or secondary that develops most often in dairy cows as a form of vagal indigestion. Impactions may have a hereditary predisposition in Suffolk sheep.

Pathogenesis: The abomasum is loosely suspended by the greater omentum and lesser omentum, thus facilitating its movement from its normal position on the ventral part of the abdomen to the left or right side, termed as Left displacement of abomasum or Right displacement of abomasum. The abomasum can rotate on its long axis, while displaced to the right and lateral to the liver, called abomasal volvulus. The abomasum can shift from its normal position to left displacement or to right displacement over a relatively short period termed as pendulous left displacement of abomasum. Abomasal volvulus can develop rapidly or slowly from a right displacement. Both LDA and RDA result in partial ileus. Abomasal volvulus leads to complete ileus and abomasal wall ischemia. Left displaced abomasum (LDA) refers to displacement of the gas-filled, distended abomasum to the left side of the abomasum, trapping it between the rumen and the left abdominal wall. Right displaced abomasum (RDA) is the displacement of the gas-filled, distended abomasum from the ventral abdominal wall into the craniodorsal right abdominal cavity.

Abomasal volvulus refers to displacement of the gas-filled, distended abomasum from the ventral abdominal wall into the craniodorsal right abdominal cavity, secondarily creating a volvulus by vertical and horizontal rotation (abomasum, wrapped in the greater omentum)

In RDA, hypomotility, gas production, and displacement of the partially gas-filled abomasum occur as it does in LDA. Mild hypokalemic, hypochloremic, metabolic alkalosis develops in both the left and right displacement of abomasum. After the dilation phase, rotation of the abomasum leads to volvulus and local circulatory impairment and ischemia. The volvulus is usually in a counterclockwise direction when viewed from the rear and the right side of the animal. The omasum is displaced medially and can be involved in the volvulus with occlusion of its blood supply, termed as omasal-abomasal volvulus, and displacement of the liver and reticulum. In rare cases, the reticulum can be involved in the volvulus, called reticular-omasal-abomasal volvulus. As much as 50L of chloride-rich fluid may accumulate in the abomasum, and hypochloremic, hypokalemic metabolic alkalosis develops. The blood supply to the abomasum, and often the omasum and proximal duodenum, is compromised, eventually resulting in ischemia of the abomasum and proximal duodenum as well as dehydration and circulatory failure. As circulatory failure progresses, a metabolic acidosis due to hyper-L-lactatemia and azotaemia can develop along with the preexisting metabolic alkalosis.

Clinical signs: The clinical deterioration with abomasal volvulus is rapid. The typical history of abomasal displacement includes anorexia and decreased milk production. The affected ruminants most commonly lack appetite for grain with a decreased or normal appetite for roughage. The decrease in milk production is not as extensive as with traumatic reticuloperitonitis or peritonitis due to other causes. In abomasal volvulus, anorexia is complete, milk production is more markedly and progressively reduced, and clinical deterioration is rapid. In abomasal displacement, temperature, heart rate, and respiratory rate are usually normal. The caudal part of the rib cage on the side of the displacement may appear "sprung." Hydration appears subjectively normal with LDA except in some chronic cases. On the contrary, dehydration occurs very early in abomasal volvulus. Rumen motility may be normal but is often reduced in frequency and strength of contraction. Faeces are usually reduced in quantity and more fluid than normal. However, normal consistency dung may be voided.

Diagnosis: The most important diagnostic finding is a ping sound on simultaneous auscultation and percussion of the abdomen, when performed in the area marked by a line from the tuber coxae to the point of the elbow, and from the elbow toward the stifle on both sides of the cattle. The ping is most commonly located in an area between 9th and 13th rib in the middle to upper third of the left abdomen in left sided displacement of abomasum. However, the ping can be more ventral or more caudal, or both. Pings associated with a rumen gas cap are usually more dorsal, less resonant, and extend more caudally through the left paralumbar fossa. Additionally, splashing sounds can be heard during the swinging auscultation of the displaced abomasum. Rectal examination can confirm a gas-filled rumen or an extremely empty rumen that correlates with the rumen ping in these cases. Pings associated with pneumoperitoneum typically are less resonant, present on both sides of the abdomen, and inconsistent in location on repeated evaluation. Frequently, secondary ketosis develops, and ketone bodies are present in blood, urine, or milk in abomasal displacement. Abomasal displacement associated secondary ketosis responds only transiently to treatment and recurs.

The ping associated with RDA is most commonly located in the area between 10th and 13th ribs on the right abdomen, where the splashing sounds can be ausculted. Differentiation between various causes of a right-side ping can be difficult in some cases, although a ping cranial to 10th rib usually indicates the presence of abomasal volvulus, as the liver is displaced medially by the distended viscus. The liver displacement can be confirmed via percussion or ultrasonographic examination.

A small, right-side ping underlying 12th or 13th rib and extending as far forward as 10th rib is common in cows with functional ileus from a number of causes. This ping is most often associated with gas in the ascending colon and resolves with correction of the underlying condition. Caecal dilatation and rotation are characterized by a right-side ping. The ping extends through the dorsal paralumbar fossa in caecal dilatation. The ping sound is usually noticed more caudally well into the paralumbar fossa in caecal rotation than the ping in RDA/ abomasal volvulus. The volvulus of the duodenal sigmoid flexure is another cause of ping on the right side. Per rectal palpation is helpful in differentiating an RDA/abomasal volvulus from caecal dilatation or rotation. Other right-side pings are produced by pneumoperitoneum or gas in the rectum, descending colon, duodenum, or uterus.

Clinical signs associated with abomasal volvulus are more severe than with simple displacements because of the vascular compromise. However, it is difficult to distinguish an early abomasal volvulus from an RDA, except by the presence of a right-side ping cranial to 10th rib indicating medial displacement of the liver by the abomasal volvulus. In contrast to cases of displacement, an abomasal volvulus causes tachycardia proportional to its severity, and the area of the ping is usually larger, extending as far forward as 8th rib, and a greater amount of fluid can be suctioned in volvulus. The animal is more depressed, and signs of weakness, toxaemia, and dehydration develop as the disease progresses. The caudal extent of the abomasum is usually palpable per rectum. In addition, measurement of serum L-lactate concentration may help differentiate between the two conditions. The animal often becomes recumbent within 48–72 hours after developing volvulus without therapy. Death occurs from shock and dehydration and is sudden following the rupture of ischemic abomasum.

Diagnosis of Abomasal Displacement Volvulus

The diagnosis of abomasal displacement or volvulus can be made by clinical evaluation after combined percussion and auscultation of a ping sound that can be confirmed from results of the laboratory testing. In case of abomasal displacement or volvulus, diagnosis is based on the presence of the characteristic ping on simultaneous auscultation and percussion and by exclusion of other causes of left- or right-side pings. Spontaneous fluid-splashing or gas-tinkling sounds may be heard on auscultation of the ping area or on simultaneous ballottement and auscultation of the abdomen. The characteristic rectal examination findings with LDA include a medially displaced rumen and left kidney. The abomasum is rarely palpable in LDA and only occasionally in right displacement. The abomasum is severely dilated in abomasal volvulus that can be more frequently reached with the tips of the fingers during rectal examination. In left sided abomasal displacement, the abomasum may become pendulous intermittently, which might be misleading during a physical examination. Ultrasonographic examination confirms the diagnosis of LDA, RDA, or abomasal volvulus, but it cannot reliably differentiate RDA from abomasal volvulus. Recent parturition, partial anorexia, and diminished milk production suggest displacement. The abomasal displacement is mostly associated with ketosis that is responsive to treatment. The typical signs during a physical examination in addition to the ping, rectal examination, and laboratory sample evaluation also support the diagnosis.

Melena or signs of peritonitis such as fever, tachycardia, localized abdominal pain, pneumoperitoneum, with left sided displacement abomasum may indicate a bleeding or perforated abomasal ulcer. In cattle with abomasal volvulus, blood L-lactate concentrations ≤ 2 mmol/L indicates a positive outcome with surgical correction, whereas cattle with blood L-lactate concentrations ≥ 6 mmol/L have a high probability of a negative outcome.

Treatment of Displaced Abomasum and Abomasal Volvulus: Several treatment options are available that allow correction of the displacement and fixation of the abomasum to the anatomically correct position. Conservative medical treatment for left abomasal displacement may be successful, especially if the diagnosis is made early, but prompt surgery is always required for right abomasal displacement due to high risk for volvulus. Improved management techniques may help reduce the frequency of all three conditions in dairy herds. Medical or surgical treatment can be successful for LDA. Abomasal volvulus and RDA require immediate surgical treatment. Antioxidative drugs are suggested prior to surgery in patients with abomasal volvulus. The time from diagnosis to the start of the treatment, including supportive care, is highly relevant to the outcome. Generally, both conservative (medical) and surgical therapy options are available for cattle with LDA. However, surgery is the only possible treatment for those with abomasal volvulus.

Cattle with LDA can be medically treated with administration of spasmolytic and analgesic drugs. In addition, rolling a cow through a 180° arc after casting it on its right side corrects most LDAs. Additionally, it has been reported that movement during transport of the patient with LDA may result in return of the abomasum to its normal position. However, the recurrence is very likely in nonsurgical approaches. Open (surgical) and closed (percutaneous) techniques can be used to correct abomasal displacements. LDA can be corrected surgically using right- or left-flank omentopexy, right-paramedian abomasopexy, combined left-flank and right-paramedian laparoscopy (a two-step procedure), or left-flank laparoscopy (a one-step procedure). Blind suture techniques (toggle-pin fixation or the "big needle" blind-stitch method, performed in the right-paramedian area, are percutaneous methods for correction of LDA.

The main disadvantage of these techniques is uncertainty of the exact location of the suture. Potentially fatal complications can develop after blind suture techniques, and the reported success rate is less than that of surgical correction by right-flank pyloric omentopexy.

The typical smell of abomasal gas or pH can be checked to confirm that the pin is in the abomasum with toggle-pin fixation, which reduces the likelihood of attaching rumen, small intestine, or omentum to the body wall rather than the abomasum. More recently, endoscopic methods (one- or two-step procedures) have become routine procedures to correct LDAs. Both RDA and abomasal volvulus are corrected surgically using right-paralumbar fossa omentopexy. The time between onset of the disease and surgery is crucial for treatment success.

Antioxidative drugs such as vitamin C, vitamin E, and dexamethasone can be administered in combination. However, any such medications should be administered before surgery. Right-paramedian abomasopexy should be performed only to correct RDA and abomasal volvulus in cattle, unable to stand.

Ancillary treatment of patients with abomasal displacement includes treatment of any concurrent periparturient diseases like metritis, mastitis, ketosis. Administration of calcium borogluconate or calcium gluconate subcutaneously or calcium gels orally help restore normal abomasal motility in many cases. Administration of erythromycin at the dose rate of 10 mg/ kg body weight intramuscularly at the time of surgery increases abomasal emptying rate and milk production in the immediate postoperative period. Because surgical correction of abomasal displacement or volvulus is frequently performed on the farm, the prokinetic effect of erythromycin suggests that it might be preferred if antimicrobials are administered to control intraoperative infection.

In simple displacement, fluid and electrolyte abnormalities correct spontaneously with access to water and a salt block. Providing electrolyte solutions (60 g sodium chloride and 30 g potassium chloride in 19 L of water) via nasogastric tube is helpful in cases of longer duration. Patients with notable dehydration and metabolic derangements require intravenous therapy, typically administered as hypertonic saline (7.2% NaCl, 5 mL/kg, IV over 5 minutes).

Occasionally, animals with abomasal displacement or volvulus have atrial fibrillation, due to concurrent hypokalemia and metabolic alkalosis. Correction of the displacement or volvulus almost always results in correction of the atrial fibrillation within 5 days.

Aggressive treatment of ketosis plays an important role in successful treatment of abomasal displacement, because most of the patients that die after surgical correction of LDA and RDA do so from the metabolic consequences of prolonged anorexia.

The prognosis after correction of simple LDA or RDA is good, with survival rates of 95%. Abomasal volvulus has a variable and less favourable prognosis with average survival rate of 70%. A high heart rate, moderate to severe dehydration, a longer period of illness, a large quantity of fluid in the abomasum, increased blood or plasma L-lactate concentration, and the presence of omasal-abomasal or reticulo-omasal-abomasal volvulus are associated with a poorer prognosis.

Prevention of abomasal displacement and volvulus: Maintaining optimal dry cow and calving management, avoiding rapid dietary changes, maintaining adequate roughage in the diet, avoiding post-parturient hypocalcaemia, avoiding endometritis/ metritis, and minimizing and promptly treating concurrent disease and ketosis contribute to effective prophylaxis of LDA, RDA, and abomasal volvulus. The incidence of abomasal displacements can be decreased by ensuring a rapid increase in rumen volume after calving and by feeding a total mixed ration rather than feeding grain twice daily.

Abomasal lesions can be classified as ulcers if the necrosis of the abomasal wall reaches into the submucosal layer, deeper than the lamina muscularis mucosae into the submucosal layer. In contrast more superficial lesions are defined as erosions.

A system of abomasal ulcer classification is based on the depth of penetration or the degree of haemorrhage or peritonitis caused by the ulcer:

Abomasal Ulcer

- Type I - an erosion or ulcer without haemorrhage
- Type II - haemorrhagic
- Type III - perforated, with acute localized peritonitis
- Type IV - perforated, with acute diffuse peritonitis
- Type V - perforated, with peritonitis within the omental bursa

This classification, however, is very difficult to assess by physical examination and may be more applicable during necropsy examination. Ulcers seem to occur less frequently in small ruminants but they are very commonly found in South American Camelids (SAC).

Abomasal ulcers affect mature cattle and calves and have several different manifestations. The common clinical signs include anorexia, bruxism, abdominal pain, occult blood in the faeces, and tachycardia. Ultrasonography and laboratory tests may aid making a diagnosis. Treatment is based on decreasing acid secretion in the abomasum. Blood transfusion may be required in severe cases.

Etiology and Pathogenesis: The causes of abomasal erosions and ulceration are not well understood except for lymphosarcoma of the abomasum that develop in viral diseases such as bovine viral diarrhoea, bovine leukaemia virus, and bovine malignant catarrhal fever. Abomasal ulcers are common in high-producing, mature dairy cows within the first 6 weeks after parturition, although they can occur any time during lactation. Although in contrast to monogastric animals, and not scientifically proven, One cause of ulcer development may be a prolonged inappetence, which results in sustained periods of low abomasal pH. The situation is very different from monogastric animals, as the abomasal volume does not vary substantially in ruminating animals. Abomasal ulcers may also arise in association with lymphosarcoma, abomasal disorders, or increased luminal pressure causing ischemia of the abomasal mucosa. The role of stress due to transport, new groups, heat, dehorning of calves in the pathogenesis of ulcers in abomasum is not completely understood and seems to vary substantially between animals. However, the stress is a likely major contributing factor.

Abomasal ulcers are very common in milk-fed calves after they have consumed milk or milk replacer for 4 -12 weeks. Most of these ulcers are subclinical and non-haemorrhagic. Occasionally, milk-fed calves < 2 weeks old are affected by acute, haemorrhagic abomasal ulcers that may perforate and cause rapid death. Well-nourished suckling beef calves 2 to 4 months old may be affected by acute abomasal ulcers. Abomasal trichobezoars are common in these calves but do not appear to increase risk of ulcer formation.

Clinical Findings: The abomasal ulcer syndrome varies, depending on whether ulceration is complicated by haemorrhage or perforation and by the severity of such haemorrhage or peritonitis. There may be only a single ulcer or many acute and chronic ulcers. Cattle with bleeding abomasal ulcers may be asymptomatic except for intermittent occult blood in the faeces. The affected cattle may die acutely from massive haemorrhage. Common clinical signs include mild abdominal pain, bruxism, sudden onset of anorexia, tachycardia (90–100 bpm), and faecal occult blood or melena that may be intermittent. It is difficult to identify melena unless the bleeding is extensive. Signs of blood loss occur with major haemorrhage and may include tachycardia (100–140 bpm), pale mucous membranes, weak pulse, cool extremities, shallow breaths, tachypnoea, and melena. More severe signs include acute rumen stasis, generalized abdominal pain with a reluctance to move and an audible grunt or groan with each breath, weakness, and dehydration. Melena may not be present in peracute cases, because only after 8 hours abomasal blood can be detected in the faeces. As the condition progresses, body temperature drops, and the animal becomes recumbent and dies within 6–8 hours.

In general, bleeding ulcers do not perforate, and perforating ulcers do not bleed into the GI tract sufficiently to produce melena. However, haemorrhage and perforation occur together occasionally, usually in chronic cases or those associated with abomasal displacement.

Calves with abomasal ulceration and hairballs may have a distended gas- and fluid-filled abomasum that is palpable behind the right costal arch. Deep palpation may reveal abdominal pain associated with local peritonitis due to a perforated ulcer. In calves, perforating ulcers are more common than bleeding ulcers.

Ulceration is most common in the fundic region in adult cattle and in the pyloric antrum in milk-fed calves. The single or multiple ulcers measure from a few millimetres to 5 cm in diameter. The affected artery is usually visible after ingesta and necrotic tissue are removed from a bleeding ulcerated area. Most cases of perforation are walled off by the omentum, which forms a cavity 12–15 cm in diameter that contains degenerated blood and necrotic debris. Material from this cavity may infiltrate widely through the omental fat. Adhesions may form between the ulcer and surrounding organs or the abdominal wall.

Clinical Diagnosis and Ultrasonography: The diagnosis of abomasal ulcer in cases with only slight bleeding and mild clinical signs is difficult and may require repeated faecal evaluations for occult blood. Other conditions that can cause partial anorexia and decreased milk production should be excluded by physical examination and laboratory test evaluation, including abdominocentesis when perforated abomasal ulcers are suspected.

Diagnosis of perforating abomasal or Type III ulcers is based on physical examination and excluding other causes of peritonitis. Rupture of a distended viscus can occur with abomasal volvulus or caecal rotation and produces similar signs. Furthermore, abomasal ulceration with perforation and local peritonitis may be indistinguishable from traumatic reticuloperitonitis. A magnet in the reticulum, confirmed by use of a compass, or an accurate history of having given the cow a magnet before the onset of signs decreases the likelihood of traumatic reticuloperitonitis. Traumatic reticuloperitonitis does not occur in SACs; however, peritonitis due to other causes may be considered. Ultrasonographic examination is a valuable tool to diagnose peritonitis and can be combined with abdominocentesis evaluation. However, ultrasonography is not suitable to diagnose non-perforating ulcers in the gastric wall.

Occult Blood: The diagnosis of abomasal ulcer in cases with melena, can be based on physical examination alone. An occult blood test of the faeces can confirm melena. Other conditions that result in blood in the faeces should be eliminated, because blood from portions of the GI tract distal to the abomasum reacts during faecal occult blood tests. It is usually bright red if from the large intestine or is raspberry-coloured if from the small intestine. Animals with abomasal lymphosarcoma can have a bleeding syndrome similar to that associated with abomasal ulcers but do not respond to therapy. Occasionally, oral, pharyngeal, and laryngeal lesions bleed, and the swallowed blood appears in the faeces. Similarly, pulmonary abscesses that form as a sequela of rumenitis by embolization to the lungs and liver can erode blood vessels and result in haemoptysis. Swallowing blood can also result in melena. Faecal occult blood may also be due to abomasal volvulus or rarely to bloodsucking helminths.

Haematology: There is neutrophilia with a marked left shift and haemoconcentration in cases of abomasal ulcer with diffuse peritonitis. In bleeding ulcers, the PCV can help to determine the extent of haemorrhage, although PCV decreases only 4 hours after an acute haemorrhage.

Abdominocentesis: In perforated ulcers, abdominal fluid is usually readily obtainable in large quantities. Evaluation of fluid obtained by abdominocentesis will confirm peritonitis if total protein or D-dimer concentration and nucleated cell count are increased. Nucleated cell count may also be normal because of dilution or utilization.

Intracellular bacteria or degenerate neutrophils are rarely seen as the infection is rapidly walled off in most cases. However, bacteria may be present, glucose concentration is close to zero, and LDH activity and D-dimer concentration are increased in septic peritoneal fluid.

Treatment: The treatment is aimed at improving dietary intake, decreasing acid secretion in the abomasum, and restoring blood volume by transfusion in severe cases. The most important treatment is to get the animal to eat, because food is an excellent buffer, and continual flow of forestomach contents (pH 6.0–7.0) into the abomasum helps increase abomasal pH. Broad-spectrum antimicrobial therapy, given for ≥5 days or until the rectal temperature is normal for 48 hours, is indicated for perforating ulcers. Antacids effectively increase abomasal pH in milk-fed calves, when administered at 4-6 hours intervals in a manner that induces oesophageal groove closure. However, their efficacy is extremely questionable in adult ruminants because of dilution by the large rumen volume. H_2-receptor antagonists effectively increase abomasal pH in milk-fed calves. However, the oral dosages required for cimetidine (100 mg/kg, 3 times a day) and ranitidine (50 mg/kg, 3 times a day) are high, making treatment expensive.

Proton pump inhibitors, such as omeprazole (2 mg/kg, IV, or 4 mg/kg, PO in calves) effectively increase luminal pH, but, again, treatment is expensive. The efficacy of administration of oral omeprazole (4 mg/kg) in adult ruminants is difficult to assess because absorption rates vary substantially.

The prognosis for localized peritonitis associated with perforating abomasal ulcers is good with medical therapy and dietary alteration. Recovery generally takes 1–2 weeks, and patients fully recover for 1-2 weeks. Surgery is indicated for perforating abomasal ulcers only when the abomasum is displaced. However, extensive abdominal contamination can occur in the process of breaking down adhesions and resecting or oversewing the ulcer.

Animals with diffuse peritonitis after perforation of an abomasal ulcer rarely respond to therapy, and the prognosis is grave. Treatment consists of rapid and continued intravenous fluid therapy based on the metabolic status and intravenous broad-spectrum antimicrobials. The surgery to close the perforation and intensive lavage of the peritoneal cavity can be performed. The few animals that recover from diffuse peritonitis usually develop massive abdominal adhesions.

Blood transfusions and fluid therapy may be necessary in addition to dietary management, stall confinement, and oral antacids for bleeding ulcers. If haemorrhage is acute, the PCV may not reflect the severity because equilibration between intravascular and extravascular fluid after blood loss takes at least 4 hours. Generally, a blood transfusion is required whenever weakness and lethargy are present. Complete recovery usually takes 1–2 weeks. The prognosis is good if weakness and lethargy have not developed before treatment is started.

Prevention: Stress factors such as transportation, regrouping, and overcrowding should be avoided to prevent abomasal ulcers. The feeding of coarse roughage to calves with an underdeveloped rumen should be avoided. Animals should be encouraged to keep eating to avoid prolonged periods of inappetence and low abomasal pH.

Abomasal Impaction

It is an infrequently diagnosed disease of cattle, characterized by drier-than-normal abomasal contents and a larger-than-normal abomasal volume. Impaction may involve the pyloric antrum alone or both the pyloric antrum and abomasal body. The initial clinical signs are typically complete anorexia, scant faeces, moderate distention of the abdomen, weight loss, and weakness. Diagnosis is based on history, clinical signs, and laboratory test results. Medical treatment with laxatives and prokinetic drugs may be effective.

Impaction of the abomasum develops in pregnant cows during cold winter months when cattle have decreased water intake and are fed poor-quality roughage. Impaction has also been seen in cattle fed a variety of mixed rations containing chopped or ground roughage (straw, hay) and cereal grains and in late-pregnancy dairy cows on similar feeds. Impaction of the pyloric antrum is an underdiagnosed condition in dairy cows in early lactation.

Etiology and Pathogenesis: The etiology of abomasal impaction can be divided into dietary (primary impaction) and nondietary (secondary impaction) causes, which can occur in various combinations.

Dietary Causes: The consumption of excess roughage low in both digestible protein and energy, in combination with decreased water availability is the cause of dietary abomasal impaction. Impaction with sand can occur if cattle are fed hay or silage on sandy soils or fed sandy or dirty root crops. The cause of nondietary abomasal impaction in post-parturient dairy cows is related to abomasal hypomotility and decreased abomasal emptying. Abomasal impactions have also been reported in cows in late gestation, caused by a mechanical interaction between the pregnant uterus and the abomasum, which is displaced by the uterus in a more cranial position. Another cause may be the decreased abomasal motility and emptying caused by vagal indigestion.

Once the abomasum becomes impacted, subacute obstruction of the upper GI tract develops. Hydrogen and chloride ions are continually secreted into the abomasum despite impaction, resulting in atony and alkalosis with hypochloraemia. Varying degrees of dehydration develop, because fluids are not moving beyond the abomasum into the duodenum for absorption. Sequestration of potassium ions in the abomasum results in hypokalemia. Impaction of the abomasum may be severe enough to cause irreversible abomasal atony.

Clinical Findings: Complete anorexia, scant faeces, moderate distention of the abdomen, weight loss, and weakness are usually the initial clinical signs of dietary abomasal impaction. Body temperature is usually normal but may be subnormal during cold weather. A mucoid nasal discharge tends to collect at the external nares and on the muzzle. The muzzle is usually dry and cracked, caused by both the failure of the animal to lick its nostrils and the effects of dehydration. The heart rate may be increased, and mild dehydration is common.

Most often, the rumen is static and distended with dry contents, but it may contain excess fluid if the cow has been fed finely ground feed. The pH of the ruminal fluid is usually normal (6.5–7). Protozoal activity in the rumen ranges from normal to a marked reduction in numbers and activity. The impacted abomasum is usually located in the right lower quadrant on the floor of the abdomen. Deep palpation and strong percussion of the right flank may indicate the presence of a large, firm mass, and elicit a grunt, probably because of distention of the abomasum and stretching of its serosa.

Severely affected cattle die in 3-6 days after the onset of clinical signs. The abomasum ruptures in some cases, and death happens precipitously in a few hours from acute, diffuse peritonitis and shock. In sand impaction, there is considerable weight loss, chronic diarrhoea with sand in the faeces, weakness, recumbency, and death in a few weeks.

Metabolic alkalosis, hypochloraemia, hypokalemia, haemoconcentration are common in abomasal impaction. The total and differential WBC counts remain within the normal range. At necropsy examination, the abomasum is commonly enlarged many times its normal size and impacted with dry, rumen-like contents. The omasum may be similarly enlarged and impacted. The rumen is grossly enlarged and filled with dry contents or fluid. The GI tract beyond the pylorus is characteristically empty and has a dry appearance. Varying extents of dehydration and emaciation are also present. If the abomasum has ruptured, lesions of acute, diffuse peritonitis are present. In dairy cattle in early lactation, typically only the pyloric antrum is impacted.

Diagnosis: Physical examination, history, ultrasonographic examination and laboratory sample diagnostics indicating decreased abomasal emptying, aids in diagnosis of abomasal impaction. Clinical diagnosis of dietary abomasal impaction is based on the nutritional history, clinical evidence of impaction, and laboratory results. The disease must be differentiated from secondary abomasal impaction as a form of vagal indigestion. Ultrasonographic examination is a valuable tool to assess size and position of the abomasum and to diagnose peritoneal lesions.

Impaction of the abomasum may be a complication of traumatic reticuloperitonitis that usually occurs in late pregnancy in a sporadic case. A mild fever may or may not be present, and there may be a grunt on deep palpation of the xiphoid. The rumen is enlarged and may have hypermotility in early stage or atonic in late stages. In many cases, it is impossible to distinguish between the two causes of impacted abomasum, and examination via right flank laparotomy may be necessary to explore the abdomen for peritoneal lesions.

Treatment: Laxatives and prokinetic drugs are used for medical treatments, and surgery, in some cases. The challenge is to recognize the cases of dietary abomasal impaction that will respond to treatment and those that will not. Cows that are weak, have a severely impacted abomasum, and have a marked tachycardia (100–120 bpm) are poor candidates for treatment, with a high risk of failure. Medical treatment usually requires a confirmed diagnosis via examination by right-side laparotomy. In treated cows, the metabolic alkalosis, hypochloraemia, hypokalemia, and dehydration should be corrected. Lubricants can be used to move the impacted material. It is necessary to empty the abomasum surgically only in cattle with severe impaction. Balanced electrolyte solutions are administered intravenously continuously for up to 72 hours at a daily rate of 80–120 ml/kg body weight. Some cows respond well to this therapy and begin ruminating and passing faeces in 48 hours.

Mineral oil should be administered at the dose rate of 4 l/day for 3 days. Additionally, the laxative dioctyl sodium sulfosuccinate (DSS) can be injected once into the abomasum during standing right flank laparotomy at 60–100 mL of a 25% solution for 450kg cattle. Dioctyl sodium sulfosuccinate (DSS) should not be administered orally as it kills rumen protozoa. A beneficial response cannot be expected in < 24 hours. The improvement is usually seen in cattle that respond to treatment, by the end of day 3 following treatment. Erythromycin (10 mg/kg, IM, once to twice daily) can be administered as a prokinetic in cattle that do not improve after surgery, provided that mineral oil has been administered and physical obstructions were not identified during surgery.

Surgery may be performed, but results are often unsuccessful, as abomasal atony may appear to worsen after surgery. An alternative may be to perform a rumenotomy to empty the rumen and infuse mineral oil directly into the abomasum through the reticulo-omasal orifice in an attempt to soften and promote evacuation of the abomasal contents. Cattle with secondary impactions that develop as a sequela of traumatic reticuloperitonitis or abomasal volvulus usually show signs of vagal indigestion, and abomasal impaction may be diagnosed during exploratory surgery.

The induction of parturition by administering dexamethasone (20 mg, IM) may be indicated in affected cattle within 2 weeks of term and in which the response to treatment for a few days has been unsuccessful. Parturition may assist recovery because of a reduction in intra-abdominal volume. For sand impaction, affected cattle should be moved off the sandy soil and fed good hay and a grass mixture containing molasses and minerals. Severely affected cattle should be treated with administration of mineral oil.

Prevention: Prevention of dietary abomasal impaction is possible by providing the necessary nutrient requirements. The low-quality roughage should be analysed for crude protein and digestible energy. Based on the analysis, grain is usually added to the ration to meet energy and protein requirements. Adequate fresh drinking water should be supplied at all times.

Abomasitis

A variety of putative aetiologies for abomasitis exist historically, including primary bacterial or fungal infection, immunosuppression, and pica, trauma from coarse feed or trichobezoars, and vitamin or mineral deficiencies. It is a sporadic disorder of neonatal to weanling age calves, lambs, and kids. This disease is characterized by diffuse, haemorrhagic to necrotizing inflammation of the abomasal mucosa, frequently involving the deeper layers of the abomasal wall in severe or chronic cases. Intramural emphysema and edema of the abomasal wall may be present. Abomasal ulceration and perforation may occur in a subset of affected animals.

In 1987, investigators at Kansas State University detected C. perfringens types A and E in stomach contents of affected calves and the following year reproduced the disease experimentally by intraruminal inoculation of C. perfringens type A in calves. The ability of this organism to produce gas is considered to contribute to the abomasal dilation and intramural emphysema evident in affected animals. Abomasal stasis and ruminal accumulation of milk have also been proposed as risk factors for the abomasitis/ulceration and the disease is further discussed in Chapter 5.

Diagnostic studies indicate that C. perfringens type A is now the most common bacterial isolate in cases of dairy calf clostridial enteritis and that affected calves typically exhibit tympany, haemorrhagic abomasitis, and abomasal ulceration. The presence of the organism as a member of the normal flora of all mammals has clouded its definitive etiologic role, but because inoculation with the organism can reproduce the disease experimentally, much interest has been focused on this organism, particularly in herds that experience repeated cases, or outbreaks, of peracute abomasitis, tympany, and death in nursing calves.

22

Acute Diarrhoeas in Ruminants and Equines

Acute diarrhoea in adult ruminants is more likely to develop due to maldigestion and malabsorption, secondary to infectious enteritis, whereas osmotic diarrhoea develops secondary to carbohydrate engorgement. Diarrhoea is defined as the passage of loose or watery stools, typically at least three times in a 24-hour period. It reflects increased water content of the stool, whether due to impaired water absorption and/or active water secretion by the bowel. Based on duration of diarrhoea, it is called acute (14 days or fewer in duration), persistent (more than 14 but fewer than 30 days in duration) or chronic (more than 30 days in duration). Invasive diarrhoea or dysentery is defined as diarrhea with visible blood or mucus, in contrast to watery diarrhea. Dysentery is commonly associated with fever and abdominal pain. Diarrhoea can have serious financial and animal welfare implications in dairy herds. It is a common complaint in young ruminants, particularly during the first few months of life. Acute diarrhoea in the pre-weaning period accounts for 75% of early calf mortality in dairy herds. The management practices and pathogens that cause diarrhoea in calves, also affect lambs and goats. Most herds are exposed to diarrhoea causing pathogens. However, the management practices largely determine the health impact of pathogens in young stocks. Diarrhoea in most young ruminants is caused by more than one factors or pathologic agents. It is important to arrive at correct diagnosis, and institute appropriate treatment against diarrhoea in livestock, and to prevent further outbreaks of disease through effective management strategies.

Etiology

Diarrhoea is a multifactorial disease entity. Most cases of acute-onset diarrhoea in equines are caused by an infectious agent. The infectious diseases causing colitis and diarrhoea include salmonellosis, clostridiosis, coronavirus and, Potomac horse fever. Most cases of acute diarrhoea are infectious, more commonly caused by viruses. However, the most cases of severe diarrhoea are caused by bacterial pathogens. Diarrhoea can be classified as infectious and non-infectious on the basis of etiology. Infectious diarrhoea is caused by pathogenic agents like virus, bacteria, parasites, mycotoxins etc., whereas non-infectious diarrhoea occurs due to poor hygiene, stress, overfeeding, indigestion, faulty diet, intestinal injury, inflammation and malabsorption. As a whole, the possible causes of acute diarrhoea include bacterial and viral infections, certain chemicals, intestinal parasites, poor diet, overfeeding on milk or lush grasses, poisonous plants and toxins, food allergies and stress. The diarrhoea results from intestinal failure to absorb fluids adequately with increased secretion into the intestine. The correct determination of the cause of diarrhoea is important in order to take effective therapeutic and preventive measures.

Bacterial Diarrhoea

Colibacillosis: *Escherichia coli* is the major cause of acute diarrhoea in young calves. Bacteria invade the intestinal and other mucous membranes and produce toxins that cause severe inflammation of the intestinal lining resulting in enteritis, and may lead to death within hours. A less severe form of the disease is usually manifested by diarrhoea accompanied by progressive dehydration. Colibacillosis lasts for 2 to 4 days and its severity depends on the age of the affected calf. *E. coli* inhabits in the intestine and is excreted in the faeces that contaminate kennels, stables, floors, paddocks and even water supplies.

Salmonellosis: Calves are usually affected at 6 days of age or older. The signs of salmonella scours include diarrhoea, presence of blood and yellow clots in the faeces, depression and elevated temperature. The bacteria multiply in the intestine and many reach the bloodstream, causing blood infection and sudden death. The presence of a membrane-like cover in the intestine during post mortem examination suggests salmonellosis.

Tick-borne diseases and underfeeding predispose calves to salmonella scours. Heavily infected animals may become severely depressed following treatment due to release of endotoxins following death and lysis cell wall of Gm -ve bacteria after administration of bactericidal drugs.

Enterotoxaemia (Pulpy kidney disease): It is usually associated with the change in weather, feed of cows, or management practices that cause the calf to nurse for a longer period of time than usual. The hungry calf may overconsume milk that establishes a conducive environment in the gut for the growth of the bacteria and production of toxins. The disease usually starts quite suddenly. Affected animals become listless, display uneasiness, and strain or kick at their abdomen. Bloody diarrhoea may or may not occur. The gut may be red in colour or have bloody, purplish areas in dead animals.

Johne's disease or paratuberculosis: Johne's disease (JD) is a chronic, progressive intestinal disease caused by *Mycobacterium paratuberculosis (Map)*. The causative agent was first identified in European cattle a century ago, and it was discovered in the United States in the early 1900s. The long incubation period of the disease, clinical signs similar to many other common diseases of cattle, expensive and relatively low sensitive diagnostic tests, and lack of accepted standards for diagnosis and control, make the disease problematic for its control. These problems are compounded by a lack of awareness about the disease. The slow progression of the disease makes financial losses not easily perceptible to the individual producer. Johne's disease is an incurable wasting disease of adult cattle. The management practices, especially close confinement in dairy operations, promote easier transmission of the organism. The causative acid-fast organism causes an infectious inflammation of the intestines with severe weight loss and diarrhoea. The infected animal becomes emaciated but continues to eat well and look bright despite persistent diarrhoea, lasting for months. The manure of typical cases is not mixed with blood or mucus.

Control of Johne's disease on farms has been difficult for several reasons. The disease has a long subclinical phase, during which animals can spread the infection without themselves exhibiting signs of illness. The biosecurity measures should in place in dairy herds for the control of the disease. Precautions should be taken to prevent introducing the disease through purchased animals. Clinically infected animals should be isolated and calves may be reared in separate calf-rearing area. Recent offsprings of clinical cases are culled as soon as possible. Calves should be removed from dams immediately upon birth before suckling. Colostrum is harvested from cows with cleaned and sanitized udders and fed to calves by bucket, and thereafter feeding only milk replacer or pasteurized milk. The contamination of calf feedstuffs, water, or bedding by effluent from the adult herd is avoided. The manure from the adult herd is only applied to cropland or pastures grazed by adult stock.

Non-bacterial Causes of Calf Diarrhoea

Coccidiosis: The disease commonly affects young animals of 3 weeks of age and older, managed in groups in unsanitary conditions. Calves as young as 16 days of age may be affected. It occurs usually following stress, poor sanitation, overcrowding or sudden change in feed. Older cattle are less susceptible to disease than younger cattle unless they are experiencing extreme stress or have depressed immune systems. The symptoms are diarrhoea, slimy and bloody faeces, emaciation, weakness and anaemia. The affected calves strain extremely during defecation. Diarrhoea with faeces smeared over the rump as far around as the tail is a typical sign of coccidiosis in young calves.

Coccidia are host-specific protozoan parasites. The oocysts are shed in the faeces of both the symptoms-showing affected animals and carriers. The oocysts sporulate, undergo maturation in moist warm environments and become infective. Coccidia oocysts are ingested by animals, while consuming contaminated feed, water, pastures or licking a dirty hair coat. The infective stage develops inside the host animal, causing damage to intestinal cells and potentially resulting in diarrhoea with blood in the faeces of infected animals. The damage to the intestinal mucosa also impairs the ability of animals to absorb fluids to compensate for the water losses in diarrhoea. The oocysts are highly resilient and can survive in moist shaded areas for several years.

Many coccidia-exposed and infected animals do not exhibit clinical symptoms. However, they develop species-specific immunity. Disease occurs when large numbers of the infective form of the protozoal oocysts are ingested. Stress of weaning, shipping or moving, and compromised immune response in animals precipitate the illness and clinical signs. Sick animals usually have acute diarrhoea with or without blood, decreased appetite and mild depression. This may progress to severe depression, dehydration, pale mucous membranes, straining and severe weight loss in more severe infections. Some cattle with coccidiosis may show neurologic symptoms. Death may occur despite therapy.

The more chronic form of the disease causes reduced growth rates, and acts as a stressor, causing increased susceptibility to other infections, such as salmonellosis, or Bovine Respiratory Disease. Coccidiosis is primarily a disease of young animals, but can affect older animals those are maintained in poor conditions or are immune-compromised. It occurs commonly in confined conditions, but can occur in free-ranging animals that have congregating areas, such as feeding shades and watering areas.

Coccidiosis may cause significant economic losses due to reduced performance, death from direct infections, and by predisposing cattle to secondary bacterial and viral infections. Additional economic losses occur because of the labour demand for care and treatment of infected cattle and medication costs.

Transmission: Coccidiosis is transmitted from animal to animal by the faecal–oral route. Infected faecal material can contaminate feed, water and soil. Cattle may suffer from the disease by eating and drinking from contaminated sources, or by licking itself or other animals. Calves may become infected by nursing contaminated udders. The severity of clinical disease depends on the number of ingested oocysts.

Diagnosis: Coccidiosis is diagnosed by clinical signs, faecal sample examination by flotation or smear technique, and by postmortem examination. Clinical signs are usually seen about 17 days after ingestion of oocysts. The damage is far advanced by the time clinical signs are observed and life cycle of coccidia is completed in the animal. The history frequently includes a preceding stressful event for the animals. It must be noted that coccidia can be found in the faeces of normal healthy cattle and diagnosis must rule out other diseases such as BVDV, salmonellosis, internal parasites and toxicities.

Treatment: The most effective treatment for the already sick animals is supportive therapy (fluids) and antibiotics to ward off secondary infections. The affected animals should ideally be isolated to prevent increased contamination of the premises. Amprolium at the dose rate of 10 mg/kg/day for five days and sulfonamides are commonly used to treat the diseased animals. The severely affected animals may need to be handled and treated individually as drugs administered in feed or water may not be consumed by sick animals.

The presence of one clinical case in a group of cattle increases the probability of other animals in the group being exposed and harbouring coccidia in the intermediate stages of development. The preventive therapies are practised in cases of coccidiosis due to the drug susceptibility of the parasite at the intermediate stages of development. Drugs useful for treatment are not necessarily useful for preventive therapy and vice versa. Anticoccidial-preventive therapies, commonly used in cattle should follow label recommendations paying careful attention to maintaining preventive levels for long enough period of 28 days or longer, to affect the life cycle of coccidia. Anticoccidial preventive therapies may be incorporated into beef cattle rations and supplements. The most effective coccidiosis control program focuses on preventive therapies before clinical signs appear.

Viral Diarrhoea

The aetiology of viral diarrhoea in ruminants is complex. A variety of viruses have been associated with the syndrome.

Rotavirus diarrhoea: It is mostly seen within 24 hours of birth. The infected calves are severely depressed with watery faeces, anorexia and drooling of saliva. The faeces colour varies from yellow to green. Calves lose appetite and the mortality rate may be as high as 50%. Post mortem changes are not prominent. However, there may be an increased volume of fluid in both the small and large intestine.

Coronavirus diarrhoea: This occurs in calves over 5 days of age. The animal is not as depressed as in rotavirus diarrhoea. The initial signs may be the same as in rotavirus infection, but later on the faeces may contain clear mucus that resembles the egg white. The mortality is as low as 1 to 25 %.

Bovine Virus Diarrhoea (BVD): Bovine Viral Diarrhoea (BVD) is a disease of cattle caused by the Bovine Viral Diarrhoea Virus (BVDV). BVD can be a serious, costly disease in a susceptible herd. Diarrhoea begins 2 to 3 days after exposure to the virus and may persist for a long period. Ulcers on the tongue, lips and in the mouth are the usual lesions found in the affected live calves. The virus is widespread, and most herds are at risk for infection.

Pathophysiology of Diarrhoea: Diarrhoea may be physiological or pathological. Physiological causes of diarrhoea are mainly an increase in intestinal peristalsis, hypersecretion of intestinal fluid, damage to intestinal mucosa, and malabsorption. The intestine fails to adequately absorb fluids, and in many cases, there is increased secretion into the intestine. Loss of fluids in diarrhoea leads to tissue dehydration and loss of certain body salts, resulting in changes in body tissue composition and severe depression of affected animals. Death from scours usually results from dehydration and loss of body salts, rather than invasion of infectious agents.

Several pathogens responsible for causing diarrhoea are zoonotic agents, such as *Salmonella* spp., *Cryptosporidium* spp., *Giardia* spp., and certain types of enteropathogenic *E. coli.* So, proper biosecurity measures must be followed while handling diarrhoeic animals, contaminated bedding, and faecal samples to avoid contracting the disease. The main cause of death in acute diarrhoea is dehydration that results from loss of fluid and electrolytes in the stool. Diarrhoea is an important cause of malnutrition, as calf with diarrhoea eats less and their ability to absorb nutrients is reduced. The nutrient requirements are also increased as a result of infection.

Nutritional diarrhoea: Nutritional scours are caused by anything that disrupts the normal nursing pattern, including storms, strong wind or the temporary absence of mother. The hungry calf may take in excessive quantity of milk from the udder that contains more milk than normal upon getting an opportunity to nurse, resulting in nutritional scours. It is typically white scours caused by undigested milk passing through the intestinal tract. Milking a cow to limit the milk intake by the calf usually clears up the problem. Oral antibiotics may be used, if the calf becomes depressed. Nutritional causes also include acidosis like grain overload, lush green feed, cobalt or selenium deficiency, copper deficiency or toxicity, There are many other diseases where diarrhoea presents as a secondary manifestation. Some of these include liver disease, peritonitis (e.g. hardware disease), heart failure, chemical or plant poisonings due to lead or nitrate.

Clinical Signs

The affected animals pass watery droppings many times a day, and show symptoms due to loss of water and salts from their body. Animals become weak and thin, and loose appetite. Loose and watery faeces contain mucous and sometimes, blood. The milk production declines and the delay in the treatment may cost the life of the animal.

The immune status of the exposed animals and the strain of the infecting virus influence the clinical signs in Bovine Virus Diarrhoea. The incubation period is about 3 to 5 days. The infection with a virulent strain of the virus in susceptible non-vaccinated population is likely to appear as an acute, severe sickness, with bloody diarrhoea, high fever (105-107°F), off-feed, mouth ulcers, and often pneumonia. Some infected animals may die, while others recover, usually within one or two weeks. Occasionally, an affected animal dies very quickly before other signs are apparent. Since BVD is a viral disease, antibiotics are not effective. Abortions, infertility, and embryonic deaths associated with BVD lead to significantly reduced reproductive performance and increased premature culling. Animals developing acute diarrhoea and fever may die or have a long and costly recovery period with decreased milk production and growth. The immune system is often depressed, making them more susceptible to other diseases during recovery. The pregnant animals may abort about 2 to 4 weeks after exposure, especially when infection occurs in the second trimester of pregnancy. The pregnant cows exposed in the first trimester may experience early embryonic death, while open cattle may fail to conceive and return to heat. Cows infected during 60 to 120 days of pregnancy may not lose their foetus, rather they deliver a persistently infected (PI) carrier calf that shed lots of BVD virus for the rest of its life infecting other animals. Persistent infection (PI) carrier only develops in utero, only where the dam is exposed to BVDV at less than 125 days of pregnancy. The disease is generally much less severe in well vaccinated cattle, and those are not heavily stressed. The only visible signs in such cases are sporadic abortions or repeat breeding.

Treatment of Calf Scours

The treatment of diarrhoea is very alike regardless of the cause. Symptomatic treatment of the diarrhoea is frequently followed to correct the fluid loss, acidosis and loss of salts. Specific treatment is often not possible, as confirmatory diagnosis is required for such healthcare. The management of patients with acute diarrhea begins with general measures such as fluid repletion and nutrition maintenance, with adjustments in diet, if necessary. Patients who have bothersome symptoms may benefit from symptomatic pharmacologic therapy. Antibiotic therapy is not indicated in most cases since the illness is usually self-limited. Nevertheless, empiric and specific antibiotic therapy may be appropriate in certain situations, mainly in patients with severe disease, with symptoms and signs suggestive of invasive bacterial infection, or at high risk for complications.

Fluid repletion: The most critical therapy in diarrhoea is rehydration, preferably by the oral route, with solutions that contain water, salt, and sugar. Diluted fruit juices and flavoured soft drinks along with saltine crackers and broths or soups may meet the fluid and salt needs in patients with mild illness.

Oral rehydration solutions (ORS), including standard World Health Organization ORS or commercial ORS, may be more appropriate in patients with more severe diarrhoea.

Calves may be given milk diluted with an equal quantity of clean water. Antibiotic or sulpha drug can be given simultaneously along with the treatment for dehydration. The electrolyte powder is mixed with water for oral administration.

Treatment of Infectious Diarrhoea

Various drugs are available in the market for treatment of infectious diarrhoea caused by bacteria, protozoa and to check secondary infections in viral diarrhoea.

Norfloxacin with Tinidazole combination has been most effective for the treatment of bacterial and protozoal diarrhoea. Norfloxacin is a synthetic broad spectrum antibacterial agent belonging to the fluoroquinolone group. It exerts its bactericidal effect by inhibiting the A subunit of DNA gyrase, an essential enzyme involved in DNA replication. Tinidazole is a nitroimidazole that has antimicrobial action against microaerophilic protozoa *Giardia lamblia*, *Entamoeba histolytica* and *Trichomonas vaginalis* and against obligate anaerobic bacteria. It acts by damage of DNA strands or inhibition of their synthesis.

Azithromycin is preferred for patients with fever or dysentery (bloody or mucoid diarrhea) and in other patients suspected to be at risk for a fluoroquinolone-resistant pathogen (in patients with diarrhea). Azithromycin can be given as a single 1 g dose (for patients without dysentery) or as 500 mg once daily for three days (for patients with or without dysentery).

Fluoroquinolones – Fluoroquinolones include ciprofloxacin (a single 750 mg dose or 500 mg twice daily for three to five days) or levofloxacin (500 mg as a single dose or given once daily for three to five days). The single-dose regimens are associated with greater nausea than the multidose regimens. Specific circumstances may warrant empiric treatment for particular pathogens.

The antimotility agent, loperamide can be used cautiously in patients in whom fever is absent or low grade and the stools are not bloody. The dose of loperamide is two tablets (4 mg) initially, then 2 mg after each unformed stool for ≤2 days, with a maximum of 16 mg/day. Antimotility agents are avoided in patients with clinical features suggestive of dysentery unless antibiotics are also given because of concerns that antimotility agents can prolong disease in such infections or lead to more severe illness. In such patients, bismuth salicylate (Pepto-Bismol, 30 mL or two tablets every 30 minutes for eight doses) is an acceptable alternative, although it is somewhat less effective and there is the potential for salicylate toxicity, especially in those who take aspirin for any reason. Another antisecretory agent, racecadotril, is an effective option for symptomatic therapy. The addition of loperamide to antibiotics also decreases the time to symptom resolution compared with antibiotics alone. Diphenoxylate (Lomotil) is an alternative antimotility agent, but it has not been as well studied and may cause central opiate or cholinergic side effects. The dose of diphenoxylate is two tablets (5 mg) four times daily for ≤2 days.

Role of herbs in treatment of non-infectious diarrhoea: Non-infectious diarrhoea largely occurs due to dietary error and intestinal disturbances. Herbs possess a great value to restore gastrointestinal functions and in repairing mucosa. Herbs form protective layer on mucosal surface of intestine and reduce the severity of diarrhoea. They regulate the peristaltic movements of intestine and provide healthy environment for growth and multiplication of microflora.

Prevention

Identifying the cause of diarrhoea in the herd will allow you to make a sound plan about how to prevent or minimise the effect. This can include reviewing rations, grazing plans, worm management and biosecurity practices.

23

Chronic Diarrhoea in Adult Ruminants and Equines

Chronic diarrhoea refers to continuous or intermittent passage of soft or watery faeces for a prolonged period of time. However, continuous passage of liquid faeces is unlikely to occur, making the definition inconsistent. Most of the causes of chronic diarrhoea are non-infectious, as opposed to acute diarrhoea, caused by infectious aetiologies.

Etiology

The causes of chronic diarrhoea could be different, and in general could be due to inflammatory causes or imbalance of the normal physiological processes. Among inflammatory causes, different aetiologies could be involved such as bacterial or parasitic infection and infiltrative disorder with involvement of the hindgut. The horse always manifests the problem in a cyclic way. The mycobacterial disease, paratuberculosis is associated with chronic diarrhoea and wasting, often without change in appetite, and occurs as a sporadic disease in adult cattle. Chronic salmonellosis and chronic bovine virus diarrhoea (BVD) are responsible for chronic diarrhoea and wasting in younger animals.

Causes of chronic diarrhoea include malabsorption disorders such as pancreatic insufficiency and celiac disease, small intestinal bacterial overgrowth (SIBO), mesenteric ischemia, and gastric bypass surgery. Inflammatory diarrhoea is caused by inflammation of the colon that affects the mucous lining of the colon. It causes swelling and impairs the ability of colon to absorb. Inflammatory diarrhoea tends to come out more frequently and in smaller amounts. It may be accompanied with stomach pain, fever or bleeding. Inflammatory conditions that can cause chronic diarrhoea include persistent infections, such as parasites or an overgrowth of the bacteria like Clostridium difficile, causing pseudomembranous colitis. Inflammatory bowel disease includes ulcerative colitis or Crohn's disease, ischemic colitis, diverticulitis, radiation colitis and cancer.

Diarrhoea can also be graded based on pathophysiologic mechanisms, such as watery, fatty or inflammatory. These three categories are further subdivided according to aetiopathogenesis. Watery diarrhoea is divided into secretory, osmotic, and functional. Fatty diarrhoea is divided based on functional disturbances such as malabsorption and maldigestion. Inflammatory diarrhoea is divided into inflammatory bowel disease, infectious diseases, caused by bacteria, virus, and parasites, neoplastic and radiation colitis. The duration of diarrhoea can last from weeks to months, and an episode of acute diarrhoea may become chronic. The faecal consistency can vary from soft to watery and there may be periods of normal faeces.

Epidemiology

Sex and breed have not been identified as specific risk factors except very few contradictions, as more common affections seen in Standardbred horses. Age is a risk factor for certain conditions that can cause diarrhoea, such as cyathostominosis, caused by pathogenic intestinal helminths of horses.

Pathophysiology

Many of the causes and mechanisms of chronic diarrhoea are poorly understood. Chronic diarrhoea is a challenge from both the diagnostic and therapeutic point of view. Diarrhoea, to be considered 'chronic', has to have been present for at least 7 to 14 days. In some cases, the diarrhoea will persist for weeks or months, with recurrent attacks of changed faeces texture from "cowpat" consistency to real watery diarrhoea, separated by periods of relatively normal faecal consistency. In the adult horse, chronic diarrhoea is almost invariably associated with large intestinal disease. It could be caused either by physical damage to the colonic wall or physiological disturbances of colon functions. The causes of chronic diarrhoea could be different, and in general could be due to inflammation.

Chronic diarrhoea could be a combination of different mechanisms including mucosal inflammation and damage, impaired digestion, increased secretion across the gastrointestinal wall, and osmotic overload of the luminal contents.

The osmotic type diarrhoea is caused by poorly absorbed nutrients that draw extra water into your colon. Secretory diarrhoea represents a wide variety of diseases that can cause watery diarrhoea. Excess fat in the faeces changes its consistency resulting in diarrhoea. This results from poor performance in breaking down and absorbing fats, or excessive fatty acids produced by gut bacteria.

Some of causes of chronic diarrhoea include bile acid diarrhoea that happens in presence of too much of bile in the gut, also called bile acid malabsorption (BAM), microscopic colitis, diabetes-related neuropathy. vagotomy, hyperthyroidism, adrenal disorders, neuroendocrine tumours, irritable bowel syndrome (IBS). Physically, chronic diarrhoea puts in risk of dehydration and its side effects. A lot of water and electrolytes are lost with chronic diarrhoea. Electrolyte losses can have serious consequences. They can affect the heart, lungs, brain and nervous system. Dehydration also stresses kidneys.

Clinical Signs

Clinical signs vary according to the inciting cause and involvement of other body systems. Faeces are soft and watery, and the excreted amount is greater than normal. Weight loss is commonly reported along with inappetence and dullness. Colic or recurrent colic is a less common clinical sign occurring in one-fifth of horses with chronic diarrhoea. Other signs include peripheral oedema and skin lesions. Fever may be seen for a transient period or persistent, and it is more common in diseases of inflammatory aetiology. However, it may also occur as a paraneoplastic syndrome. Low or decreased body condition and muscle mass are a frequent finding and subcutaneous oedema of the limbs, ventral aspect of the body, mandible and throat latch may be present. Increased gastrointestinal sounds may be auscultated. Physical examinations frequently reveal vital parameters as normal, even with longstanding disease. Performing an examination per rectum is recommended in all cases to establish the presence of masses, enlarged lymph nodes, increased thickness of the intestinal wall, and adult or larval cyathostomins.

Clinical Pathology

Complete blood count and serum biochemistry should be performed in all cases of chronic diarrhoea. Clinicopathological findings are likely affected by the cause of the chronic diarrhoea and the stage of the disease at the time of sampling. Anaemia may be present due to malabsorption of essential nutrients, iron sequestration, whole blood loss, or immune-mediated diseases. Abnormalities in the white blood cell count or differential count may be present. Both neutropenia and neutrophilia are the haematological findings in gastroenteritis. Horses with other types of eosinophilic intestinal disease uncommonly have changes in the white blood cell count. A consistent clinicopathologic finding in horses with chronic diarrhoea is the presence of hypoproteinaemia, characterised by hypoalbuminemia. The globulin concentrations are variable and may be reduced, normal or elevated. Significantly lower albumin levels and significantly higher alpha-2-globulin levels are observed in non-survivors compared to survivors. The hypoalbuminemia occurs due to leakage of albumin from the intestinal mucosa into the lumen, whereas the high alpha-2-globulin levels are most likely a reflection of the degree of inflammation. Acute phase proteins, including fibrinogen and serum amyloid A protein (SAA) concentrations, can be normal or elevated. Serum electrolyte concentrations are usually within normal limits. However, hyponatraemia, hypokalaemia, hypochloraemia, and hypocalcaemia are seen in chronic diarrhoea. Acid-base abnormalities are present in severe cases, secondary to electrolyte derangements and increased lactate concentration. The hepatocellular and hepatobiliary enzymes, such as sorbitol dehydrogenase (SDH), aspartate aminotransferase (AST), alkaline phosphatase (ALP) and γ-glutamyl transferase (GGT) may get elevated. Prerenal azotaemia may occur uncommonly in dehydrated horses.

Intestinal absorption tests are used to assess the absorptive capacity of the small intestine, specifically the efficiency of sugar absorption from the intestinal lumen. Oral D-xylose or glucose absorption (OGAT) tests are indicated in chronic diarrhoea irrespective of weight loss and hypoproteinaemia. The OGAT can be performed at stall side with minimal equipment. The results are immediate, and it is inexpensive. Factors such as dietary history, gastric emptying rate, intestinal transit, age, and hormonal effects can influence the glucose peak and curve shape. The D-xylose absorption is not affected by hormonal or metabolic factors and, therefore, may be a better test option. Other factors like gastric emptying, intestinal motility, intraluminal bacterial overgrowth, and renal clearance can affect the curve shape. The length of fasting markedly affect the D-xylose absorption and abnormal xylose absorption curves has been detected, even in horses with no small intestinal histopathological abnormalities. One of the major drawbacks of D-xylose testing includes the biochemical analysis and turnaround time.

Diagnosis

Diarrhoea that persists for weeks or months is a frustrating problem from the diagnostic point of view. The diagnosis of the chronic diarrhoea warrants a detailed clinical examination of patients. Anamnesis should include a chronological history including onset, duration, consistency, seasonality, instigating factors, clinical progression, additional clinical signs, diet and dietary changes, supplements provided, environment, travel, previous medical history and medical intervention in the past.

Abdominocentesis is another diagnostic tool that can be used for investigating cases of chronic diarrhoea. The peritoneal fluid analysis includes total protein concentration, total nucleated cell count, red blood cell count, and cytology. Additionally, lactate, fibrinogen, glucose, serum amyloid A, total bilirubin, amylase, and lipase may be measured. Gastroenteritis in horses typically shows normal peritoneal fluid. Eosinophilic gastrointestinal disease occasionally results in increased eosinophils in the abdominal fluid. The total nucleated cell count and protein may be elevated in cases with abdominal neoplasia, and exfoliated neoplastic cells may be detected depending upon the underlying neoplasm. Neoplastic cells are detected on cytology in horses with lymphoma.

Visual inspection of the faeces may provide some clues regarding abnormalities in digestion or transit time of digesta in the gastrointestinal tract. Dental or masticatory disorders, colonic microbial dysbiosis or abnormal colonic transit times may result in increased faecal water content or poorly digested feed particles. Adult cyathostomins and its 4th stage larvae may be observed in the faeces, but microscopic examination of wet faecal smears may be more helpful in identifying the larvae. However, the absence of larvae does not rule out cyathostomins as a cause of clinical disease, but antemortem identification of cyathostomin larvae in the faeces supports the diagnosis of these cases. Faecal flotation for parasites should be performed in horses with hypoalbuminemia, weight loss, and chronic diarrhoea. The detection of sand or gravel in the faeces can be easily accomplished with a simple sedimentation test and is highly suggestive if significant quantities are present in the manure.

The detection of single causative agent in each case in horses is often negative. However, the faecal samples are submitted for microbiological testing for common enteropathogens. Faecal culture and molecular test panels are used to screen multiple enteropathogens associated with horse diarrhoea.

The granulomatous enteritis and lymphadenitis in horse are caused by *Mycobacterium branderi*. *Lawsonia intracellularis* is a well-characterised cause of chronic diarrhoea in young horses, and faecal PCR is the current method of choice for its confirmatory diagnosis. Serological testing is also commercially available. Immunoperoxidase monolayer assay (IPMA) confirms the presence of circulating antibodies to *L. intracellularis* and is indicative of active or recent infection. Assessment of the equine faecal microbiome is predominantly using 16S gene sequencing in horses. Clinical interpretation of these differences is difficult as there are differences in the normal equine microbiome depending on geographic location, housing and diet.

Diagnostic imaging: Transcutaneous ultrasound is one of the most widely used imaging modalities both in the field and at referral centres to assess internal organs. Ultrasound evaluation can be very useful for choosing a site for peritoneal fluid sampling, assessing parenchymal organ appearance, detecting abnormal structures like masses and abscesses, and measuring intestinal lumen distension, motility, and thickness.

Equine proliferative enteropathy caused by *Lawsonia intracellularis* in young horses causes enterocyte proliferation resulting in markedly thickened small intestinal mucosa and occasionally large intestinal mucosa. Increased small intestinal wall thickness is a common finding on abdominal ultrasound examinations of affected foals.

Ultrasonographic measurement of mural thickness and the appearance of the right dorsal colon can be supportive of the diagnosis of right dorsal colitis. The right dorsal colon can be imaged at the right 11th, 12th, and 13th intercostal spaces, below the margin of the lung and axial to the liver. Horses with encysted cyathostomins can develop marked caecal and colonic oedema, and therefore marked intestinal wall thickness, which can be seen with transabdominal ultrasonography. Since there is currently no confirmatory antemortem diagnosis of larval cyathostomins, ultrasonography can be used in combination with the signalment, and deworming history to aid in the diagnosis.

Abdominal radiographs have limited diagnostic value due to the large size and mass of the average adult horse abdomen. Chronic diarrhoea is commonly reported in horses diagnosed with sand enteropathy, and radiography is considered the reference standard for diagnosis of sand accumulation. A grading system has been developed to categorise sand accumulations based on height and length of the accumulation visible in the radiographs. However, sand accumulation in the intestine can be found in horses with no apparent clinical significance, and therefore, the clinical applicability of the grading system is questionable.

Endoscopic examination and biopsy: Adequate assessment of the pylorus and proximal segment of the duodenum can be performed with the advent of endoscopic equipment designed for horses. In suspected IBD cases, duodenal or rectal biopsies should be taken to assess the infiltration of the intestinal mucosa by different cell types. Gastric ulcers have been observed in horses with chronic diarrhoea. Such lesions are associated with the underlying cause of the diarrhoea or due to the inappetence, commonly seen in gastric ulcers.

Pinch biopsies can be obtained from the stomach and duodenum. These may be obtained in cases where there are macroscopic mucosal abnormalities, ultrasonographic abnormalities, or when infiltrative disease is suspected based on other diagnostics. This procedure is performed through non-invasive endoscopy. However, biopsy samples obtained by this method contain the mucosa and occasionally the superficial submucosa. This method does not allow for the acquisition of samples containing the deep submucosa or muscularis layers. Superficial biopsy segments may only consist of villi or small segments of crypts which limits the ability to interpret the sample. Detection of one or more infiltrating cell types provides supportive evidence for the diagnosis of IBD. However, there are resident mucosal cells including lymphocytes, plasma cells, eosinophils, and macrophages that reside in the GI tract of normal horses. Differentiating resident mucosal cells from pathologic infiltration is difficult and can result in unreliable interpretations. Diseases that do not impact the mucosa or are sporadic in their distribution are unlikely to be detected by these biopsies.

Evaluation of cytokine gene expression and number of T regulatory cells in rectal biopsies from horses affected with inflammatory bowel disease were compared to healthy controls in a prospective study. Horses with IBD have increased expression of IL-17A, IL-12p40, and TLR4. T regulatory cells are increased in horses classified as having chronic active simple proctitis. Increased TLR4 and TLR3 expression is thought to play a role in gastrointestinal immune hyperresponsiveness. Specific TLR4 polymorphisms are associated with an increased risk of IBD, giving some support for a genetic predisposition.

The findings in both duodenal and rectal mucosal biopsies, may not represent the entire gastrointestinal tract since only a small section of the mucosa is examined, and this section may be remote from the site of pathology. Pathological findings in such samples are clinically relevant to make a diagnosis. The presence of neoplastic cells and neutrophils within the surface epithelium or crypts on biopsy histopathology is always considered abnormal.

Abdominal exploratory laparotomy/laparoscopy: These diagnostic evaluations are usually recommended as a last resort due to the expense and invasive nature of the procedures. The most common reason for exploratory surgery is to collect full-thickness intestinal biopsies in cases of suspected infiltrative bowel disease. Other reasons include further investigation of suspected or confirmed intra-abdominal masses, abscesses, and neoplasia for diagnostic and treatment purposes. The advantages of surgical biopsies include full thickness sampling, often includes multiple crypts and villi, and better preservation of tissue architecture and orientation. Exploratory laparoscopy is a minimally invasive technique that may be considered as an alternative to midline celiotomy. However, there are several limitations when laparoscopically exploring the abdomen, including potentially poor visualisation of targeted organs/ structures, limited access, and inability to collect adequate samples. Additional limitations include access to laparotomy equipment and comfort with the laparoscopic equipment and suturing. Incisional healing has been a concern in horses that are systemically compromised with hypoproteinaemia.

Diarrhoea that persists for weeks or months is a frustrating problem from the diagnostic and therapeutic point of view, both for the owner and the clinicians. In this condition, faces are soft and watery, and the amount excreted is greater than normal. Horses not only have the production of watery faeces, but also can have "free faecal water" where horses could even produce normal faeces. However, before, after, or during, or independent from defecation, faecal water is produced and pollutes the tail and legs, and may even cause skin lesions. Diarrhoea should be present for at least 7 to 14 days to be considered chronic. In some cases, the diarrhoea will persist for weeks or months, with recurrent attacks of changed faeces texture from "cowpat" consistency to real watery diarrhoea, separated by periods of relatively normal faecal consistency. Chronic diarrhoea in the adult horse is almost invariably associated with large intestinal, caecal and colonic diseases. It could be caused either by physical damage to the colonic wall or physiological disturbances of colonic function.

Treatment

Sulfasalazine is administered @10-20 mg/kg PO once a day, when a definitive cause is not proven. Nutritional management could be of higher importance for horses that suffer from this problem because it improves the well-being of the hindgut environment.

In the case of chronic diarrhoea, especially those caused by motility abnormalities without other physiological disturbances of colonic function with no decrease in body weight, it is necessary to consider the following:

(1) Rebalancing of the diet to the real nutritional requirements of the patient

(2) Avoiding excess carbohydrate from both forage and concentrate,

(3) Selecting the right proportion between good quality first-cut meadow hay and grounded and pelleted meadow hay that helps to reduce the mechanical and physical load of the colon and could help in the formation of more homogeneous digesta,

(4) Reducing the stress with adequate feed consumption time, meal size, and time spent in a overgrazed paddock, and

(5) Increasing the proportion of omega-3 fatty acids in the diet.

Taken together, these factors help reducing the inflammation and influence the motility pattern, absorption, and secretion of nutrients in the colon. Good quality hay is necessary to promote the activity of the microorganism in the hindgut. Structural carbohydrates (chemically defined as neutral detergent fibre or NDF) are fermented to varying degrees by large intestinal microorganisms. NDF can be further partitioned into acid detergent fibre and hemicelluloses components. Fermentability of hemicelluloses and cellulose components of NDF is proportional to the amount of lignin in the cell wall (ADL). Higher lignin content results in lower total digestibility. Certain foods and drinks can cause diarrhoea. Isolating the offending food or drink and reducing it or eliminating it from the diet may solve the problem.

Many medications can cause chronic diarrhoea as a side effect. They can cause it by several different means, and include NSAIDs (nonsteroidal anti-inflammatory drugs), such as aspirin and ibuprofen, Proton pump inhibitors (PPIs), Antibiotics, such as amoxicillin and cephalexin, certain herbal supplements and tea ingredients, such as senna, high doses of certain vitamin and mineral supplements, such as vitamin C and magnesium, antacids containing magnesium, antidepressants (selective serotonin reuptake inhibitors) such as fluoxetine and sertraline, Beta-blockers, and metformin. Removal of such medication can bring cessation of chronic diarrhoea.

24

Calf Diarrhoea

Increased frequency of passing liquid faeces is commonly referred to as diarrhea. Because of the high mortality rate, high cost of treatment and potential for stunting in the wake of diarrhoea episode in calves, neonatal diarrhoea poses a serious financial risk to the cattle breeding enterprise.

Origin

There are both infectious and non-infectious causes of calf diarrhea. A variety of enteric infections, comprising bacteria, viruses, and protozoa, contribute to the development of this illness. When calves have diarrhoea, co-infection is often seen, yet in certain instances, a single main pathogen may be the cause. Diarrhea is caused by multiple infectious pathogens that can coexist or even succeed one another in the same calf, making it difficult to identify each one with precision.

- Bacteria: mostly Escherichia coli, but also Salmonella, Clostridium, Yersinia, and Campylobacter genera are known to exist.
- Viruses: primarily BVD, coronavirus, and rotavirus. Occasionally, isolates of Calici viruses and Toga viruses are also found.
- Parasites: including Strongyloides sp., Eimeria sp., Giardia sp., and Cryptosporidium parvum. The parasites that are most commonly found are cryptosporidia.

Dietary factors, such as excessively fat-containing milk, excessively oversized or cold drinks, can be additional contributing factors to these infectious reasons. Food-related diarrhoea is plaster-textured and white in colour. The calf may be somewhat or completely dehydrated, but by becoming weaker, they run the risk of encouraging the onset of parasite diarrhea.

Prevention

Young calves with enteritis episode are susceptible to a number of risk factors. Controlling these elements is therefore necessary for prevention.

Regulation of colonic intake

Although calves have a functional immune system from birth, active immunological defense does not occur for several weeks. Therefore, the transfer of passive immunity via colostrums is a crucial phase that needs to happen in the first few hours of life. Through colostrum, a newborn calf absorbs antibodies that are meant to shield him from infectious agents that cause diarrhea. However, it is frequently the case that the calf does not receive enough protection from these agents, either because of inadequate colostrum consumption or low-quality colostrum. The quality of colostrum, particularly against these agents, is improved when women are vaccinated against neonatal diarrhea agents during pregnancy.

Hygiene protocols

The coexistence of calves of varying ages and the cleanliness of the calving and lodging areas are important factors in the infectious pressure, as the older animals excrete germs that could infect the younger ones. Using individual niches during the first few weeks of life is a useful way to prevent the excessive transmission of germs in dairy calves, especially in the case of the most vulnerable newborns.

Control of food and rationing

Calf health is greatly influenced by the diets of mothers, particularly during the final third of pregnancy. The ideal nutritional balance between energy and nitrogen content must be achieved; additional focus will be placed on vitamin and macro and trace element supplementation.

25

Foal Diarrhoea

Foal diarrhoea is very common in neonatal foals aged between 5 days to 15 days, and occurs in 75% to 80% of foals. It is one of the important diseases in young foals and may occur in more than half of foals until weaning age.

Etiology

Several infectious and noninfectious underlying causes have been implicated for foal diarrhoea and more etiologies are being evolved with scientific advancements and investigations. There are established and putative infectious and noninfectious causes of foal diarrhea. Risk factors include suboptimal deworming of mares and extensive use of antibiotics. Noninfectious causes include foal heat diarrhoea, perinatal asphyxia syndrome, necrotizing enterocolitis (NEC), dietary imbalance, equine gastric ulcer syndrome, luminal irritant diarrhoea, and secondary lactose intolerance. The etiology of foal heat diarrhoea remains speculative and includes changes in mare's milk composition during foal heat. However, orphan foals on milk replacer also develop diarrhea at similar ages, making this theory less probable. More recently, maturational changes in bacterial intestinal flora during that early life period have been elucidated and this adaptation of the microbiota is considered as the cause of foal heat diarrhoea.

The most common infectious agents include rotavirus (RV), *Clostridium perfringens* types A and C, *Salmonella* spp, *C difficile*, *Cryptosporidia*, and *L intracellularis*. Less common causes are *Coronavirus*, *Rhodococcus equi*, and *Strongyloides westeri*. Other very uncommon causes of foal diarrhoea include *Aeromona hydrophyla*, *Neorickettsia ristici,* Candida and *Listeria monocytogenes*. Co-infections may be the cause of foal diarrhoea in large number of cases as seen with *C perfringens* and *C difficile*. Rota viruses belong to the Reoviridae, subfamily Sedoreovirinae, genus *Rotavirus*, and are the most frequently detected infectious agents in foals with diarrhoea, and accounts for 35% to 90% of cases up to 3 months of age. The virus has a ubiquitous world distribution in the horse population, is highly contagious, replicates rapidly, and is shed in high concentrations in feces of affected animals.

C perfringens is a gram-positive anaerobic spore-forming bacterium, associated with enterocolitis in foals and horses. *C perfringens* was long considered part of normal microbiota of the large colon. In neonatal foals, *C perfringens* was isolated from the feces of 90% of 3-day-old foals. The most common genotype identified (85%) is type A; *C perfringens* type C was isolated in less than 1% to 3% of neonatal foals.

The less common infectious and parasitic agents responsible for foal diarrhoea include Corona virus, Cryptosporidium spp, Lawsonia intracellularis, Rhodococcus equi. enterotoxigenic Escherichia coli, Enterotoxigenic Bacteroides fragilis, and sepsis. Diarrhoea is common in septic foals, due to hemodynamic alterations leading to gastrointestinal (GI) mucosal hypoperfusion, inflammatory mediators associated with SIRS, and dysmotility. Cryptosporidium spp can infect adult horses and foals, Cryptosporidium parvum is the most common genotype identified. Infection occurs with ingested oocysts that reach the ileum. It affects mainly foals with severe combined immunodeficiency (SCID), but outbreaks in immune-competent foals have also been reported. Self-limiting diarrhoea can occur for up to 8 days and in older foals; chronic and intermittent diarrhea may also occur

Equine coronavirus is considered an enteropathogen in foals. However, the importance of this agent is not known. *L intracellularis* is a gram-negative obligate intracellular bacterium affecting mainly foals aged 4 months to 7 months (range 2–13 months) causing proliferative enteropathy. It is transmitted via the faecal-oral route. *R equi*, a gram-positive facultative intracellular pathogen, is a common etiologic agent in foal pneumonia. It has been reported that up to 50% of foals with *R equi* bronchopneumonia also had intestinal lesions on necropsy, and 4% of foals had intestinal lesions without pneumonia. *Strongyloides westeri* infects foals early in its life, but its role in diarrhoea is questionable. *N. risticii* was found in 4% of foals with diarrhoea.

The causes of the noninfectious diarrhea include necrotizing enterocolitis, hypoxic gastrointestinal damage, equine gastric ulcer syndrome, luminal irritant diarrhoea, lactose intolerance and dietary diarrhoea. Primary lactose intolerance in foals is rare. Secondary lactose intolerance is produced by small intestinal enteritis associated with Rota virus and *C difficile*. Dietary intolerance in foals is more commonly seen in foals receiving enteral nutrition with milk replacers. Foals may ingest large amounts of sand or other abrasive materials due to pica that can be extremely irritating and damaging to intestinal mucosa resulting in diarrhoea with colic. NEC is a disease primarily of premature infants and consists of necrotic injury to the mucosal and submucosal layers of the gastrointestinal tract. The distal small intestine and proximal colon are most commonly affected. Equine gastric ulcer syndrome occurs in foals from 2 days to 9 months of age and has been associated with diarrhoea. The changes in mare's milk composition during foal heat may induce diarrhoea. However, orphan foals on milk replacer also develops diarrhea at similar ages, making this theory less likely. More recently, maturational changes in bacterial intestinal flora during the early life period have been elucidated and this adaptation of the microbiota may lead to foal heat diarrhoea.

Pathogenesis

The transmission occurs through the faeco-oral route of contaminated feces or fomites, resulting in outbreaks. The incubation period is 1 day to 4 days. Virus shedding can start before the onset of diarrhoea that persists during the clinical phase up to12 days post resolution of diarrhoea. Foals with RV diarrhoea presented to a referral hospital had a survival rate of 94%. The lactose intolerance may result in foal diarrhoea. Additional virulence factors include β2 toxin, an enterotoxin produced by all *C perfringens* but most commonly by type A. Recently, an additional toxin named *netF* has been identified in foals with NEC and in vitro cell cytotoxicity has been proved. The role of these additional virulence factors in pathogenesis of diarrhea is still debated. Lactase is mainly produced by the brush border epithelium of the villous enterocytes. The two major toxins, toxin A (TcdA) an enterotoxin and toxin B (TcdB) a cytotoxin, are responsible for causing increased paracellular permeability of mucosal surfaces, cell rounding, and eventually cell death or apoptosis. The toxins also cause inflammation resulting in increased fluid exudation and mucosal damage, leading to diarrhea or pseudomembranous colitis.

Clinical Signs

Diarrhoea is one of the most important clinical signs in young foals and may occur in more than half of foals until weaning age. Foal diarrhoea is usually a self-limiting condition that lasts 3 to 4 days. The diarrhoea may not have systemic involvement, and the foal may continue to suckle well. However, a few foals may be affected by more severe and prolonged diarrheic episodes. Diarrhoea is characterized by increased frequency of defecation with increased water content in faeces. Foals showing clinical signs of diarrhoea usually have enteritis, which is associated with systemic inflammatory response syndrome (SIRS). More than 50% of foals experience one or more bouts of diarrhoea in the first 6 months of life. Outbreaks can occur depending on the etiology, or single foals might be affected. The other observations include depression, decreased suckling reflex, weakness, colic, weight loss, and occasionally sudden death

Diarrhoea commonly occurs within the first 6 months of life. Vaccination and worming status, information on number and ages of the affected foals, and hygienic conditions on the farm are important to determine the etiology and therapeutic outcome. Failure of passive transfer of immunity contributes to the pathogenesis of the disease. The common infectious agents of foal diarrhoea sorted by age range of foals

Age Range	**Common Causes**
< Two weeks	Foal heat diarrhea, *C perfringens, Rotavirus, C difficile,* Septicemia, *Cryptosporidium,* Neonatal asphyxia, NCE, *Rotavirus, Cryptosporidium*
2 weeks to 2 months	*Salmonella* spp, *S westeri, L intracellularis*
> 2 months	*S westeri, N risticii, Salmonella* spp
All ages	Lactose intolerance, Luminal irritants

Rota virus diarrhoea is primarily malabsorptive but also includes a toxin-mediated secretory component. It preferentially infects the mature absorptive villous enterocytes of duodenum, jejunum, and ileum-sparing crypts, causing villous damage resulting in malabsorptive diarrhoea

The initial clinical signs include colic, hypermotility or amotility, and abdominal distension. Faecal consistency can range from watery to pasty, with different colors, and may contain blood or casts. Foals tend to dehydrate quickly and can reach severe dehydration as evidenced by sunken eyeballs and a prolonged skin tent. Depression, weakness, anorexia, salivation, and bruxism are frequently observed. Body temperature varies from fever to hypothermia. Cases of enteritis commonly have signs of SIRS, including congested mucous membranes, tachycardia, weak pulses, tachypnoea, and cool extremities, leading to recumbence, coma, and death. Bacteremia is seen up to 50% of diarrheic foals less than 30 days of age with diarrhea

C perfringens was the second most significant organism isolated from foals with diarrhea, and may cause sudden death without prior clinical manifestations. Foal diarrhea associated with *C perfringens* occurs commonly early in life; 96% of affected foals had adequate transfer of passive immunity. Diarrheic foals less than 1 month of age are more likely to be positive for *C perfringens*. Mortality ranges from 12% to 83%, and the highest mortality has been reported for infection with *C perfringens* type C. *C perfringens* type A produces α toxin; type B produces α, β, and ε toxins; type C produces α and β toxins; type D produces α and ε toxins; and type E produces α and ι toxins.

Salmonella spp are rod-shaped, gram-negative Enterobacteriaceae. There are more than 2200 serovars in 8 subspecies, with subspecies I responsible for 99% of clinical mammalian diseases. Diarrheic foals older than 1 month of age are more likely to have fecal *Salmonella* isolated than younger foals. *Salmonella* colonizes the intestine using the invasion-associated type III secretion system causing inflammation and massive mucosal damage of the ileum and colon.

Clostridium difficile, an anaerobic, spore-forming, gram-positive, bacterium commonly associated with diarrhoea and colitis in mammals and humans, is considered as one of the important agents of enteric disease of foals and adult horses. It has been isolated in 11% of diarrheic foals.

Clinical Pathology

Dehydration results in higher hematocrit value. Signs of inflammation may be present in the leukogram, including leukopenia or leukocytosis, commonly due to abnormal neutrophil counts. A left shift might also be present, particularly in cases with SIRS. Plasma total proteins can be increased or decreased, as in cases with sepsis, protein-losing inflammation (salmonellosis), or *Lawsonia intracellularis* infections. Acute phase proteins, such as fibrinogen and serum amyloid A, are usually elevated. Electrolyte derangements are common and should be evaluated. Up to 66% of the foals have been reported as hyponatremic and 33% hypokalemic. Hyperchloremia or hypochloremia also can occur. Foals with diarrhea usually have a strong ion metabolic acidosis with low blood pH, low plasma bicarbonate levels, and a decreased strong ion difference (strong ion difference = Na + K-Cl lactate = 36–44) due to electrolyte derangements. Increased lactate concentrations are also common and contribute to the low SID. Urea and creatinine levels are often increased, indicating a prerenal and occasionally also a renal azotemia.

Diagnosis

Establishing a diagnosis of diarrhea is easily achieved whereas establishing the cause of the diarrhea is often difficult. The minimum database of a foal with diarrhea should include a thorough history, physical examination, complete blood cell count, biochemistry profile, and determination of serum immunoglobulin concentrations

It is important to investigate all known potential causes and identify infectious agents to avoid outbreaks, evaluate the level of systemic compromise, and establish adequate therapy. It is also important to differentiate foals that require referrals from those who can be managed in field conditions.

Several infectious and noninfectious underlying causes have been implicated but scientific evidence of pathogenesis is constantly evolving

Practically it is important to investigate all known different potential causes and to identify infectious agents to avoid outbreaks, as well as to evaluate in individuals the level of systemic compromise and establish an adequate therapy.

Abdominal ultrasound can be useful to determine signs of NEC and to assess intestinal wall thickness. Abdominal radiography can be used to determine presence of sand and gas accumulations. A major differential in diarrheic foals is sepsis and if available a sepsis score and blood culture should be performed.

Fecal testing is recommended and tests should be based on the most likely causative agent of a given case based on age and clinical signs. A positive diagnostic test does not always confirm that this agent is the etiologic cause of the diarrhea. Several pathogens can be found in feces of healthy as well as diarrheic animals.

Treatment

The treatment is mainly symptomatic but some specific treatments of various causes are available.

Crystalloids, colloids, or combinations of these 2 can be used to correct hypovolemia. Replacement crystalloid fluids for foals include lactated Ringer solution, Normosol R, and Plasma-Lyte 148. Hypertonic solutions are not commonly used in foals, because foals may not be tolerant of large sodium loads. Synthetic colloids can be used as 3-mL/kg to 5-mL/kg. Synthetic colloids have been associated with renal failure and a worse outcome in human medicine and their use is currently questioned in horses as well. An attempt should be made to correct hypovolemia without synthetic colloids if deemed feasible. Plasma is a better choice of colloids than synthetic colloids in foals. The risks for anaphylactic reactions during colloids administration can be minimizing by infusing at a constant rate infusion (CRI) of 1 to 2mL/kg/h.

The speed and volume of infusions depend on the degree of hypovolemic, cardiovascular status and plasma protein levels. However, there are no studies in foals to suggest accurate fluid rates. The most recent recommendation based on pediatric patients suggests an initial fluid bolus of 20 mL/kg crystalloid solution over 10 minutes to 20 minutes. Perfusion should be reassessed after the bolus and, if still inadequate, this bolus can be repeated up to 3 times in total. Clinical signs of adequate perfusion include normalization of tachycardia, improvement in mentation, return of suckle reflex, and urine voidance. If possible, laboratory parameters should be re-evaluated. After 3 boluses, continuous rate infusion of fluids is recommended. If response to fluid therapy is inadequate (perfusion does not normalize), inotrope therapy should be added. Dobutamine diluted in isotonic saline, 5% dextrose, or lactated Ringer solution is used and the dose should be titrated from 1 μg /kg/min to 3 μg/kg/min up to 5 μg/kg/min CRI. The remainder of the dehydration should be corrected in within 12 hours to 24 hours and ongoing losses added to the fluid regimen. Maintenance amounts of fluid should be added if the foal is not allowed to nurse. The maintenance fluid needs in foals should be calculated using the Holliday-Segar formula for body weight (BW). Correction of electrolyte imbalances during maintenance fluid therapy is crucial for recovery. This also aids in balancing the acid base status of foals. Hyponatremia should be corrected at a rate not exceeding 0.5 mEq /h to avoid central pontine myelinolysis. Hyponatremia can be partially corrected by increasing the plasma sodium concentration by 2 mEq/L to 4 mEq/L to abolish seizure activity. Hypokalemia is treated with empiric supplementation of fluids with 10 mEq/L to 40 mEq/L of potassium chloride, and it depends on the existing plasma potassium concentration. Hyperkalemia can be empirically treated with administration of 20 mL/kg of 0.9% sodium chloride containing calcium (eg, 0.4 mEq/L to 1 mL/kg of 23% calcium borogluconate) and 4 mg/kg/min to 8 mg/kg/min of dextrose. Isotonic sodium bicarbonate (1.3%) can be used with added dextrose as an alternative treatment. In refractory and severe cases of hyperkalemia, insulin treatment can be tried.

Sodium bicarbonate therapy should be reserved for foals with severe refractory acidosis, often due to hyperchloremia. In these cases, isotonic sodium bicarbonate (1.3%) intravenously (IV) can help ameliorate signs of acidosis by supplying sodium while decreasing the chloride load. Alternatively, oral sodium bicarbonate can be administered.

All foals with compromised systemic status usually benefit from glucose administration. A starting dose is 4 mg/kg/min. The rate can be increased to 6 mg/kg/min to 8 mg/kg/min if needed. Glucose should be administered separately from fluid therapy if possible, using a fluid pump. This allows adjusting fluid and glucose therapy separately.

Nonsteroidal anti-inflammatory drugs (NSAIDs) should be used cautiously in foals less than 1 month old due to renal and gastric side effects. Polymyxin B (6000 U/kg IV) can be used to treat endotoxemia. Plasma transfusion should be considered for foals with failure of passive transfer of immunity. Because foals with diarrhoea may use their globulins despite initially adequate levels, assessment of immunoglobulin concentration should be repeated periodically to assess the need for further plasma transfusion.

Adsorptive agents, such as kaolin/pectin and bismuth subsalicylate, have been used but efficacy studies are lacking. These agents are used at 0.5 mL/kg to 4 mL/kg, 1 to 4 times daily. Di-tri-octahedral smectite can has been shown to neutralize *C difficile* and *C perfringens* toxins *in vitro*. It is possible that *in vivo* neutralization of bacterial toxins in the gut could also occur.

The use of probiotics in neonatal foals cannot be recommended based on currently available data. The use of fecal microbiota transplant has been proposed based on its remarkable success in treating recurrent *C difficile* infection in humans.

The prophylactic use of antiulcer medications is controversial in neonatal foals but indicated in older foals. It has been shown that neonatal foals on antiulcer medication had a higher risk of developing diarrhea. Omeprazole is the only approved drug for the treatment of EGUS, and 4 mg/kg orally, every 24 hours, is the recommended dose. Gastro-protectants should be considered if NSAIDs are given for prolonged periods of time.

In foals with signs of SIRS or sepsis, the use of broad-spectrum antibiotics is warranted. The antibiotics commonly used are a combination of aminoglycosides (mainly amikacin 20–25 mg/kg IV every 24 h) and penicillin (22.000–44.000 IU/kg IV every 6 h). Metronidazole (15–25 mg/kg body weight every 8 h) is widely used for clostridial diarrhea, but there are no current objective data supporting its use. Vancomycin has been used in cases of resistance and in humans is used in severe cases of *C difficile*–associated diarrhea. Bacitracin was used in the past but is no longer available.

Antibiotic treatment in foals with known *Salmonella* infection is indicated due to the risk for bacteremia and sepsis. The treatment should be continued beyond clinical recovery to prevent secondary seeding. Antimicrobials effective against *Salmonella* spp include extended-spectrum cephalosporin or ampicillin-sulbactam alone or in combination with an aminoglycoside (gentamicin or amikacin) or fluroquinolones. Antimicrobial choices recommended for treating *L intracellularis* include oxytetracyclines, macrolides, or chloramphenicol. Treatment of Rhodococcal enterocolitis is similar to pulmonary *R equi* infection using macrolide antimicrobials combined with rifampin. Therapy for cryptosporidiosis is largely supportive in foals, but therapy with the aminoglycoside paromomycin could be attempted. In cases of *S westeri*–associated diarrhoea, anthelmintics like oxibendazole is given at the dose rate of 15 mg/kg or ivermectin at the dose rate of 0.2 mg/kg.

Foals with mild to moderate diarrhea without evidence of colic should be allowed to continue nursing. In contrast, any foal with abdominal discomfort will likely benefit from a period of brief GI rest (12–24 h). Fluid therapy has to be administered if a foal less than 4 weeks old is not nursing for more than 6 hours. In cases of suspected lactose intolerance, exogenous lactase (Lactaid, 3000–6000 FCC Unis orally every 6 h) can be used. Reintroduction to feeding should be slow and gradual. Feeding of 10% of body weight per day divided into hourly or 2-hour feeding intervals through a nasogastric tube or feeding tube is an adequate starting level.

Control

Foal infectious diarrhea prevention should focus on 3 important aspects of disease prophylaxis: minimize exposure to pathogens (disinfection and isolation), increasing immunity (vaccination), and optimization of management practices. Detection of a sick foal demands that in-contact animals should not be moved to other locations, because they may be a source of infection. People traffic should be curtailed by ensuring that healthy animals are attended to first, followed by in-contact foals, and finally clinical cases

26

Gastritis in Dogs and Cats

Gastritis refers to inflammation of the gastric mucosa, and is principally characterised by a syndrome of acute or chronic vomiting. It may occur as a short episode, called acute gastritis, or is chronic in nature that shows manifestation for a long duration, and may be associated with underlying tissues making the condition more serious.

Etiology

Irritation, infection, antigenic stimulation, or injury caused by physical or chemical agents, erosion, and ulceration of the gastric mucosa are the common causes of gastritis in dogs and cats. The causes of acute gastritis in dogs and cats also include dietary indiscretion such as ingestion of foreign material or novel, spoiled or contaminated foods, ingestion of toxin or drug, such as antibiotics, NSAIDs, corticosteroids, plants, and chemicals. Systemic illness like pancreatitis, uremic gastropathy, hypoadrenocorticism, and endoparasitism such as Physaloptera species in dog and cat, *Ollulanus* sp. in cat or viral infection like canine parvovirus gastroenteritis, feline panleukopenia causes gastritis.

Acute gastritis occurs frequently in dogs. Dogs are particularly prone to dietary indiscretion, including ingestion of spoiled or raw food, garbage, cat litter, foreign objects, plants, toxins, moulds, and poisonous mushrooms. Feeding of inappropriate foodstuffs, such as table scraps or leftovers, or being fed large quantities of food can also cause gastritis. Some of the common causes or conditions associated with gastritis in dogs include antibiotics oral adminstration, fungal infection, liver disease, nonsteroidal anti-inflammatory drugs (NSAIDs), fungal toxins, mast cell tumour, bacterial infection, gastrinomas, neuroendocrine tumours characterized by the secretion of gastrin with resultant excessive gastric acid production, or other cancers, moulds, chemical irritants, granulomatous gastric disease, overeating, chemotherapy, heavy metal poisoning, pancreatitis, corticosteroids, hypoadrenocorticism called Addison's disease, poisonous plants, diabetic ketoacidosis, idiopathic gastritis, spoiled food, dietary indiscretion, immune-mediated disease, stress, endocrine disease, inflammatory bowel disease, toxins, food allergy, intestinal parasites, uterine infections, foreign body, kidney disease and viral infections.

Epidemiology

The older, male, small-breed dogs such as Lhasa Apso, Shih Tzu, Maltese, Miniature Poodle, more frequently suffer from gastritis. A unique, breed-associated form of atrophic gastritis not linked with *Helicobacter pylori* infection, in Norwegian Lunde hounds has been noticed with gastric adenocarcinoma.

Pathophysiology

Damages to the gastric mucosa stimulates the release of inflammatory and vasoactive mediators with subsequent disruption of gastric epithelial cells, increased gastric acid secretion, and impaired gastric barrier function. Visceral receptors sensitive to gastric distention, gastric inflammation, and tonicity of gastric contents send impulses via vagal and sympathetic nerves to the vomiting centre of the medulla oblongata, thereby stimulating the vomiting reflex.

Chronic hypertrophic gastropathy is characterized by diffuse or focal hypertrophy of the gastric mucosa, muscularis layer, or both, with variable inflammatory infiltration. The lesions are often most pronounced in the pyloric region, with resultant obstruction to gastric outflow, and projectile vomiting of food within hours of eating. Hypergastrinemia due to exaggerated secretion as seen in gastrin-secreting neoplasia, Basenji gastroenteropathy or inadequate clearance in hepatic or renal disease, achlorhydria may initiate mucosal hypertrophy. The role of *Helicobacter* infection in the development of atrophic gastritis is unknown.

Clinical Signs

Vomition and anorexia are the most common clinical signs in gastritis. Other clinical symptoms may include dehydration, lethargy, depression, abdominal pain, increased thirst, and presence of blood in the vomitus or faeces. Vomiting of sudden onset is characteristic of gastritis. The vomitus may contain bile, food, froth, frank or digested blood, or an ingested substance like grass, bones, foreign material, etc. The severity and frequency of vomiting as well as the underlying cause determine the additional clinical signs. Dogs suffering from gastritis mostly experience extensive vomiting. The vomitus may contain yellowish, foamy bile, especially seen if the stomach is empty. Blood or food is frequently seen in the vomitus, particularly following ingestion of inappropriate foodstuffs. The dogs may breathe heavily after eating or drinking and may have tenderness around the stomach area. The affected dogs have no energy for movement, possibly, due to loss of appetite.

The intermittent or persistent vomiting lasting longer than 14 days is considered as the manifestation of chronic gastritis in dogs and cats. The most common clinical sign is intermittent vomiting of food or bile. Systemic illness, weight loss, and gastro-intestinal ulceration are infrequent and should raise suspicion of a more serious condition or diffuse gastrointestinal inflammation, like inflammatory bowel disease. Chronic atrophic gastritis is often characterized by marked mononuclear cell infiltration, thinning of the gastric mucosa, and atrophy of the gastric glands.

Diagnosis

Diagnosis is arrived at thorough history-taking, clinical findings, and the response to symptomatic treatment. A specific diagnosis should be explored in cases of dog having access to foreign objects or toxins, and non-recovery within 2 days of symptomatic therapy and continuance of hematemesis or melena, and the dog is systemically unwell. Dogs may signal the presence of cranial abdominal discomfort by adopting a "praying" posture with raised hindquarters and holding the chest and forelegs close to floor in an attempt to get some relief. A complete blood count, serum biochemical profile, and urinalysis followed by more specific clinicopathologic tests such basal serum cortisol concentration, ACTH stimulation test, evaluation of vomitus for specific toxins are indicated but are frequently unremarkable in dogs with chronic gastritis. Diagnostic imaging, including plain or barium contrast abdominal radiographs and abdominal ultrasound, may be indicated. Histologic evaluation of endoscopic or surgical gastric biopsies is required for definitive diagnosis and classification of chronic gastritis. Lymphocytic-plasmocytic gastritis and eosinophilic gastritis are characterized by diffuse infiltration of the gastric mucosa and lamina propria with lymphocytes and plasma cells, or eosinophils, respectively. Similar cellular infiltrates may be seen in the small intestine. Concomitant lymphoid hyperplasia, mucosal atrophy, or mucosal fibrosis is infrequently seen. Dietary allergy or intolerance or hyperimmune response to normal antigens have been proposed as possible causes. Eosinophilic gastritis with eosinophilia and skin lesions should raise suspicion for dietary sensitivity or hyper-eosinophilic syndrome.

Tests for gastritis may include blood tests, urinalysis, faecal tests, abdominal radiographs (X-rays), abdominal ultrasound, and endoscopy. In acute cases, only minimal diagnostics such as blood and urine tests are required. However, several other tests are performed to rule out the conditions like tumours, foreign bodies, systemic infections, poisoning, pancreatitis, parvovirus, neoplasia, dietary indiscretion, and to determine the exact cause of chronic gastritis in dogs and cats with clinical signs of vomiting.

Treatment

Acute gastritis is typically of short duration that last for less than twenty-four hours. The disease is mostly self-limiting as the clinical signs usually resolve before diagnostic testing are performed to establish the exact cause of the disease. Most dogs with acute gastritis recover in one to three days with supportive treatment including a short period of withholding food. The prognosis is usually good.

A small amount of water is offered frequently during the first 24 hours. Intravenous fluid infusion is resorted to, if fluids cannot be given orally without inducing vomiting. A small amount of highly digestible, low-fat, and low-fibre food is offered, if there is no vomiting after 24 hours. The feeding is resumed with frequent small meals, usually about half the normal daily amount of food, divided into four to six portions. Gradually, the quantity of food is increased over the next two to three days.

Medical treatment for dogs with gastritis includes anti-emetic drugs, such as maropitant, ondansetron or metoclopramide, fluid therapy for dehydrated dogs, Proton pump inhibitors, such as omeprazole, gastrointestinal protectants, such as sucralfate, H_2 receptor antagonists, such as famotidine or ranitidine.

The prognosis for chronic gastritis is based on the underlying cause, and the severity of vomiting. In addition to symptomatic care and dietary modification, animals with moderate to severe conditions, and histologically confirmed diseases generally require immunosuppressive therapy. Prednisone (or prednisolone in cats) is started at 2 mg/kg/day, PO (cats and small dogs), or 40–50 mg/m2 (medium to large dogs), and tapered to the lowest dosage that controls clinical signs. Assuming continued clinical remission, prednisone therapy is ultimately discontinued and strict adherence to dietary therapy maintained.

If clinical signs persist despite dietary modification and prednisone therapy, treatment with an additional immunosuppressive agent (dogs: azathioprine at 2 mg/kg, PO, every 24–48 hours, or cyclosporine at 5 mg/kg, PO, every 12 hours; cats: chlorambucil at 2 mg [total dose], PO, every 48–72 hours for 2–4 weeks, then tapered to 2 mg every 72–96 hours) can be considered.

27

Diarrhoea in Dogs and Cats

Diarrhoea is defined as an increase in frequency, fluidity or volume of faeces. It occurs in dogs and cats can occur due to a number of reasons, including a sudden change in diet or eating something inappropriate, characterised by loose faeces. The dietary origin diarrhoea usually resolves on its own after a few days.

Etiology

Besides dietary origin of diarrhoea, the other causes are described as follows

Stress: Stress from changes in routine or environment

Infections: There are many potentially pathogenic organisms which may cause or contribute to the development of acute diarrhoea, and include bacterial, viral, or parasitic infections of the gut. Parasitic infections tend to be more common in younger animals. Identification of worms or their eggs in the faeces of diarrhoeic animals is an indication for treatment with an appropriate endoparasiticide. While protozoal infections can be subclinical, diarrhoea may follow heavy infestations, infections in young or immunocompromised animals or infections alongside other pathogens or concurrent disease.

Bacterial – *Salmonell*a spp., pathogenic E. coli, Campylobacter jejuni, Clostridium difficile, Clostridium perfringens, Yersinia spp

Parasites – Parasites such as Giardia, hookworms, and roundworms, more specificaly *Toxocara* spp., Taenia spp., *Dipylidium caninum*, *Trichuris vulpis*, *Uncinaria stenocephala*

Protozoa – Giardia, Cryptosporidium, Coccidia, Isospora, *Tritrichomonas foetus*

Algal – Prototheca

Viral – Various viruses have been associated with acute diarrhoea in dogs and cats. The most commonly detected are parvoviruses, coronaviruses, paramyxoviruses, rotaviruses and caliciviruses, FeLV, FIV, distemper, rotavirus, circovirus, norovirus, canine adenovirus, feline torovirus, reovirus, enteroviruses, feline astrovirus. Parvovirus and paramyxovirus have been shown to be strongly associated with acute haemorrhagic diarrhoea. Parvoviral infection may also lead to permanent changes in the cytoarchitecture of the gut. Severe disruption to the intestinal mucosal barrier may facilitate bacterial translocation, which could result in septicaemia

Foreign bodies: Eating something poisonous or a foreign body

Other diseases: Internal diseases such as kidney failure, or failure of digestive enzyme production such as pancreatic insufficiency

Diarrhoea may occur secondary to disorders affecting the small or large intestines, or both. Localisation to the small or large bowel may not be possible in diffuse disease. It may also be less relevant in acute diarrhoea as biopsies are unlikely to be required and symptomatic treatment remains almost the same.

Sudden diet change, indiscretion (scavenging), hypersensitivity (allergy), intolerance, poor quality diet, mouldy food. Dogs which have experienced a recent change of diet by scavenging or through the owner, and are fed a home-cooked diet that live in a multi-dog household or have recently stayed in kennels, are at increased risk of diarrhoea.

Epidemiology

Acute diarrhoea in dogs and cats is a common complaint at first-opinion small animal veterinary practice institutions. Diarrhoea lasting for less than two weeks is generally described as acute that commonly resolves without the intervention of veterinary practitioner. In terms of prevalence, 14.9 per cent of experience an episode of diarrhoea is very common in dogs The prevalence of diarrhoea in a rescue cat population was 11.9 percent. There are potential causes of diarrhoea which may initially present as acute diarrhoea, but then develop into chronic diarrhoea.

Pathophysiology

A healthy gastrointestinal microbiota consists of a diverse mix of microorganisms, including some which are potentially pathogenic, it can be difficult to determine whether the presence of certain microorganisms is of clinical significance in cases of diarrhoea.

Healthy dogs and cats should normally produce firm stools that are well-formed and are brown in colour. A dog or cat that is suffering from diarrhoea will produce faeces that are soft, or sometimes even liquid in consistency. There may also be a change in normal colour of faeces of the pet, or the pet may pass stool with mucous or blood in it.

Clinical Signs

The signs of diarrhoea in dogs include passage of loose faces, increased frequency of defecations, and increased bowel movements than usual. Signs of diarrhoea in cats include loose and soft faeces in their litter tray and matting of hairs on their back with faeces.

If the diarrhoea is severe, pet can rapidly become dehydrated. Pets may also suffer from vomiting, loss of appetite, or blood in their faeces. All of these factors may also contribute to loss of fluid from the body, so it is important to recognize the symptoms of dehydration. Signs of dehydration include sunken eyes, pale, 'tacky' or sticky gums, sleepiness or lack of energy, loss of appetite, loss of skin elasticity.

Diagnosis

The aim of faecal analysis in cases of acute diarrhoea is to identify the presence of infectious agents to guide treatment or to provide prognostic information. There is no doubt that detection of parasites or protozoa will often indicate that certain treatments are required, or that isolation of certain viruses (eg parvovirus) will help the clinician to provide prognostic guidance for the owner. However, given that *Salmonella*, *Campylobacter*, *Yersinia* and *E. coli* O157 have been isolated from healthy animals, detection of these bacteria in faecal samples does not necessarily confirm the cause of the diarrhoea. All of these bacteria are zoonotic, so testing may be indicated if there is concern regarding zoonotic disease within a household (eg presence of children, immunocompromised people or owners with significant symptoms themselves). However, the use of antibiotics solely to try to achieve elimination of these bacteria from the pet's stool samples is questionable.

Treatment

Mild cases of diarrhoea in both cats and dogs can be treated at home by feeding bland diets that are easy to digest. Only rest to the gastrointestinal is adequate for the recovery. The regular food can be gradually reintroduced once the pet feels better. Despite the fact that it is often self-limiting and likely to resolve with symptomatic treatment alone, antibiotics are frequently prescribed as part of the therapeutic management plan. Antibiotics are ineffective in the treatment of viruses. However, when considering the use of antibiotics in acute diarrhoea, viral infections become particularly relevant if they are likely to increase the risk of sepsis, for which antibiotics would then be required. This may be through immunosuppression or disruption to the mucosal barrier which could enable translocation of commensal bacteria. Antibiotics are rarely required to treat diarrhoea in dogs and cats and can sometimes even make diarrhoea worse. This is because antibiotics destroy the good bacteria in pet's gut, along with any bad bacteria. Generally, antibiotics will only be prescribed in cases where pets are very ill.

Probiotics and Prebiotics

Probiotics and prebiotics may also help to aid pet's recovery from diarrhoea. Probiotics are the 'good bacteria' that live in the digestive tract and prebiotics are a source of food for these live bacteria which helps them to multiply. Products containing probiotics and/or prebiotics may also contain natural binders which can help the pet to produce firmer stools.

Home remedies for Diarrhoea

It is important to ensure that the cat or dog has plenty of fresh water available so that it can replenish the fluid loss due to diarrhoea. Proper rest should be given to the animal without any disturbance.

Control: Avoiding completely scavenging of inappropriate food items can prevent any unwanted diarrhoea in the pet. The change in the regular diet should be done slowly and gradually by feeding increasing amounts of the new food and decreasing amounts of the old food, over a period of 7 days. It is also important to ensure that the pet has regular faecal tests and is up to date with parasite treatment.

28

Colic in Horses

Colic in horses refers to clinical signs of abdominal pain due to gastrointestinal diseases. It is the most important and frequently recorded disease of horses and the principal cause of death, characterised by clinical signs of rolling or thrashing, pawing frequently and shock.

Etiology

The etiology-based classification of disease includes (1) obstructive (2) obstructive and strangulating (3) non-strangulating infarctive (4) inflammatory like peritonitis, and enteritis. Colic cases can also be classified on the basis of the duration of the disease: acute (« 24-36 h), chronic (> 24-36 h) and recurrent (multiple episodes separated by periods of > 2 days of normality. Another system classifying colic is based on anatomic location of the origin of pain.

Epidemiology

Equine colic has been reported worldwide. However, there are regional variations in respect to the types of colic. Mortality due to colic ranges between 0.5 and 0.7 deaths per 100 horse per year, representing 28% of overall horse deaths. The case fatality rate is 6-13%. Risk factors for colic are categorized as 1) intrinsic horse characteristics, 2) feeding practices, 3) management, 4) medical history, and 5) parasite control.

Horse characteristics: Horses, 2-10 years of age are 2.8 times more likely to develop colic than horses less than 2 years. Foals less than 6 months of age have an incidence of 0.2 cases of colic per 100 horses per year, while horses more than 6 months of age had incidence of approximately 4-6 colic affected horses per 100 horses in a year. The confounding factors that alter with the age of the horse, such as its use, feeding and management practices mask the age effects. There is no overall effect of sex as a risk factor for colic, but certain diseases are limited to a particular sex. For instance, inguinal hernias occur only in males, whereas entrapment of intestine in the mesometrium is restricted to mares.

Breed: Arabian horses are at increased risk of colic, but the reason for this has not been explained. Thoroughbreds are reported to be at increased risk of colic, independent of their use.

Diet and feeding practices: Horses at pasture are at a lower risk of developing colic than stabled horses, fed concentrate feeds only. The risk of colic increases with the quantity of concentrate feed consumed, such that a horse fed 5 kg of concentrated feed per day has 6 times more risk of developing colic.

Management: Watering horses without constant access to water are at increased risk of developing colic, whereas horses with access to ponds or dams have a reduced risk of colic compared to horses provided with water from buckets or trough. This might represent a confounding effect of pasturing, in that horses with access to dams are probably at pasture and benefit from the lower risk of colic associated with that management practice. Alternatively, horses provided with water from buckets may be at greater risk of having periods when water is not available.

Housing: Increased duration of indoor housing per day is associated with an increased risk of colic. Horses cared for by their owner and horses in stables with large numbers of horses are less likely to develop colic. Exercise appears to be an increased risk of colic among horses that are undertaking physical activity or that have a recent change in the amount of physical activity.

Medical history: Horses with a history of suffering from colic are more likely to have another episode, and horses that have had colic surgery are approximately five times more likely to have another episode of colic than are horses that have not had colic. There is no association between dental care and incidence of colic or recent vaccination and colic.

Parasite control: Inadequate parasite control programs have been estimated to put horses at 2-9 times greater risk of developing colic. The presence of tapeworms is associated with a 3 times greater risk of ileal impaction. The important losses caused by colic in horses are due to death of the patient, cost of treatment and the emotional trauma to the owners.

Pathophysiology

The pathogenesis of equine colic is variable depending on the cause and severity of the inciting disease. A horse with a strangulating lesion involving 50% of its small intestine has a much more rapidly evolving disease, with severe abnormalities, than does a horse affected with mild spasmodic colic or impaction of the pelvic flexure of the large colon. The equine colic often involves changes in many body systems, notably the gastrointestinal, cardiovascular, metabolic and endocrine systems. The features common to severe and milder colic are pain, gastrointestinal dysfunction, intestinal ischemia, endotoxemia, compromised cardiovascular function and metabolic abnormalities. Pain is the hallmark of gastrointestinal diseases in horses and is attributable to distension of the gastrointestinal tract and stimulation of stretch receptors in the bowel wall and mesentery, stretching of mesentery by displaced or entrapped bowel, and inflammation and irritation of the bowel, peritoneum or mesentery. The intensity of the pain depends on the severity of the inciting disease. Horses with mild impaction of the large colon of short duration (<24 h) often have very mild pain, whereas a horse with a strangulating lesion of the small intestine will have very severe pain. Gastrointestinal pain has an inhibitory effect on normal gastrointestinal function, causing a feedback loop in which the pain inhibits normal gut motility and function, allowing accumulation of ingesta and fluid, resulting in distension and further pain. Horses can respond very violently to abdominal pain and may injure themselves when rolling or thrashing. Colic is almost invariably associated with impaired gastrointestinal functions, usually alterations to motility or absorptive function. Gastrointestinal motility may be increased, as is presumed to be the case in spasmodic colic, altered in its character or coordination, as in some cases of impaction colic, or absent, such as in ileus secondary to inflammation or ischemia of the bowel or to the presence of endotoxemia. Increased or uncoordinated gastrointestinal motility causes pain through excessive contraction of individual segments of bowel or distension of bowel because of the loss of normal propulsive activity. Ileus is associated with fluid distension of the small intestine and stomach and fluid and gas distension of the large colon, both of which cause severe pain and can lead to gastric or colonic rupture. The absorptive function of the intestine may be decreased by inflammation or ischemia, which results in distension of the small intestine or large colon, pain and potentially rupture of the stomach or colon. Impairment of the barrier function of the gastrointestinal mucosa by inflammation or ischemia can result in leakage of endotoxin into peritoneal fluid and endotoxemia. Most cases of lethal colic involve some degree of ischemia of the intestine, with subsequent loss of barrier function, evident in its most extreme form as rupture of the viscus, endotoxemia, bacteraemia, cardiovascular collapse and death. Ischemia may result from impaired blood flow to or from the intestine, because of torsion or volvulus of the intestine, entrapment of the intestine and associated mesentery, strangulation such as by a pedunculated lipoma, or thromboembolic disease. Ischemia may also result from severe gastrointestinal distension, as occurs in the terminal stage of severe colon impaction. Mild ischemia impairs normal intestinal motility and function. Endotoxemia death in fatal cases of colic is due to the absorption of endotoxins from the gut lumen into the systemic circulation. Rupture of the stomach or intestine is also a characteristic termination of distension of the intestine in the affected horse. The resulting deposition of large quantities of highly toxic ingesta or faecal contents into the peritoneal cavity causes profound shock and death within a few hours. The usual cause of death in severe colic is cardiovascular collapse secondary to endotoxemia and hypovolemia. In less severe colic, hypovolemia and cardiovascular dysfunction may contribute to the development of the disease, and rapid correction of hypovolemia is central to the effective treatment of colic. Hypovolemia is due to the loss of fluid and electrolytes into the lumen of the gastrointestinal tract or loss of protein from the vascular space with subsequent reduction in the circulating blood volume. Hypovolemia impairs venous return to heart and therefore cardiac output, arterial blood pressure and oxygen delivery to tissues. Measures of circulatory status are good predictors of the outcome of colic. Cardiorespiratory function is impaired if there is severe distension of gut, such as in large-colon torsion, because of restricted respiration by pressure on the diaphragm and reduced venous return to the heart because of pressure on the caudal vena cava.

Severe colic, especially that involving ischemia or necrosis of intestine, is associated with abnormalities in coagulation and fibrinolysis, characterized by hypercoagulation of blood and decreases in rate of fibrinolysis. Disseminated intravascular coagulation is common among horses with ischemia or necrosis of the gut and is a good prognostic indicator of survival. Changes in coagulation and fibrinolysis include decreases in antithrombin activity and fibrinogen concentration and increases in prothrombin time, activated partial thromboplastin time and concentration of thrombin-antithrombin complexes in plasma.

Simple obstructive colic is associated with obstruction to the aboral passage of ingesta without any ischemia or strangulation of bowel. In the terminal stages there is often ischemia caused by distension of the intestine. Small-intestinal obstructive lesions include ileal hypertrophy, ileocecal intussusception and foreign body obstruction of the lumen. The course of the disease is often 24-72 hours and sometimes longer depending on the extent of the obstruction, partial obstructions having much less severe signs and disease of longer duration. The principal abnormality is reduced aboral flow of ingesta, with subsequent distension of intestine cranial to the obstruction, causing pain and, if the distension is severe, gastric rupture. Large intestinal obstructive lesions include impaction and simple (non-strangulating) displacements of the large colon. The course of disease is prolonged, often more than 72 hours. Signs of abdominal pain are due to distension of the bowel. There is progressive distension with fluid and gas and ultimately ischemia of the bowel and rupture. Obstructive and strangulating Diseases that cause both obstruction and strangulation as an initial event, such as torsion of the small intestine or volvulus of the large colon, result in severe and unrelenting pain that is little relieved with analgesics. Obstruction causes distension and strangulation causes ischemia, loss of barrier function and endotoxemia. These diseases have a short course, usually less than 24 hours and sometimes as short as 6 hours, and profound clinical signs. Endotoxemia and cardiovascular collapse are characteristic of these diseases. Infarctive diseases, such as thromboembolic colic, are characterized by ischemia of the intestinal wall with subsequent alterations in motility and absorptive and barrier functions. Ileus causes distension of the intestines and stomach and altered barrier function causes endotoxemia. The course of the disease is usually less than 48 hours and is terminated by cardiovascular collapse and death. Inflammation of the intestine or peritoneum alters gastrointestinal motility and absorptive function leading to accumulation of fluid and ingesta, distension and abdominal pain.

Clinical Signs

Visual examination of behaviour: Pain is manifested by pawing, stamping or kicking at the belly or by restlessness evident as pacing in small circles and repeatedly getting up and lying down, often with exaggerated care. Other clinical signs include looking or nipping at the flank, rolling, and lying on the back. Often the penis is protruded without urinating or with frequent urination of small volumes. Continuous playing with water without actually drinking, called sham drinking, is common. Pain may be continuous or, more commonly, intermittent with bouts of pain lasting as long as 10 minutes interspersed with similar periods of relaxation. The intensity of the pain is of about the same severity for the duration of the illness. Sudden exacerbations may indicate a change in the disease status or the development of another abnormality.

Horses in the terminal phase of the disease may have a marked diminution of pain associated with relief of pressure after rupture of distended bowel and depression caused by toxaemia and shock. Pain responses in colic may be so severe and uncontrolled movements are so violent that the horse may do itself serious injury. The posture is often abnormal, with the horse standing stretched out with the forefeet more cranial and the hindfeet more caudal than normal, 'saw-horse' stance. Some horses lie down on their backs with their legs in the air. Distension of the abdomen is an uncommon but important diagnostic sign. Symmetrical, severe distension is usually caused by distension of the colon, sometimes including the cecum, secondary to colon torsion, or impaction of the large or small colon and subsequent fluid and gas accumulation. Maximum distension of stomach or small intestines does not cause appreciable distension of the abdomen. Vomiting Projectile vomiting or regurgitation of intestinal contents through the nose is very unusual in the horse and is a serious sign suggesting severe gastric distension and impending rupture. Defecation patterns can be misleading. It is often mistakenly assumed that there is no complete obstruction because faeces are still being passed. But in the very early stages of acute intestinal obstruction there may be normal faeces in the rectum, and the animal may defecate several times before the more usual sign of an empty rectum with a sticky mucosa is observed.

Heart and respiratory rates: The heart rate is a useful indicator of the severity of the disease and its progression but has little diagnostic usefulness. Horses with heart rates less than 40/min usually have mild disease whereas horses with heart rates above 120/min are usually in the terminal stages of severe disease. Horses with obstructive, non-strangulating disease often have heart rates between 40 and 60/min, whereas horses with strangulating disease or necrotic bowel will usually have heart rates over 60/min. However, heart rate is not an infallible indicator of disease severity, as horses with torsion of the colon can have heart rates of 40-50/min. The respiratory rate is variable and may be as high as 60/min during periods of severe pain.

Mucous membranes of normal horses and of horses without significantly impaired cardiovascular function are pink, moist and regain their normal colour within 2 seconds after firm digital pressure is removed. Dehydrated horses have dry mucous membranes, although the capillary refill time and colour are normal. Horses with impaired cardiovascular function have pale, dry mucous membranes with delayed capillary refill (> 2sec). Endotoxemia horses will often have bright red mucous membranes with normal or delayed capillary refill. As the disease becomes more severe the mucous membranes develop a bluish tint and capillary refill is longer than 3 seconds. Terminal stages of disease are associated with cold, purple, dry mucous membranes with a capillary refill time of more than 3 seconds. Cool extremities may be indicative of compromised cardiovascular function but should be interpreted with caution and only in the context of the rest of the clinical examination. Sweating is common in horses with severe abdominal pain and, when present in a horse with cool extremities and signs of cardiovascular collapse, is indicative of a poor prognosis.

Auscultation of the abdomen can provide useful diagnostic and prognostic information and should be performed thoroughly. All four quadrants (dorsal and ventral, left and right sides) of the abdomen should be examined for at least 1 minute at each site. Attention should be paid to the intensity, frequency and characteristics of the spontaneous gut sounds (borborygmi). Repeated observations are often necessary to detect intermittent or rapid changes in the character of the borborygmi. Continuous, loud borborygmi distributed in all or most quadrants are indicative of intestinal hypermotility and consistent with spasmodic colic, impending. diarrhoea or the very early stages of a small-intestinal obstructive strangulating lesion. The absence of sounds, or the presence of occasional high-pitched, brief sounds, sometimes with a splashing character, is consistent with ileus. These sounds should not be mistaken for the rolling, prolonged sounds of normal peristalsis. Combined percussion and auscultation are a valuable procedure for defining the presence of extensive gas caps. The detection of such sounds indicates the presence of tightly gas-distended bowel near the body wall. Such bowel is almost always large colon or cecum and is consistent with gas distension secondary to ileus, small l or large colon impaction, gas colic or colon displacement, including torsion.

A careful rectal examination is probably the most important part of the clinical examination in colic. The horse should be restrained so that the examination can be performed with minimal risk to both the examiner and patient. Rectal examination in small or unruly horses should be approached with caution. Only approximately

40% of the abdomen can be examined in a mature horse, the cranial and ventral structures being outside the reach of the examiner. The cecum is readily palpable in the right caudal abdomen, with its ventral band running from the dorsal right quadrant ventrally and slightly to the left. The base of the cecum may be palpable as a soft, compressible structure containing fluid and gas. The caudal border of the spleen is readily palpable in the left side of the abdomen against the body wall. Gas and fluid distension of the cecum and colon, fluid distension of the small intestine, impaction of the large and small colon, and displacement of the large colon can be diagnosed during the per rectal examination. Small intestinal distension is evident as loops of tubular structures of up to 10-15 cm diameter that may extend as far caudally as the pelvic canal. The presence of distended small intestine is an important sign suggestive of a small-intestinal obstructive lesion or anterior enteritis. Gas and fluid distension of the large colon is evident as large (> 20 cm) taut structures often extending into the pelvic canal. The distended bowel may extend into the pelvic canal, preventing examination of the caudal abdomen. Impaction is evident as columns of firm ingesta in the large or small colon.

The course of the disease depends on cause and the severity of the associated lesions. Spasmodic and gas colic usually resolves within hours of onset. Strangulating lesions have severe clinical signs and usually the patient die within 24 hours of the onset. Obstructive lesions have longer courses, often 48 hours to 1 week, and the patient die when distension causes paralysis and rupture of the bowel. The intestinal rupture results in sudden onset of toxaemia and shock. The horse may be recumbent but most continue to stand until the last few minutes, when they literally drop dead. The respiration is sobbing. There is muscle tremor and abundant sweating, and there is staggering wandering.

Clinical Pathology

Analysis of clinico-pathological parameters is useful in assessing the severity of the deviation as a consequence of colic than arriving at a definite diagnosis. Some of variables have prognostic implications and are monitored repeatedly in severe cases. Haematology and serum biochemistry indicate the extent of alterations and severity of the disease. The measurement of haematocrit and plasma total protein concentration is useful in assessing hydration status. Haematocrit increases as a consequence of splenic contraction releasing the erythrocytes or dehydration, making the use of either of the two variables alone as an indicator of hydration status unreliable. However, increases in both haematocrit and total protein concentration indicate dehydration, and these variables can be used as basic evaluations of response to fluid therapy. Plasma total protein concentrations may decline with significant loss of protein into the gut lumen or peritoneal cavity. Measurement of the blood leukocyte count has little diagnostic implication, except in endotoxemia accompanied with devitalized bowel, enteritis or peritonitis, characterised by leukopenia along with a left shift. Horses with severe colic often have anomalies in coagulation in respect to antithrombin activity and prolonged prothrombin and activated partial thromboplastin times, with higher deviation in horses with strangulating lesions. The serum electrolytes concentration is important in providing an assessment of the electrolyte status and assessing the effects of fluid therapy. The nature of the abnormalities depends to some extent on the cause and severity of the disease. Mild hyponatremia is common but may not be clinically significant. Hyperkalaemia is common in severe acidosis and horses with large sections of devitalized intestine. Hypokalemia is common in horses with more long-standing colic, for instance impaction of the large colon, that have not eaten for several days. Hypocalcaemia and hypomagnesemia are common in horses with colic, especially horses with severe colic. Measurement of total concentrations, both ionized and nonionized, can be misleading in that reductions in concentration of the physiologically important ionized component can be present in horses with normal concentrations of the total ion. Horses with colic or diarrhoea are more likely to have hypomagnesemia. Serum enzyme activities are rarely useful in supporting the diagnosis or treatment of horses with colic, with the exception that serum gamma glutamyl transferase (GGT) activity is elevated in approximately 50% of horses with right dorsal displacement of the colon, whereas such elevations are rare in horses with left dorsal displacement. The elevated GGT, and less commonly serum bilirubin concentration, in horses with right dorsal displacement is attributable to compression of the common bile duct in the hepatoduodenal ligament by the displaced colon. Semm and peritoneal alkaline phosphatase activities are higher in horses with ischemic or inflammatory bowel disease than in horses with other forms of colic. Semm urea nitrogen and creatinine concentrations are useful indicators of hydration status and renal function. Prerenal azotemia is common in horses with colic, and may progress to acute renal failure in severe cases of colic. Most horses with severe colic have metabolic acidosis that is attributable to L-lactic acidosis. However, respiratory acidosis and metabolic alkalosis may also occur. An estimate of the plasma lactate concentration can be obtained by calculating the anion gap that is = (sodium + potassium) - (bicarbonate + chloride). If bicarbonate concentrations are not available, total serum carbon dioxide can be substituted. Anion gaps of less than 20 mEq/L (mmol/L) are associated with 81 % survival, 20-24.9 mEq/L (mmol/L) with 47% survival, and 25 mEq/L (mmol/L) or more with 0% survival. Analysis of peritoneal fluid is an important component of the complete examination of a horse with colic. The total protein concentration increases when there is an insult to the gastrointestinal tract that compromises the serosal surface of the bowel, for instance strangulating lesions of the small intestine or in the terminal stages of an impaction colic in which the bowel wall is devitalized. The presence of intracellular bacteria, plant material and degenerate neutrophils is indicative of gastrointestinal rupture provided that one is certain that the sample came from the peritoneal space and not from the bowel lumen.

Deep congestion (dark red) or cyanosis (purple) and capillary refill times much longer than 2 seconds are indicators of peripheral circulator failure. The development of a 'ping' on auscultation-percussion indicates accumulation of gas under pressure.

Diagnosis

Accurate diagnosis of the cause of the colic has some prognostic usefulness, but assessment of the physiological state by measurement of heart and respiratory rates, mucous membrane colour and refill time, arterial blood pressure, haematocrit and serum total protein concentration, and other parameters allows more accurate prognosis.

Radiology: The large size of the adult horse precludes detailed radiographic examination of intra-abdominal structures. However, enteroliths and sand accumulation can be detected with reasonable accuracy, provided suitable radiographic equipment is available. Diaphragmatic hernias can be detected on radiographic examination of the thorax.

Arterial blood pressure: Arterial blood pressure is a very good indicator of the degree of shock in colic, and the availability of a simple technique makes it a practical aid in assessing prognosis in a clinical case. If normal systolic pressure is about 100 mmHg (13.3 kPa), a pressure below 80 mmHg (10.6 kPa) indicates a critical situation. It may fall as low as 50 mmHg, 6.6 kPa. In horses with very severe pain but not associated shock, the systolic pressure becomes too is likely to be very high, up to 250 mmHg (33.3 kPa).

Treatment

The specific treatment of each case of colic varies depending on the nature of the lesion and the severity of the disease. However, several principles are common to the treatment of most colic that include analgesic administration, correction of fluid, electrolyte and acid and base abnormalities, oral administration of gastrointestinal lubricants or administration of faecal softeners.

Analgesic administration is important to relieve discomfort, minimizes the physiological consequences of pain, including the pain-induced reduction in gastrointestinal motility, permits a thorough clinical examination and reduces the likelihood of the horse injuring itself by rolling or thrashing. Analgesics can be classified into non-steroidal anti-inflammatory drugs (NSAIDs), sedating analgesics and spasmolytics. The analgesic and its dose rate should be chosen such that the pain is relieved but signs of progressive cardiovascular compromise are not masked. Non-steroidal anti-inflammatory drug such as Flunixin meglumine is a potent, long-acting analgesic with the ability to mask signs of surgical disease. It is used routinely in horses being monitored for progression of disease unless such monitoring is frequent and thorough. Similar to flunixin meglumine, ketoprofen and phenylbutazone, which has relatively weak analgesic effects in colic patients are used to control pain. Dipyrone is a weak analgesic that is useful in treatment of mild cases of colic. The alpha-2 agonists (xylazine, detomidine, romifidine) provide potent analgesia, especially when combined with the opiate butorphanol. Duration is relatively short (up to 90 min for detomidine), which means that signs of progressive disease are readily detectable. The effect of alpha-2 agonists in reducing gastrointestinal motility is not clinically Opiates, including butorphanol, meperidine (pethidine), morphine and pentazocine, are potent analgesics useful in the management of abdominal pain in the horse. Other agents, Acetyl promazine has almost no analgesic properties, although it is a potent sedative, and should not be used in the routine treatment of colic. Acetyl promazine is a potent hypotensive agent and should not be administered in dehydrated patients. Hyoscine butyl bromide, a parasympatholytic drug, is widely used as the drug of choice in the initial treatment of field cases of colic. It is often combined with dipyrone and is effective in the treatment of mild, uncomplicated colic.

Atropine causes gastrointestinal stasis in horses and should not be used in the routine treatment of colic. Lidocaine is a potent analgesic when administered systemically, but must be given by constant intravenous infusion. Overdosing results in central nervous system excitement. Prophylaxis and treatment of endotoxemia Administration of plasma from horses hyperimmunized with *Salmonella typhimurium* or *E. coli* reduces the severity of clinical signs and shortens the duration of disease in horses with endotoxemia secondary to enterocolitis or colic.

Polymyxin attenuates the effect of endotoxin @ 5000 IU/kg intravenously every 8-12 hr and is used for the prevention and treatment of endotoxemia in horses. Aspirin @10 mg/kg orally every 48 h is administered to diminish platelet aggregation around intravenous catheters. Flunixin meglumine (1 mg/kg intravenously every 8-12hr or phenylbutazone (2.2 mg/kg intravenously every 12hr, is given for analgesia and to prevent endotoxin-induced increases in plasma prostaglandins. The antibiotics of choice should have a broad-spectrum activity against Gram-negative and positive and anaerobic bacteria. Fluid and electrolyte therapy, preferably a balanced, isotonic, poly-ionic fluid such as lactated Ringer's solution is provided in cases of dehydration, compromised cardiovascular function or electrolyte imbalances. Intestinal lubricants and faecal softeners are given only through a nasogastric tube as its aspiration is associated with severe and usually fatal pneumonia. Mineral oil is useful in cases of mild impaction colic and is often administered when the cause of the colic is not known, provided that there is no reflux of gastric contents through the nasogastric tube. Dioctyl sodium sulfosuccinate (DSS) is a faecal softener with the potential to be toxic at therapeutic doses and its use is now not generally recommended. Magnesium sulfate is an effective faecal softener useful in the treatment of impaction colic. However, it can cause hypermagnesemia and toxicity characterized by depression and signs of central nervous system dysfunction. Sodium sulfate is a safe and effective faecal softener, although it may induce mild hypernatremia and hypokalemia. Promotility agents may be used in cases of ileus or large colon impaction.

Heparin and low-molecular-weight heparins have been recommended for the treatment and prevention of coagulopathies associated with severe colic. The use of heparin or low-molecular-weight heparin is associated with increased risk of haemorrhage and heparin use causes a decrease in haematocrit.

It may be necessary to relieve the gas distension of the colon or caecum by trocarisation in severe cases of flatulent (gas) colic or in cases of colon torsion in which the abdominal distension is impairing respiration. Trocarisation is usually performed through the right paralumbar fossa immediately caudal to the last rib. The area of loudest ping will indicate the point of insertion of the trocar. The trocar is kept in position as long as gas is escaping.

The only definitive treatment for many causes of equine colic is surgical correction or removal of the lesion. Gastrointestinal surgery should be performed by trained and experienced surgeon. Horses with severe pain refractory to treatment with analgesics should have an exploratory laparotomy even if no other significant abnormalities can be detected. Minimization of colic episodes depends on management factors, including ensuring adequate parasite control, feeding large quantities of forage and minimizing the amount of concentrate feed, and providing dental care. However, most cases of colic not attributable to parasites or dietary factors cannot be prevented.

29

Diseases of Liver and Pancreas in Bovine

Affections of Liver

Liver is the largest exocrine gland located in right side of the abdominal cavity obliquely from lumbo-costal angle downward and forward to intercostal space, about 3-4" (10 cm) above the abdominal floor.

Manifestation of hepatic dysfunction

1. Icterus or jaundice
2. Gastrointestinal abnormalities
3. Intermittent pyrexia
4. Ascites
5. Abdominal pain
6. Alternation in liver size
7. Oedema of body parts due to hypoproteinaemia and emaciation
8. Photosensitization
9. Haemorrhagic diathesis
10. Hepatic encephalopathy.

(A) Jaundice / Icterus

Yellowish discolouration of skin and visible mucous membrane (sclera and vaginal mucosa) occurs due to hyperbilirubinemia.

Types of Icterus

1. Haemolytic jaundice- without impairment of bile flow
2. Obstructive jaundice- impairment of bile flow

Metabolism of bilirubin:

Production:

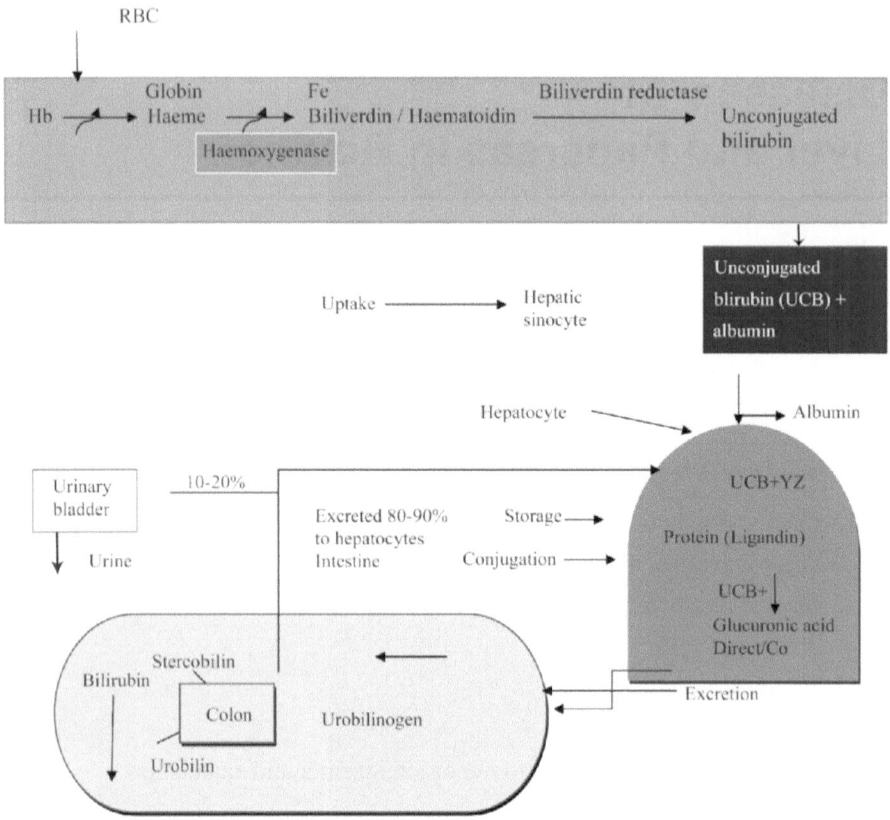

(B) Gastrointestinal abnormalities

- Constipation followed by diarrhoea, nausea, vomition and regurgitation and distension of GI tract due to gas accumulation.
- Stretching of nerve ending of distended part leading to reflex nerve spasm resulting in colic or abdominal pain.
- Distension causes drawing of more fluids to the part, hence causing further distension. However, there shall be no distension, pain and colic following attaining the critical limits and it causes paralytic ileus.
- When more fluids are drawn to a par,t then dehydration occurs resulting in hypovolemia, arrhythmia and lastly shock.

(C) Intermittent pyrexia

- Infection is associated with liberation of toxin. These toxin causes stimulation of immune response and there is release of cytokines and interleukins, which cause pyrexia.
- When detoxifying capacity of liver is reduced, there is accumulation of toxic substances which causes toxaemia and increases in temperature.

(D) Ascites

- Portal hypertension occurs due to dysfunction of liver. It leads to increase in hydrostatic pressure of plasma as it draws fluid from the system.
- When there is disruption to protein metabolism, there is hypoproteinaemia or hypoalbuminemia, which reduces osmotic pressure of plasma and more fluid is drawn from vascular system resulting in ascites.

(E) Alteration in liver size and shape

- Infection of liver leads to hepatomegaly. In terminal stage, liver fibrosis leads to decrease in liver size.

(F) Emaciation

- Liver dysfunction impairs protein metabolism. Hence, anabolism of amino acids in liver is hampered leading to emaciation.

(G) Photosensitization

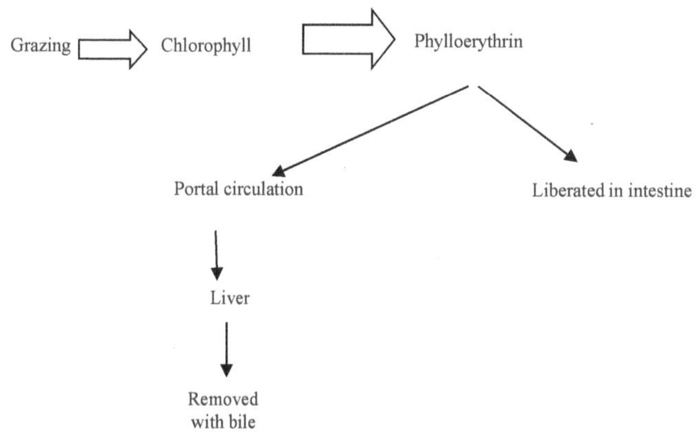

In liver diseases, phylloerythrin cannot be removed out and get accumulated in skin throughout the body (photo-dynamic substances). It reacts with radiant energy of U.V ray and there is peroxidation of cellular lipid and formation of free radicals. These radicals cause cell injury at whitish non-hairy part and there is dermatitis and sloughing of skin.

(H) Haemorrhagic Diathesis

Liver disease ⟶ decrease vitamin-K absorption ⟶ increase in blood clotting time ⟶ continuous haemorrhage by leakage of erythrocytes

(I) Hepato-encephalopathy/ Hepatic encephalopathy

Liver disease ⟶ accumulation of amines, NH and phenols ⟶ affects brain and CNS ⟶ Nervous disorder

Diagnosis of Liver Diseases

1. Palpation at abdominal area immediately next to last costal arch only possible in dog and cat.
2. Liver biopsy- cumbersome and highly invasive procedure.
3. Ultrasonography.
4. Liver function test - sulpho-bromophthaline test, indole cyanine green excretion test, urea tolerance test, takatara test.
5. Enzyme test: ALT, AST, SDH and LDH.
6. Bile acid estimation mostly cholic acid, lithocholic acid and deoxycholic acid.

Treatment

1. Glucose solution intravenously to counter hypoglycaemia.
2. Amino-acid mixture containing methionine to counter hypo-proteinemia.
3. Provision of diet rich in carbohydrate, protein and calcium.
4. Choline citrate to repair hepatocyte.
5. Liver extract injections.
6. Anthelmintics to reduce parasitism.
7. Operative surgery.

Affections of Pancreas

Pancreatitis: It refers to inflammation of pancreas. It is important in case of small animals like dog and cat. It may occur in two forms:

1. **Acute pancreatitis**: No irreversible damage to pancreas and can be recovered.
2. **Chronic pancreatitis**: damage to pancreas and persist for long periods.

Etiology

1. **Infectious agents** like bacteria, viral and mycoplasmal agents. It starts with generalized systemic infection, later on causing pancreatitis.

2. **Dietary origin:** Diet containing more fat and less protein causes hyperlipidemia. Triglyceride concentration in pancreatic capillaries increases leading to increased breakdown of triglyceride by lipase which increases the production of toxic fatty acids eventually causing pancreatitis.
3. **Drugs:** Diuretic like thiazide and frusemide and long-term use of sulfonamide and tetracycline cause pancreatitis.
4. **Duodenal reflux:** The obstruction or intestinal overload, leads to fluid distension and highly acidic duodenal fluid returns to pancreas causing pancreatitis.
5. **Hypercalcemia:** Increased calcium level in blood stimulates pancreatic activity and may lead to pancreatitis.
6. **Scorpion venom:** Experimentally, scorpion venom causes pancreatitis by stimulation.
7. **Obstruction of pancreatic duct:** Due to neoplasm, surgical wound, hyperplasia and parasites may cause pancreatitis.

Pathogenesis

The most important way in which pancreatitis occurs is auto-digestion due to failure of excretory, process of pancreas by any means. Excretory failure causes unwanted effect on lysosomes and zymogen granules. Lysosomes present in cytoplasm causes lysis of bacteria. But when fusion occurs, zymogen enters into lysosomes and acts as trypsin inhibitors. There will be activation of trypsinogen which digests the same cells. This phenomenon is called auto- digestion.

Clinical Signs

1. Anorexia, depression, dullness and weakness
2. Nausea and vomition are regular features
3. Sometimes diarrhoea
4. Vomition and diarrhoea leads to dehydration
5. Congestion of visible mucous membranes
6. Abdominal pain and softening of abdomen
7. Peculiar position of dog (Prayer position)
8. Mild rise in body temperature
9. Secondary symptoms like jaundice, respiratory distress and cardiac arrhythmia may be seen.

Clinical Pathology

1. Blood

- Leucocytosis.
- Hb and PCV % increases
- Hyperglycaemia due to lack of insulin.
- Hyperglucagonemia.
- Azotemia.
- Increased concentration of pancreatic enzymes in serum

Treatment

1. Feeding should be withheld for 3-5 days in suspected cases and fluid therapy should be given with electrolytes and plasma expanders.
2. Antiemetic like metoclopramide and ondansetron may be given to counteract vomition.
3. Insulin treatment subcutaneously.
4. Start with proteinaceous diet or carbohydrate diet.
5. Hepatotonic and digestive enzymes are recommended
7. Vitamin-K injection is given if there is bleeding.

30

Jaundice

Introduction

Icterus, otherwise known asjaundice,refers to a yellow discolouration of tissues resulting from the excessive accumulation of a yellow pigment. The discolouration is most easily seen in the gingivae and sclerae. Jaundice signifies the abnormal accumulation and distribution of bilirubin pigments into the systemic circulation, causing yellow discoloration of plasma and nonpigmented soft tissues including skin, sclera, mucous membranes of oral cavity, vagina, penis, and urine. Jaundice is often the primary clinical sign in animals with disorders involving the gallbladder or extrahepatic biliary structures.

Etiology

The three basic pathological types of jaundice are prehepatic (associated with haemolysis), hepatic (associated with primary or secondary intrinsic liver disease), and post-hepatic (associated with biliary obstruction or gall bladder diseases). The parenchymal changes in the liver are basically characterized as inflammatory, vacuolization, or neoplastic. Post-hepatic jaundice is associated with cholangiostasis, which occurs from the bile flow stagnation. The various causes of such condition include biliary obstruction, various hepatic inflammatory diseases, drugs, toxins, parasites, neoplasia, or systemic disease, especially acute pancreatitis and sepsis.

The pathologic processes leading to jaundice include overproduction of bilirubin due to increased haemolysis, reduced uptake in liver, impaired conjugation due to lack of enzymes, impaired intrahepatic secretion due to hepatic damage, intrahepatic cholestasis due to biliary obstruction, impaired extrahepatic secretion due to obstruction of the bileduct. Therefore, jaundice is classified into (1) haemolytic or prehepatic jaundice, (2) toxic or intrahepatic jaundice, and (3) obstructive or post-hepatic jaundice.

Causes of haemolytic or prehepatic jaundice

Bacteria: Leptospirosis icterohaemorrhagica,Clostridium haemolyticum

Virus: Equine infectious anaemia (EIA)

Protozoa: Babesiosis, Anaplasmosis, Haemobartonellosis, Trypanosomiasis

Nutritional: Phosphorus deficiency - Post parturient haemoglobinuria

Phytotoxins: Resin, Saponin, Kale, Rape and Onion poisoning

Animal toxin: Snake venom

Chemicals: Copper, selenium toxicity in sheep

Immune-mediated: Icterus neo-natrum, incompatible blood transfusion, Tick fever (ehrlichiosis), Leptospirosis, Parasitic infections, Viral infections (adenovirus), Fungal infections (histoplasmosis)

Prehepatic jaundice reflects enhanced bilirubin formation. The most common cause is accelerated RBC turnover, usually due to haemolysis. Jaundice develops when released heme exceeds the hepatic capacity to process and excrete bilirubin. The most common cause is RBC haemolysis associated with an immune-mediated haemolytic anaemia in dogs or Heinz body -related haemolytic anaemia in cats. Prehepatic jaundice may be aggravated by shortened RBC lifespan in transfused blood, especially in recipients of numerous repeated transfusions where immune reactions hasten RBC removal. This form of jaundice can also evolve in patients needing a large blood volume exchange transfusion. The least common cause of prehepatic jaundice is heme released from blood sequestered in large hematomas (bleeding disorders, trauma) or haemorrhagic effusions, such as intra-abdominal bleeding as encountered with a ruptured spleen (eg, hemangiosarcoma in dogs).

Causes of Toxic or intrahepatic jaundice

The most common causes of liver disease include viral or bacterial infections, ingestion of toxic plants or chemicals, certain drugs and medications, cancer, autoimmune diseases, and certain breed-specific liver diseases, and include the followings.

Bacteria: Leptospirosis, Salmonellosis

Virus: Infectious canine hepatitis

Phytotoxins: Senecio, crotalaria

Chemicals: Phosphorus, chronic copper poisoning, chloroform, carbon tetrachloride.

Hepatic jaundice reflects impaired hepatocyte bilirubin uptake, conjugation, or canalicular excretion, which are encountered in animals with critically decreased functional hepatic mass or disrupted sinusoidal circulation (e.g. liver fibrosis, regenerative remodelling). However, impaired bilirubin conjugation and canalicular transport are also often encountered secondary to endotoxemia, sepsis, or other infections. Animals with impaired canalicular transport usually have canalicular cholestasis.

Jaundice associated with bacterial toxins, especially endotoxin, also known as lipopolysaccharide (LPS), may not be associated with commensurate increases in liver enzyme activities. All potential pathological mechanisms causal to hepatic jaundice can emerge in animals with severe critical hepatocyte dysfunction (ie, fulminant hepatic failure, acute liver failure, acute-on-chronic liver failure, advanced-stage-liver disease [formerly categorized as cirrhosis]).

***Causes of obstructive or post-hepatic jaundice*:** The most common causes of bile duct obstruction in dogs include pancreatitis, abdominal trauma, abdominal or liver cancer, gall bladder stones (gallstones), and severely thickened bile duct. These include *b*locking of bile duct from within, *Ascaris lumbricoides* in swine, *Thysanosoma astiniodes* (fringed tape worm), *Fasciola gigantica* in cattle, Gall stones, pressure on bile ducts from outside by tumours, abscesses, granulomas, fibrosis, enlarged pancreas or lymph nodes, inflammatory processes in biliary system.

Cholangitis, cholecystitis – fascioliasis, *Dicrocoelium dendriticum*

Closure of bile duct orifice in duodenum

Duodenitis – thickening of mucosa

Post-hepatic jaundice reflects cholestatic disorders involving larger ductal elements, notably the extrahepatic or common bile duct or gallbladder or rupture of the biliary tree, causing bile peritonitis.

Although cholestasis can occur at the canalicular level due to disrupted canalicular bile flow (canalicular cholestasis), associated jaundice is better considered as hepatic jaundice.

The least common cause of post-hepatic jaundice is bile-induced chemical peritonitis (ie, bile peritonitis), in which there is spillage of bile from the biliary tree as seen in ruptured common bile duct or gallbladder, typically associated with abdominal effusion.

Epidemiology

Any disease that affects red blood cells, the liver, and/or the gallbladder can significantly affect bilirubin creation, thereby leading to jaundice. Though it's often referred to in the same context as liver disease, it is important to understand that jaundice can occur due to other diseases.

Unfortunately, jaundice is a common occurrence in dogs and a common cause of humane euthanasia. While jaundice itself is not contagious, there are some underlying causes that can be contagious, or even zoonotic, meaning they can be transmitted from animals to people. This is seen in diseases such as leptospirosis.

Pathogenesis

Jaundice is caused by the build-up of bilirubin in the tissues due to high blood levels of bilirubin (hyperbilirubinemia) in the blood. Hyperbilirubinemia occurs either due to increased red blood cell destruction, called haemolysis, or an inability of the liver to process or excrete bilirubin normally.

Haemolysis can be caused by toxic plants, drugs, parasites on the red blood cells, heartworm disease, autoimmune diseases, and cancer. Several additional tests may be needed to determine the cause of haemolysis in a dog. Excessive haemolysis results in production of greater amount of unconjugated bilirubin. Since conjugation of bilirubin is a rate limiting metabolic process, all unconjugated bilirubin cannot be converted to conjugated bilirubin. Hence, some amount is left in the blood. Since large amount of conjugated bilirubin is formed, it stains faeces yellow. The absorption of conjugated bilirubin from the gastrointestinal tract increases its excretion in urine, as urobilinogen, giving an intense yellow colour to the voided urine. The liver fails to convert normally produced unconjugated bilirubin, when there is necrosis of the hepatocytes. Since the degenerated cells are swollen and disorganised and biliary capillaries are blocked, unconjugated bilirubin escapes into sinusoids and enters general circulation. Hence, blood contains both conjugated and unconjugated bilirubin. The destruction of red blood cells, the process known as haemolysis, can occur within blood vessels (intravascular haemolysis) or in the spleen and liver (extravascular haemolysis). Haemolysis causes a type of anaemia called haemolytic anaemia, which is different from blood loss anaemia due to bleeding from an injury. Any disease or condition that damages or destroys liver cells can cause icterus. Bile is stored in the gall bladder and transported into the small intestine by the bile duct. If bile becomes abnormally thick, if gallstones form in the gall bladder, if the gall bladder or bile ducts become inflamed, or if the liver becomes swollen so that the bile ducts are constricted, bile flow will become obstructed.

The obstruction to normal flow of bile results in regurgitation of bile. In this case, the production of conjugated and unconjugated bilirubin is normal. Biliary stasis of extra hepatic may occur due to cholestasis pressure, worms, inflammation and duodenitis. No urobilinogen is formed since bile is not entering to intestine. Faeces greasy and grey colour due to failure of fat emulsification and lack of faecal pigment. Urine is not containing urobilin. Clotting defects will occur due to failure of obstruction of vitamin K which is required for prothrombin formation.

Clinical Signs

The diverse range of conditions are responsible for jaundice. The symptom in a dog with jaundice include yellowing of mucous membrane,fever, decreased appetite and weight loss, vomiting, diarrhoea, lethargy, increased thirst and urination, bruising or pinpoint haemorrhages, abdominal swelling, pain, weak pulse, increased heart rate, behavioural changes, seizures, head pressing and circling. However, the signs and symptoms of jaundice in dogs varies depending upon the actual cause of icterus.

Clinical Pathology

The tests for jaundice in dogs include Complete blood count (CBC),biochemistry panel, and urinalysisto further define the cause of jaundice. Further, ultrasound, X-ray, FNA, liver biopsy and/or bacterial culture from fluid found in or around the liver are performed for pin pointing the cause of jaundice. More specific tests for infectious diseases include FeLV, feline coronavirus, and fungal testing.

Determining the reason for the icterus can be a challenge and requires a systematic approach. There are several potential causes for icterus within each category.The first step is to perform screening tests to determine which of the three categories is involved and to narrow down the list of likely diseases. Serological tests include Liver function test (LFT) by estimating activities of Alanine Transaminase (ALT), Aspartate Transaminase (AST), Alkaline Phosphatase (ALP), Gamma-Glutamyl Transferase (GGT), level of serum bilirubin, prothrombin Time (PT), total Protein, albumin, Kidney function test (KFT), Ultrasonography (USG) of the abdomen and urinalysis.

Van den Bergh test: The test was pioneered by the Dutch physician, Abraham Albert Hijmans van den Bergh (1869–1943) of Utrecht, which helps identify the type of bilirubin, andis highly useful in understanding the nature of jaundice. The method relies on colorimetric estimation of serum bilirubin by coupling with diazotized sulfanilic acid. In the Van den Bergh reaction, diazotised sulfanilic acid reacts with conjugated bilirubin and gives purple colour immediately within 30 second, called direct positive van den Bergh reaction. Diazotised sulfanilic acid plus methanol as dissolving agent, reacts with unconjugated bilirubin and gives purple colour normally within 30 minutes, termed indirect positive van den Bergh reaction. Serum sample containing both unconjugated and conjugated bilirubin in high concentrations, the purple colour is produced immediately which is further intensified by the addition of alcohol by a biphasic van den Bergh reaction.

Jaundice results from high levels of bilirubin in the blood. Bilirubin is the normal breakdown product from the catabolism of haem, an iron containing fraction of the haemoglobin, released by the destruction of erythrocytes. Bilirubin undergoes conjugation in the liver under normal circumstances, making it water-soluble. The conjugated bilirubin is e via the bile into the gastrointestinal tract, the majority of which is excreted in the faeces as urobilinogen and stercobilin, the metabolic breakdown product of urobilinogen. Around 10% of urobilinogen is reabsorbed into the bloodstream and excreted through the kidneys. The disruption of this metabolic pathways results in elevation in the level of unconjugated, conjugated bilirubin or both, giving yellowish discolouration called jaundice.

Jaundice is yellowish discoloration of the skin and mucous membrane due to increase in serum bilirubin. Serum bilirubin exists in two forms. Conjugated bilirubin is soluble whereas unconjugated bilirubin is insoluble in water. Haemolytic Jaundice is characterized by the elevation in unconjugated bilirubin, obstructive Jaundice with elevation in conjugated bilirubin, and hepatic jaundice results in the elevation in both unconjugated and conjugated bilirubin levels.

Fine-needle aspiration:A small gauge needle attached to a syringe is inserted through the skin into the liverto perform this procedure, and cells are withdrawn or aspirated. The cells are placed on a glass slide, stained, and studied under a microscope by a veterinary pathologist. This is the least invasive and quickest test, but it has certain limitations. Because only a few cells are obtained, it is possible that a representative sample from the liver will not be obtained. It is also not possible to view the cells in their normal relationship to each other.

Needle biopsy: This procedure is similar to fine needle aspiration except a much larger needle is used. For this technique, dog requires either heavy sedation or general anaesthesia. The needle biopsy gives a core of liver tissue, not just a few cells. The sample is fixed in formaldehyde and submitted to a pathology lab for histopathological analysis. The core biopsy allows to view the cells in their normal relationship to each other. Often, this test is performed during an ultrasound examination and is called an ultrasound-guided needle biopsy.

Surgical wedge biopsy: The dog is placed under general anaesthesia, and the abdomen is opened. This permits direct observation of the liver and the surgeon chooses the exact site for biopsy. A piece of the liver is surgically removed using a scalpel. This approach gives the most reliable biopsy sample.

An ultrasound examination is a more accurate non-invasive technique to evaluate the gall bladder and bile duct. Sometimes, exploratory surgery is necessary to properly evaluate the dog for biliary obstruction or gallstones.

Diagnosis

The diagnosis of icterus is usually simple and can be attained by means of a physical examination. Occasionally, a blood sample collected for some other reason and the serum is noted to be yellow before the dog becomes visibly jaundiced. In these cases, the yellow serum is usually an indication of impending problems.The medical evaluation begins with a detailed history and physical examination and often expands to include biochemical, serological testing and diagnostic imaging. The extent of the evaluation, at times,require a liver biopsy, and sometimes even after an extensive diagnostic evaluation, the results provide a pathological description without an exact etiology.The tests are useful in detecting and diagnosing the underlying cause of jaundice. Since jaundice is a symptom that is easily detected during a routine physical examination. The list of tests can be narrowed down to expedite a diagnosis.

The samples obtained can be submitted for culture to detect evidence of bacterial infection. The histopathology is done to examine cell and tissue structure.

The differential diagnosis for jaundice is broad as it can reflect a multitude of hepatic and hematologic diseases, including diseases directly impacting or obstructing bile flow in the biliary tree, disorders causing hepatocellular dysfunction impairing elimination of bilirubin, or disorders involving haemolytic or hematopoietic abnormalities that escalate bilirubin formation exceeding the rate of hepatic elimination.The prognosis depends upon the underlying causeand the timeliness of appropriate treatment measures. Some diseases that cause icterus are ultimately fatal, such as cancer, while others are treatable, with a good prognosis for full recovery.

Treatment

Since the cause and progress of jaundice depends on the actual cause of the disease, early diagnosis and treatment are the only ways to improve the prognosis. In the early stages of the disease, medications and a special diet are the recommended first line of treatment. The most common supplements suggested are high levels of silymarin, S-Adenosine Methionine, liver detoxifying and supportive vitamins and gallbladder supportive enzymes.

Treatment of the icteric patient is usually supportive, but any specific measures, such as administering antidotes for various intoxicants and parasiticides for liver parasites, should be done as soon as possible. Various emergency measures should be done when needed, and these might include managing hepatic encephalopathy and coagulopathy, or surgical correction of biliary tract obstructions or gall bladder disease when indicated. General supportive care should include maintaining hydration with balanced physiological crystalloid solutions, correcting any blood acid-base and electrolyte disorders, providing nutrition, vitamins, and liver support drugs such as SAM-e, and antibiotics for common bacterial aetiologies.

Jaundice requires intense supportive care, including intravenous fluid, antibiotics, appetite stimulants, anti-nausea medication, antacids, digestive aids and pain medication. In some cases, such as immune-mediated haemolytic anaemia, dogs may require careful administration of immunosuppressants such as cyclosporine. It is common if the patient has received a blood transfusion recently. Other conditions such as cancer may require surgery, radiation and chemotherapy. Gallbladder stones and bile duct blockages may require immediate surgery to prevent further damage to the liver and kidneys of the pet.

Prevention of Jaundice in Dogs

Most diseases that lead to jaundice in dogs are not preventable. However, some tips may be followed for reducing your dog's risk to suffer from jaundice. The administration of any human medications or improper medications should be avoided. The body of the dog should be checked regularly for prevention of tick infestation, and dogs should be vaccinated against common diseases, particularly for leptospirosis. The possible bacterial contamination should be avoided. The contact with wildlife should be avoided and should not be given access standing water and garbage.

31

Acute and Chronic Hepatitis in Horses

The liver performs over hundred different functions, and many of those are used for digestion and processing of nutrients. Its weight ranges between 5 and 9 kilograms, depending on the size of the horse. It is situated right behind the diaphragm and in front of the stomach, located in the place by six ligaments and the pressure of the surrounding organs. The liver is a fairly simple organ, consisting of mostly identical cells called hepatocytes at the cellular level. These cells are formed continually and remain metabolically active. The liver is supported by two other glands, such as the salivary gland and the pancreas, for the digestive functions. The damage to this largest gland in a horse, as caused by ragwort poisoning, has extensive health consequences. The functions of the liver are divided into five categories, such as digestive and secretory function, metabolic function, detoxification, synthetic function, and storage function.

Digestive and secretory function: The liver produces bile that helps in fat digestion, and pH neutralisation of the contents in the small intestine. Horses do not have a gall bladder, and the bile duct is wide to compensate it. Therefore, the bile is continually produced in a diluted form and carried through the liver via the bile duct, which branches through the liver tissues. In addition to carrying bile, it also transports various waste products from the metabolic system, including hormones and enzymes.

***Metabolic function*:** The liver is involved in the metabolism of carbohydrates, proteins and fat. A large amount of fat appears in the liver during starvation or when energy requirements exceed energy intake. The accumulation of fat in the liver is a sign of rapid fat mobilisation for its metabolism to provide energy. The liver also plays an important role in the effective utilisation of protein in the body. A well nurtured horse often takes in more protein than it needs. The liver breaks down the excess non-essential amino acids, removes the amino group, which is excreted as urea.

***Detoxification*:** The detoxification function is initially directed at possible toxic substances absorbed from the intestinal tract those are delivered to the liver by the portal blood circulation system, before entering into the general circulation. The liver filters the toxins out, so that they cannot damage other tissues. Some of these toxins can cause damage to the liver itself, for example ragwort, but other organs are spared from the damages due to the plant's toxins.

Synthetic function: The liver is a major player in the production of almost all the major plasma proteins and low molecular weight proteins, such as clotting factors.

Storage function: The liver is the prime storehouse of utilisable energy, vitamins and minerals. The absorbed glucose is stored as glycogen in the liver and muscles, and glycogen is released with lowering blood glucose levels, and glycogen is stored when blood glucose rises. The process is mediated by several hormones, including insulin. The liver is a vital storage organ for minerals such as iron and copper, as well as vitamins including vitamin A, D, and B_{12}, and other proteins.

Etiology

The etiology of acute hepatitis includes infectious pathogens, toxic substances, or an undefined cause. Chronic active hepatitis (CAH) in horses is an idiopathic, chronic, progressive hepatopathy, characterized by histopathological changes like biliary hyperplasia, with concomitant periportal or biliary inflammation, or both, and associated hepatocellular damage. The exact cause of chronic active hepatitis in horses is not ascertained in most of the cases. A similar syndrome occurs in human beings and that has been linked to autoimmune disease, chronic hepatitis b virus infection, non-a and non-b viral hepatitis, called Wilson's disease, α1-antitrypsin deficiency, and drug allergy. The extrahepatic signs of autoimmune disease, including dermatitis, arthritis, and glomerulonephritis, may be present along with liver affections. Some horses with chronic active hepatitis have polyclonal gammopathy, a condition where both B cells and plasma cells produce too much of immune proteins called immunoglobulins or antibodies, that help fighting infections and illnesses, but this is a nonspecific finding in many types of chronic hepatic diseases and is not necessarily an indication of autoimmunity. Occasionally, horses with chronic active hepatitis have predominantly plasma and mononuclear cell infiltration in the liver, suggesting heightened humoral activity. The presence of dermatitis in the coronary band of horses with chronic active hepatitis may be a manifestation of autoimmune disease, although this has not been confirmed by immunohistological staining. Wilson's disease and α1-antitrypsin deficiency as established in human beings have not been documented in horses. Idiosyncratic drug hypersensitivity has been reported in horses although it is not a consistent feature of chronic active hepatitis.

In addition to the possibility of autoimmunity, chronic viral or hypersensitivity reaction may result in chronic active hepatitis in horses or it may be a manifestation of chronic cholangitis, a chronic inflammation of bile duct. Many horses with chronic active hepatitis have a suppurative inflammatory response involving the biliary system, in addition to periportal inflammation and hepatocellular necrosis. Cholangiohepatitis is often accompanied by biliary hyperplasia, fibrosis, and cholestasis. In some cases, coliform organisms have been isolated from the liver, suggesting an ascending infection from the gastrointestinal tract. In some horses, histopathologic evidence of acute inflammation, hepatocellular necrosis and fibrosis are noticed, and a diagnosis of specific cause is not attained. Although iron toxicity (hemochromatosis) is frequently incriminated as a cause of liver failure in horses, it has rarely been documented in adult horses.

Hepatitis A and E are the most common infectious causes of acute hepatitis and usually have a self-limiting clinical course, resolving in 2 to 4 weeks with supportive treatment that includes intravenous fluids, antiemetics, and symptomatic treatment. Liver disease is common in horses and foals suffering from toxic, septic, hypoxic, neoplastic, vascular or metabolic conditions, but progression to liver failure is rare.

Pathophysiology

An acute poisoning from ragwort is rare. The small but frequent ingestion of ragwort causes repeated and minor damage to the liver over a period of time. Ragwort contains a pyrrolizidine alkaloid, which passes to the liver via the hepatic portal vein. The toxin damages the hepatocytes, eventually leading to shrinking and progressive fibrosis of the liver.

The pathophysiology of hepatic encephalopathy is complex and likely involves several gut-derived neurotoxins, cerebral and systemic inflammation, cerebral vascular dysfunction, and neuroendocrine abnormalities. High concentrations of ammonia in the blood and cerebrospinal fluid (CSF) have been most commonly incriminated as the cause of pathophysiologic events of hepatic encephalopathy. Ammonia (NH_3) is thought to pass the blood brain barrier (BBB) by diffusion, while the less toxic or non-toxic ionized ammonia (NH_4^+) is limited to the transcellular route for translocation. Ammonia is believed to pay a key role in the development of hepatic encephalopathy, with glutamate, NMDA (n–methyl-d-aspartate) receptor activity and increased glutamine, all playing a central role in ammonia metabolism. Systemic inflammation and bacterial translocation are often part of the syndrome of acute liver failure and hepatic encephalopathy. Breakdown products of injured hepatocytes in addition to systemic inflammatory cytokine production increases free radical and metalloproteinase concentrations which can have systemic effects in addition to increasing permeability of the blood brain barrier (BBB). In addition, serum ferritin and iron, which may approach 100% saturation in horses with liver diseases, play the role as proinflammatory mediators.

Clinical Signs

The onset of clinical signs in chronic and acute hepatitis is insidious. Signs are compatible with progressive liver failure and include depression, exercise intolerance, weight loss, anorexia, colic, icterus, and fever. The signs may be intermittent. Although uncommon, some horses with chronic active hepatitis have moist exfoliative dermatitis at the coronate. Photosensitization, diarrhoea, and clotting abnormalities may also occur. Horses with acute liver failure are more likely to have central nervous system signs as the initial and predominant signs. Besides, it is manifested by weight loss and photosensitivity. Unlike other organs, damage to the liver is succeeded by repair and regeneration of hepatocytes to a reasonable level, and the liver generally continues to play most of its functions. Therefore, limited damage to liver cells is rarely observed clinically. A lowered blood protein level is an early sign of liver damage, that remains unnoticed without a blood biochemical test. Damage becomes clinically visible only with extensive or complete destruction of hepatocytes. The most prominent symptoms of liver damage are weight loss, lethargy, depression, ascites, peripheral oedema, haemorrhagic diathesis, an increased tendency to bleed or bruise, jaundice, soft faeces, sensitivity to sunlight, hepatic encephalopathy, failure of anabolic and synthetic functions. These symptoms can occur as a consequence of ragwort poisoning.

The clinical signs of liver failure can vary depending upon duration of damages, acute or chronic, predominant biliary versus hepatocellular injury, and the specific causes of the disease. Horses with acute liver failure are more likely to have central nervous system signs as their initial and predominant sign. Horses with chronic liver disease leading to failure commonly show weight loss and photosensitivity as the clinical findings. Gastric impaction and bilateral laryngeal paralysis are two of many complications that can occur with equine hepatic failure.

Clinical Pathology

Biochemical tests are imperative in the diagnosis of both liver disease and liver failure, and the results can be helpful in narrowing the differential diagnosis for the liver failure, and can help predicting the prognosis, when evaluated over time. Liver specific enzymes include sorbitol dehydrogenase (SDH-cytosol), glutamate dehydrogenase (GLDH-mitochondrial) and gamma Glutamyl transferase (GGT), which respectively reflect hepatocellular (SDH, GLDH) and biliary injury (GGT). Aspartate aminotransferase (AST) and alkaline phosphatase (ALP) also reflect hepatocellular and biliary injury, respectively but are not specific for liver damage. The activity of SDH and GLDH may be increased in the serum with even mild hepatocyte injury like endotoxemia. These enzymes have a short half-life ($t½$) which can be very helpful in determining resolution or progression of the hepatic insults. GGT activity in the serum increases following biliary disease, and often continues to get elevated for few days, presumably due to biliary hyperplasia, even after the hepatic insult is no longer present in a horse. The haematocrit (HCT) and serum iron may be abnormally high in horses with severe liver disease. A small percentage of racehorses may have mild but persistent increases (50-140 IU/l) in GGT without other evidence of liver disease. The elevation in GGT is supposed to occur with oxidative stress, associated with maladaptation to training/ racing.

An increase in direct bilirubin of more than 0.2 mg/dl above normal maximum is a highly sensitive and specific marker of liver failure in equines more than 3 months of age. The increase in direct bilirubin by 25% or more of the total bilirubin suggests a predominant biliary disease. Septic foals with intestinal ileus sometimes have elevations in direct bilirubin with minimal evidence of hepatocellular dysfunction. This may be due to biliary stasis or resorption of the bilirubin from the intestinal tract. Foals with severe haemolysis may also have marked increases in direct bilirubin. There may be a decrease in BUN and albumin with chronic diseases. Abnormally high serum or plasma bile acid (>20 μmol/l) is an early predictor of liver failure. Concentration of bile acids can be an accurate predictor of prognosis with chronic liver disease and marked hepatic fibrosis, but they are not good predictors of prognosis with acute hepatic disease. Some healthy foals may have values above 30 μmol/l until several weeks of age. Serum triglycerides are increased in hepatic lipidosis in horses. Hypoglycaemia is often present in foals with hepatic failure, while in adult horses, blood glucose is generally normal or elevated. Ammonia may be high in horses with liver failure but it is inconsistent. The other causes of hyperammonemia without liver disease include portosystemic shunts, when ammonia and bile acids are almost always high but serum liver enzymes are usually normal.

Diagnosis

The diagnostic criteria for chronic and acute hepatitis include abnormally increased serum transaminase activity, abnormal immunologic findings, and characteristic histopathologic changes. The serum SDH and AST activity may be mildly increased, but GGT and ALP are often markedly increased in chronic active hepatitis. The serum bile acid, total protein, and bilirubin, especially the direct-reacting conjugated fraction of total bilirubin concentrations, may be increased. Bilirubinuria may be present, and the Bromsulfthalein, also called bromsulphthalein, bromosulfophthalein, and BSP) half-life is usually prolonged. Haematology may reveal an inflammatory leukogram, with or without a left shift. Immunodiagnostics, including antinuclear antibody titre and anti-immunoglobulin immunofluorescent staining of skin lesions, may help confirm involvement of an autoimmune phenomenon. The definitive diagnosis of chronic active hepatis depends on the presence of characteristic histopathologic findings in the liver biopsy specimen. Progressive periportal hepatocellular necrosis obscures and distorts the limiting plate, the cord of hepatocytes that surrounds the portal triad. As this destructive process continues, bands of necrotic hepatocytes and inflammatory cells connect one liver lobule to another or extend from the portal tract to the central vein, a feature called bridging necrosis, and is succeeded by fibrosis and cirrhosis. Mononuclear cells may be the predominant inflammatory infiltrates. However, infiltration of neutrophils predominates in cholangiohepatitis. Biliary hyperplasia may signal concurrent cholangiohepatitis. Ultrasound examination for diagnosis of liver diseases may reveal dilated bile ducts, biliary sludge, biliary stones, hepatic fibrosis, hepatomegaly, hepatic lipidosis and hepatic masses. Liver biopsy is best performed after the liver has been visualized.

Prognosis

The prognosis for chronic active hepatitis is variable, depending on the cause and duration of the disease. The presence of cirrhosis warrants a grave prognosis. Prognosis is variable in horses with liver failure depending upon the cause. Theiler's disease in horses is usually either ends in death of the horse or on the way to recovery within 5 days after onset of clinical signs and beginning treatments. Chronic active hepatitis cases often respond to medical treatments that are continued until GGT is normal. Relapse of chronic active hepatitis has been recorded in approximately 50% horses after cessation of therapy. Cholangiohepatitis cases often have a good prognosis, if medical treatment begins to work in 7-10 days, and there is no severe fibrosis, and multiple stones seen on ultrasound examination. Toxic causes such as pyrrolizidine alkaloid toxicosis and liver failure following multiple blood transfusions in foals, generally have a poor prognosis. Hepatic lipidosis has a good prognosis, if the predisposing factor is resolved and there is a rapid and dramatic response to medical therapy. Portosystemic shunts have been surgically corrected in approximately 50% of the cases.

Treatment

The horses with affection of acute and chronic hepatitis, requires extra support, which can be given amongst others by changing their diets. Treatments of liver failure or liver disease will vary depending on cause. Supportive treatments for most cases of equine liver failure include crystalloid therapy with supplemental dextrose and potassium.

General supportive care is indicated for hepatic failure with apparent clinical signs. Specific therapy for chronic active hepatitis depends on the histopathologic findings. Corticosteroid therapy may be beneficial if the liver biopsy reveals abundant plasma cells, or other diagnostic tests suggesting autoimmune or hypersensitivity disease. In horses, initial treatment with dexamethasone at the dose rate of 0.05 to 0.1 mg/kg/day for 4 to 7 days, followed by a gradual reduction in dose over 2 to 3 weeks, has been suggested. Additional treatment with prednisolone (1 mg/kg/day) may be necessary for several weeks. However, the decision for corticosteroid therapy should be carefully considered in cases of chronic acute hepatitis. Long-term antimicrobial therapy for 4 to 6 weeks is indicated in cases of cholangitis, confirmed by liver biopsy examination. The choice of antimicrobial is best determined by culture and sensitivity results. Antimicrobials excreted in bile, such as ceftiofur, ampicillin, or broad-spectrum antimicrobials such as penicillin and gentamicin, are recommended for use.

Treatments for hepatic encephalopathy revolve around decreasing enteric-derived neurotoxins, primarily ammonia, decreasing cerebral edema, correcting glucose, electrolyte and acid-base abnormalities, maintaining perfusion and oxygenation to the brain and other vital organs. Horses with hepatic encephalopathy can have propulsive cortical signs, which may require sedation in order to properly attend the horse and to prevent injury to the horse or attendants. A low dose of Detomidine @5-10 μg/kg intravenously may suffice. It is important not to overly sedate a horse as that might cause excessive lowering of the head in the standing horse, thus promoting cerebral edema in addition to potential negative effects on perfusion of the brain, liver, and other organs. Phenobarbital administration is preferred, if additional seizure or propulsive behaviour control is required. Diazepam should not be used in horses with hepatic encephalopathy, as it can induce astrocyte swelling and worsen hepatic encephalopathy.

Metronidazole has equal effects on decreasing intestinal ammonia production but it is well absorbed, metabolized by the liver and may have neurotoxic effects. It may be preferred to combine neomycin with lactulose initially, if ammonia level is very high. Neomycin decreases ammonia production through its effect on microbial population and decreasing ammonia producing bacteria. Lactulose, a poorly absorbed carbohydrate, decreases ammonia producing bacterial degradation in the large bowel that results in increased H^+ production and conversion of some ammonia ions (NH_3^+) to poorly absorbed NH_4^+ salts. Orally administered antibiotics should not be prolonged beyond 1-3 days to lower the risk of antibiotic-associated diarrhoea. It would be ideal if these treatments soften the stool without causing diarrhoea. If the horse is eating feeds, moderate amounts of carbohydrates and high branch chain amino acid should be administered every 2-4 hours. Treatments to decrease neuroinflammation and disruption of the BBB are mostly unproven but include N acetylcysteine, minocycline, neuro-steroids such as progesterone or allopregnanolone. Horses with acute liver failure should receive supportive treatment for systemic disorders. Pentoxifylline is often recommended for inflammatory and/ or fibrosing liver diseases. Ursodeoxycholic acid (ursodiol) @15 mg/kg q 24 po should be considered in any horse with liver disease with laboratory or ultrasonographic evidence of decreased bile flow.

Antibiotics are administered in cases of septic cholangiohepatitis. Intravenous penicillin, ampicillin, gentamicin and metronidazole are reasonable choices as the combinations of enrofloxacin and metronidazole, trimethoprim sulpha and metronidazole. Parenteral nutrition is very important and regular insulin (0.1- 1.0 IU/kg) or Zn insulin (0.4 IU/kg q24h) is needed in cases of hyperglycaemia.

In horses with chronic progressive liver disease, not associated with infection, prevention of fibrosis development or progression is key for successful management. The activation of hepatic stellate cells in response to liver injury is considered as the key cellular event that drives the liver fibrosis. Pentoxifylline (7.5 mg/kg p.o.

Q12h), milk thistle or silymarin is used. Colchicine @ 0.03 mg/kg p.o. Q24h is an excellent treatment to inhibit fibrosis but expensive. Zinc supplementation and steroids, like prednisolone @0.5-1.0 mg/kg per os Q24h are all used to inhibit fibrosis. Although S adenosylmethionine, silymarin, and vitamin E are popular hepato-protectants, the oral administration of S adenosylmethionine as an intact capsule or tablet may not be clinically effective in horses. Supplementation with vitamin A should be avoided in horses with chronic liver disease as Vitamin A can promote fibrosis.

Additional B-complex vitamins, including folic acid and vitamin K, are recommended for horses with liver damage. While a healthy liver usually processes these vitamins, an unhealthy liver may do so less successfully. Zinc supplementation may also be beneficial to horses with liver failure, as it helps process amino acid breakdown products. In humans with liver disease, zinc deficiencies have been noted to cause mental abnormalities. Any additional supplementation of extra vitamins or minerals should be avoided unless specifically required. Vitamins such as A, D and K are stored in the liver and can become toxic when ingested in excess. Overload of vitamin D can lead to renal failure, and excess vitamin K may cause haemolytic anaemia.

Good quality forage is essential for any horse to ensure a healthy digestive system and to satisfy the horse's need to chew. The grazing of horse should be encouraged, as pasture is a rich source of vitamin E, which helps to support immune function. Access to hay should be provided at all other times. Horses with liver damage require forage of low to medium quality protein, but the hay must be very clean. Legume forages, such as alfalfa, should be avoided. Horses with liver damage often have poor appetite, so hay can be dampened with some diluted molasses to improve consumption. It is sometimes better to feed small hay nets on a frequent basis, rather than providing the horse with one large portion.

32

Chronic Hepatitis and Ascites in Dogs, Hepatic Encephalopathy

Chronic hepatitis is a syndrome in dogs that can result from many different disease processes together with inflammation and necrosis of liver parenchyma in the past, and is characterised by clinical manifestations of anorexia of variable degrees, lethargy, weakness, vomiting, diarrhoea, weight loss, jaundice, polyuria and polydipsia. Severe or advanced cases of chronic hepatitis in dogs are manifested by coagulopathies, ascites, and hepatic encephalopathy. The word chronic refers to going on damaging process for some time, at least for weeks. In contrast, acute hepatitis has most likely occurred for just a few days. Some cases of acute hepatitis are cured, but many transform to chronic hepatitis. The chronic hepatitis is preceded by inflammation in the liver and, possibly, necrosis associated with infiltration of the liver with various white blood cells. The invasion of white blood cells and cell death may result from previous damage to the liver by infectious agents, such as viruses or bacteria, or as a result of toxic damage. The toxic damage can be caused by poisons ingested by the dog or by abnormal accumulation of substances, natural to the body. Idiopathic chronic hepatitis is defined as chronic necrotic, and inflammatory self-perpetuating liver disease associated with a non-suppurative inflammatory infiltrate, where the underlying cause is not discovered even after rigorous investigation.

Etiology

A primary attack by the immune system against liver cells can also cause inflammation and cell death, as seen in an autoimmune disease, where own cells of the body are recognised as foreign. Cancer can result in similar liver damage, but such diagnosis precludes using the term chronic hepatitis.

Drugs responsible for inducing chronic hepatitis include Amiodarone, Carprofen, Diethylcarbamazine, oxibendazole, Phenobarbital, Trimethoprim/ sulfadiazine. Infectious disease pathogens causing chronic hepatitis are Canine adenovirus type 1, Helicobacter species, Herpes virus, and Leptospira. Besides, genetic disease/ inherited defects such as Alpha-1-proteinase inhibitor deficiency, autoimmune mechanisms, copper hepatotoxicosis, episode of acute hepatitis and various toxicities are responsible for chronic hepatitis.

Epidemiology

Chronic hepatitis is the most common liver disease in dogs. Middle-aged to older adult dogs are more commonly affected. There are no breed or sex predilections.

Pathophysiology

Chronic hepatic failure is the primary cause of ascites, a common manifestation of liver cirrhosis in dogs. The other causes of ascites include congestive heart failure, nephritic syndrome, malnutrition, ancylostomiasis and protein losing enteropathy. Ascites is the accumulation of abnormal quantities of fluid in the peritoneal cavity, resulting in visible distension of abdominal cavity called ascites. Ascites is actually not a disease by itself, but a secondary syndrome.

Ascites develops secondary to portal hypertension, with or without hypoalbuminemia. Liver disorders associated with ascites trigger physiologic responses that sustain euvolemia and splanchnic perfusion leading to sodium and water conservation.

The hepatic causes of portal hypertension include processes upsetting craniad flow of portal circulation within the liver. These disorders are subclassified relative to hepatic sinusoids, such as presinusoidal, sinusoidal, and post-sinusoidal causes. Presinusoidal causes include diffuse flow occlusion of intrahepatic portal veins by thrombi, portal vein phlebitis or sclerosis, and diffuse severe ischemic cholangiopathy due to arterial thrombi. The latter injury leads to portal vein atrophy and disappearance. However, the most common cause of severe presinusoidal portal hypertension is diffuse portal tract fibrosis associated with a proliferative-like ductal plate malformation (DPM). Posthepatic causes of portal hypertension include pathologies external to the liver that thwart sinusoidal circulatory egress via centrilobular sinusoids and hepatic veins, such as obstruction of blood flow from the liver via the hepatic vein.

Ascites is characterised by the increase in the amount of fluid in the peritoneal cavity that normally contains approximately 150 ml of fluid. Small increases in the quantity of peritoneal fluid can be detected by ultrasonographic examination, as 1.5 to 3.0 litres of peritoneal cavity is required for its detection on physical examination by perceiving shifting dullness. Demonstration of a fluid thrill or fluid wave indicative of tense peritoneal wall necessitates accumulation of at least 10 litres of fluid in the peritoneal cavity.

Hepatic encephalopathy (HE) is a neurobehavioral syndrome affiliated with either critical liver failure such as fulminant hepatic failure or cirrhosis, or portosystemic shunting that is congenital or acquired portosystemic shunting with decreased functional hepatic mass and intrahepatic shunting of blood around regenerative nodules.

Pathophysiologic mechanisms underpinning hepatic encephalopathy (HE) are complex and yet to be completely understood. Synergistic effects between failure of the liver to detoxify ammonia and accumulation of other endogenous substances has been attributed for hepatic encephalopathy. The integrated concept of HE explains episodic variability and heterogeneous precipitating factors that correlate with diverse clinical scenarios.

Ammonia, a key player in the pathogenesis of HE, is thought to sensitize neurons to other encephalogenic factors. Ammonia can influence multiple brain neurotransmitter systems directly by chemical influence, and indirectly through alteration in substrate availability for transmitter formation. Nevertheless, blood and cerebral ammonia concentrations are often discordant, disqualifying blood ammonia as a simplistic reliable measure of HE. A weak nonlinear and nonexponential correlation exists between serum ammonia levels and severity of chronic HE. A more direct correlation exists in acute HE.

Most ammonia is derived in the gastrointestinal tract from enterocyte conversion of glutamine to glutamate (energy metabolism). Smaller ammonia contributions derive from urease-producing bacteria, mucosal ureases, and bacterial protein degradation. In healthy animals, most enteric ammonia circulates directly to the liver in the portal vein whereupon it is promptly removed by hepatocytes and used for amino acid interconversions or detoxified in the urea cycle or in formation of glutamine. Approximately 25% of systemically circulated urea accesses the alimentary canal (in saliva, bile, or via enteric diffusion), whereupon it is digested to ammonia by bacterial and mucosal ureases.

Ammonia also is eliminated by renal tubular secretion and via use for glutamine synthesis in skeletal muscles (temporary ammonia detoxification). The latter pathway underscores the importance of maintaining lean body mass (muscle) in hepatic insufficiency where this process transiently buffers hyperammonemia. In animals with portosystemic shunting or liver failure, blood ammonia concentrations escalate predominantly because of hepatofugal portal circulation or secondarily because of insufficient hepatocyte detoxification. Portal regions (zone 1 hepatocytes) perform high rates of ammonia detoxification via the urea cycle whereas centrilobular regions (zone 3 hepatocytes) detoxify ammonia by synthesis of glutamine from glutamate.

Glutamine catabolism in astrocyte mitochondria further increases regional ammonia, contributing to oxidative stress and mitochondrial dysfunction (increased anaerobic metabolism with lactate accumulation). Because glutamine and lactate each impart osmotic effects, their pathological accumulation promotes cell swelling, contributing to brain edema, disturbed cerebral microcirculation, and increased intracranial pressure. These changes may lead to irreversible brain herniation, brainstem compression, and death.

A number of clinical conditions can escalate blood ammonia concentrations or precipitate HE including dehydration (prerenal/renal azotemia), alkalemia and hypokalemia (increase renal ammonia production and CNS ammonia translocation), hypoglycaemia, catabolism (increase amino acid and ammonia turnover), infection (catabolism, sepsis), PU/PD (risk for dehydration and electrolyte loss), anorexia, constipation (increased absorption of colonic ammonia and other HE toxins), widespread heme loading (ie, haemolysis, blood transfusion, GI haemorrhage, rhabdomyolysis), high dietary protein intake (especially red meat, fish, and in some animals, eggs), various drugs (eg, benzodiazepines, tetracyclines, antihistamines, methionine, barbiturates, organophosphates, phenothiazines, diuretics [overdosage], metronidazole [overdosage], and certain anaesthetics)

The most encephalogenic protein recognized is heme (haemoglobin/ myoglobin), with causal syndromes having consequences similar to consumption of raw red meat meal.

Clinical Signs

Dogs with chronic hepatitis frequently present with nonspecific clinical signs. However, more specific clinical signs are exhibited in advanced stages. Some affected dogs are asymptomatic at presentation, and clinical signs may be inapparent in patients with compensated advanced disease.

In the later advanced stages, overt clinical signs of hepatic encephalopathy may be seen.

Clinical signs of hepatic encephalopathy (HE) are variable, involving disturbed sensorium ranging from mild dullness, anxiety, and inability to respond to basic commands to overt abnormalities, including propulsive circling, head pressing, aimless wandering, weakness, ataxia, rare myelopathy (chronic HE), rare signs suggestive of polyarthropathy, amaurosis (unexplained blindness), ptyalism, rare fine tremors (with chronic recurrent HE), behaviour change (e.g. aggression), hallucinations or dementia, collapse, seizures, and coma preceding death. Early HE recognition requires serial neurological evaluations by an experienced clinician. Hepatic encephalopathy is first characterized as acute or chronic. Acute HE is encountered in animals with fulminant hepatic failure (FHF), that is closely combined with biochemical and metabolic evidence of acute and massive loss of hepatic function and is often severe and unremitting. Chronic hepatic encephalopathy is more common and reflects delayed access of portal blood to hepatocytes as seen in either congenital or acquired portosystemic shunting.

User-friendly clinical classification schemes subdivide HE into three categories: type A, reflecting acute liver failure or FHF; type B, predominantly reflecting portosystemic shunting; and type C, reflecting cirrhosis where proto systemic shunting and decreased hepatic mass contribute. Clinical signs of types B and C are similar, whereas type A is usually severe and may be associated with increased intracranial pressure, brainstem compression, and risk of cerebral herniation. According to the time course, HE is also described as episodic or recurrent (bouts of HE every 6 months or less) or persistent (denoting persistent behavioural alterations interspersed with relapses of overt HE).

Otherwise, complicated grading schemes used to classify HE in human medicine consider psychometric assessment of cognitive functions.

Using the four-part classification scheme noted previously:

- Grade 1 HE is associated with mild vacillating lethargy and decreased mental alertness such as confusion or disorientation.
- Grade 2 HE is associated with accentuation of grade 1 clinical signs, with increasing lethargy, drowsiness, personality change, mild ataxia, disorientation, and inappropriate behaviours resulting in house soiling, loss of appropriate social interactions.
- Grade 3 HE is typified by worsening incoordination or ataxia, mental confusion, ptyalism, head pressing, aimless wandering, circling, amaurotic blindness, marked personality change (sometimes to aggression), and increasing somnolence with slow arousal.
- Grade 4 HE is severe and is characterized by unresponsive recumbency or obtundation, seizures, and coma, and is associated with a risk for impending death.

Clinicopathological Findings

The laboratory findings include persistent or cyclic increases in activity of Alanine amino transferase (ALT), Aspartate amino transferase (AST), Alkaline phosphatase (ALP), and Gamma-glutamyl transferase (GGT). At a later stage, increased total serum bile acid (TSBA) concentrations are followed by hyperbilirubinemia. Other findings may include a nonregenerative anaemia, leucocytosis, and hyperglobulinemia.

In the advanced stages, portal hypertension causes development of acquired portosystemic shunts (APSSs) and the associated laboratory markers including RBC microcytosis, hypocholesterolaemia, hypoalbuminemia, activated partial thromboplastin time (aPTT) and prolonged prothrombin time (PT), and ammonium biuret crystalluria. In early disease, liver size is normal and there may be no demonstrable ultrasonographic lesions. In late-stage disease, radiographs may demonstrate a small liver with nodular lesions detected on ultrasonography. Ultrasonographic evaluations may also disclose ascites and acquired portosystemic shunts (APSSs) in dogs with advanced liver injury.

Chronic, sustained, unexplained increases in liver enzymes usually indicate for liver biopsy for definitive diagnosis and to detail acinar distribution of liver injury, the type of inflammatory infiltrates, the presence of lobular remodelling and fibrosis, and accumulation of copper or iron. Biopsy specimens should be submitted for both aerobic and anaerobic bacterial cultures and quantification of copper, iron, and zinc. Copper stains must be reconciled with quantitative copper measurements to avoid erroneous interpretations.

Diagnosis

The history taking should include possible exposure to drug and toxin along with dietary, environmental, and family history. Physical examination findings for chronic hepatitis are nonspecific, often normal and, in most cases, do not provide any information to identify the affected system unless the patient has jaundice. Laboratory investigation is required to further evaluate patients for chronic hepatitis. The World Small Animal Veterinary Association Liver Standardization Group has published standards for diagnosis of the various forms of chronic hepatitis. An antinuclear antibody test, tests for titre or antigen and investigations for exposure are performed for the diagnosis of endemic infectious diseases.

Liver biopsies must be large enough to detail at least 15 contiguous portal triads, and must be taken from several different liver lobes. Samples collected only from apparent "mass lesions" can lead to erroneous diagnoses. Application of special stains will disclose the acinar location of liver injury (reticulin staining), presence of collagen deposition of fibrosis (Masson trichrome staining), extent of iron accumulation in macrophages, Kupffer cells, and hepatocytes (Prussian blue staining) and the extent and location of hepatocellular copper accumulation.

Treatment

The use of specific treatments is recommended to slow down inflammation and fibroplasia and to restore liver antioxidant status. Antimicrobials are initially prescribed empirically until availability of the biopsy and tissue cultures results and then it is adjusted or discontinued based on culture results. The following additional treatments are recommended.

Ursodeoxycholic acid (ursodiol) is used as a hepatoprotectant and anti-inflammatory choleretic at the dose rate of 10 - 15 mg/kg/day, orally with food, every 24 hours or divided every 12 hours. Polyunsaturated phosphatidylcholine is used as an at the dose tare of 25–50 mg/kg, orally with food, every 24 hours. vitamin E, as an antifibrotic and antioxidant, is used @10 U/kg, orally with food, every 24 hourly. The bioavailable S-adenosylmethionine (SAMe) as an antioxidant is prescribed at the dose rate of 20–40 mg/kg through oral route in an empty stomach, every 24 hourly.

Immunosuppressive drugs are used only after careful consideration and exclusion of infectious or toxic causes and when an active disease process is characterized on liver biopsy. Prednisolone or prednisone is usually started at a dosage of 2–4 mg/kg, PO, every 24 hours for 7–10 days, and titrated downward to a maintenance level of 0.5–1 mg/kg, PO, every 24 hours or on alternate days, depending on patient response.

Treatment of ascites is based on identification of its cause. Drugs used to attenuate portal hypertension ideally should not enhance systemic vasodilation. No drugs can fully arrest hemodynamic imbalances initiated and sustained by splanchnic portal hypertension. Nonselective beta-blockers remain the cornerstone of treatment for splanchnic hypertension with APSSs. Propranolol is the most commonly used drug for reducing splanchnic hypertension and risk of haemorrhage from APSSs.

Therapeutic goals include a modest decrease in resting heart rate with preservation of systemic blood pressure (mean arterial blood pressure > 90 mm Hg). Propranolol has been used in dogs with decompensated cirrhosis and diuretic-resistant ascites at a low dosage of 0.05 mg/kg, PO, every 12 hours. However, investigative studies of portal hypertension in dogs, secondary to chronic biliary cirrhosis induced by chronic EHBDO did not demonstrate a decrease in splanchnic pressure with propranolol. Nonselective beta-blockers antagonize the β1 adrenergic receptors in the heart and the β2 adrenergic receptors in the periphery. The β1 effect decreases heart rate and cardiac output and subsequently decreases flow into the splanchnic circulation. The β2 effect leads to unopposed α1-adrenergic activity, which causes splanchnic vasoconstriction and further decrease of portal inflow. Propranolol should not be tried in animals with chronic obstructive pulmonary disease, reactive airway disease, asthma, or congestive heart failure because it may worsen bronchoconstriction and decreases output. Another nonselective β blocker with antioxidant properties, carvedilol, provides a greater attenuation of portal pressure in human beings.

Diet

The dietary intake of sodium should be as low as possible because a low intake makes the goal of a negative sodium balance easier to achieve. If ascites is difficult to control, the sodium intake should be further restricted to 60 mEq/day or, occasionally, to 45 mEq/day. Sodium restriction is necessary not only in the treatment of ascites but also in the prevention of re-accumulation of ascitic fluid.

Therapeutic strategies for control of ascites include dietary sodium restriction and administration of diuretics to increase urinary sodium elimination. When needed, therapeutic abdominocentesis is also used. The first step is dietary sodium restriction to an intake of ≤ 100 mg sodium/418.4 kJ (100 kcal) diet (25 mg/kg/d; < 0.1% dry-matter basis in food). However, sodium restriction alone is usually insufficient and too slow in onset for efficient management. Thus, diuretics are usually also recommended. The two most widely used diuretic agents in the management of ascites are furosemide and spironolactone. Although other agents such as chlorothiazide, bumetanide, triamterene, and amiloride have been used, but their routine use in the management of ascites cannot be recommended. Spironolactone has a milder natriuretic action in compared to loop diuretics. In contrast with furosemide, spironolactone increases sodium excretion by only 2% of the filtered amount. It acts by competitively inhibiting aldosterone and is most effective in patients with hyperaldosteronism. The intracellular changes may occur after administration of spironolactone that accounts for its relatively long duration of action. These intracellular changes also explain why a delay of 2 to 4 days is noted between the initiation of spironolactone therapy and the onset of its natriuretic effect. The loop diuretics Furosemide inhibits sodium and chloride reabsorption in the ascending limb of the loop of Henle without affecting the distal and collecting tubules. It increases sodium excretion up to 30% of the filtered sodium as a substantial amount of sodium is absorbed in the loop of Henle. The onset of action is rapid, usually within 30 minutes after oral administration. The effect peaks in 1 to 2 hours, and most of the natriuretic activity is completed in 3 to 4 hours and is insignificant after 6 hours. As loop diuretics increase renal synthesis of prostaglandins, administration of nonsteroidal anti-inflammatory agents may decrease the efficacy of loop diuretics.

The depletion in intravascular volume may occur as a consequence of imbalance between the intravascular fluid loss caused by the diuretics and the passage of fluid from the peritoneal cavity to the general circulation. Hyponatremia is another problem associated with diuretic agents. Patients with cirrhosis of the liver normally have a low serum sodium concentration because of the effect of antidiuretic hormone. Total paracentesis is preferred approach for treating tense ascites. A catheter is inserted into the left lower quadrant of the abdomen, and ascitic fluid is drained by low pressure suction applied to canisters connected in series

Diuretic-resistant ascites is a condition in which natriuresis is inadequate despite use of a diuretic dosage of spironolactone, up to 400 mg/day, and furosemide, 160 mg/day. In diuretic refractory ascites, adequate natriuresis is possible by increasing the doses of diuretics, but side effects such as hyponatremia, renal dysfunction, and encephalopathy.

Treatment of acute HE involves stabilizing vital signs, providing supportive care, curtailing HE-related neurotoxin formation, ameliorating the pathological impact of these toxins, and ensuring that metabolic complications masquerading as HE are recognized and correctly managed (ie, thiamine deficiency, severe hypophosphatemia or hyponatremia, drug-induced neurologic complications, and non-hepatic encephalopathies such as necrotizing or granulomatous meningoencephalitis common to small-breed dogs).

Animals, interrupted with severe HE. have risk for cerebral edema and brain herniation that should be addressed with mannitol (0.5–1.0 g/kg, IV, over 20 minutes). Such patients should be reclined on a slant board in head up position (25°–35° angle) to decrease gravity-related worsening of cerebral edema.

In the event of seizure activity, IV levetiracetam is recommended. Benzodiazepines (interact with GABA receptors) should be avoided. Alfaxalone, a neuro-steroid anaesthetic interacting with GABA receptors, should also be avoided. Use of any sedatives should be cautiously considered, deferring to gas anaesthesia if painful procedures requiring anaesthesia/analgesia are undertaken.

In animals with FHF due to toxic hepatopathy, there is potential for mitochondrial dysfunction. In these, propofol infusions should be avoided because this drug may critically suppress mitochondrial beta-oxidation. Oesophageal or gastric feeding should be withheld in obtunded patients with dulled or reduced level of alertness or consciousness until neurologic status improves and enteric motility (gastric and intestinal) is confirmed. Parenteral alimentation is usually not undertaken in animals with acute severe HE.

Fluids should be judiciously supplemented with dextrose (2.5%–5.0% dextrose) to avoid neuroglycopenia, potassium chloride (sliding scale), and vitamin B complex including thiamine. Thiamine should be given before starting glucose infusions. Electrolyte and acid-base imbalances should be sequentially monitored until stabilized.

33

Pancreatitis, Exocrine Pancreatic Insufficiency in Dogs

The pancreas is a vital organ located on the right side of the abdomen adjacent to the stomach. It plays two distinct roles within the body, referred to as the exocrine and endocrine functions. It produces hormones such as insulin and glucagon that regulate blood glucose metabolism after entering in to the bloodstream.

Pancreatitis refers to inflammation of the pancreas, that may be acute or chronic, characterized by abdominal pain, inappetence, and vomiting. Exocrine pancreatic insufficiency (EPI) is a condition of maldigestion and malabsorption, and it does not usually involve the endocrine component of pancreas.

Etiology

The exact cause of pancreatitis is unknown, although it may be triggered in some cases by a fatty meal or corticosteroid therapy. However, in many cases, it appears spontaneously. The causes of exocrine pancreatic insufficiency (EPI) include pancreatic acinar atrophy, chronic pancreatitis, pancreatic hypoplasia, and neoplasia. The most common cause of EPI in dogs is pancreatic acinar atrophy. The severity of this condition ranges from subclinical disease to a complete absence of secretory capacity. Pancreatic acinar atrophy is thought to be an immune-mediated condition that begins with lymphocytic pancreatitis. Selective destruction of acinar cells with replacement by atypical parenchyma, ductal structures, and adipose tissue is seen in the late stages of the disease. The prevalence of cases of EPI that develop secondary to chronic pancreatitis in dogs is still unclear. A potential cause of EPI is congenital hypoplasia of the pancreas, more commonly seen in puppies in particular. Pancreatic neoplasia can occlude the pancreatic duct, hindering release of pancreatic enzymes into the duodenum, although the pancreatic acinar cells continue to retain the ability of producing pancreatic enzymes.

Epidemiology

Pancreatitis is not uncommon in dogs, and age, sex, or breed are not the predisposing factors. Pancreatic acinar atrophy is the most common cause of exocrine pancreatic insufficiency in German Shepherds, Rough Collies, and Eurasians, whereas chronic pancreatitis is the most common cause in dogs of other breeds. The two breeds most commonly affected by pancreatic acinar atrophy, German shepherds and rough-coated collies, are thought to have inherited an autosomal recessive gene. Other breeds overrepresented among dogs with EPI irrespective of cause include Cavalier King Charles spaniels, chow chows, and English setters

Exocrine pancreatic insufficiency has a worldwide distribution. The most well-documented breed predisposition is the German Shepherd Dog (GSD). However, any breed can be affected. Rough-coated Collies, Cavalier King Charles Spaniels, Cairn Terriers, Chows, Cocker Spaniels, Eurasian dogs, and West Highland White Terriers are reported to be at increased risk of EPI. Dogs can develop EPI at any age, but the underlying etiology may affect the age of onset of clinical signs. Breeds affected by pancreatic acinar atrophy (PAA) are often affected at a younger age than those affected by EPI secondary to chronic pancreatitis. Female dogs may be overrepresented, but this is not seen in all EPI studies.

The increased risk of EPI in certain breeds makes a heritable component likely possibility, hence a search for the genetic markers of the disease. There is also a suspicion for a polygenetic mode of inheritance. Genome-wide association studies in GSDs revealed several single nucleotide polymorphisms associated with PAA, many of which were located on chromosome 12. Three single nucleotide polymorphisms were located on the same locus, which maps to the dog leukocyte antigen (DLA).

Exocrine pancreatic insufficiency due to pancreatic acinar atrophy is most frequent in young adult German Shepherds but has also been described in Rough Collies and Eurasians. Dogs and cats with EPI due to other causes are usually middle-aged to older and can be of any breed.

Pathophysiology

The exocrine pancreas secretes many different zymogens that supports digestion of carbohydrates, fats, and proteins These enzymes are produced and secreted by the pancreatic acinar cells, and secreted into the small intestines through pancreatic duct. Protein digestion is catalysed by the enzymes, such as trypsin, chymotrypsin, and carboxypeptidase. These proteolytic digestive enzymes are initially released from the pancreas as their precursor, called zymogens, such as trypsinogen, chymotrypsinogen, and procarboxypeptidase, respectively, and are activated in the small intestine. In the presence of chyme, enterocytes release enteropeptidase, which activates some of the trypsinogen to trypsin. Furthermore, the newly formed trypsin activates all three zymogens. Thus, gut enzyme-dependent delayed activation prevents autodigestion of pancreatic proteins. The activated trypsin and chymotrypsin break down proteins into smaller peptides, while carboxypeptidase further processes some of these peptides into amino acids. The carbohydrate digestion is facilitated by pancreatic amylase secretion, which hydrolyses most carbohydrates into disaccharides and some trisaccharides. The fat digestion is facilitated by the enzymes, pancreatic lipase, cholesterol esterase, and phospholipase and the zymogen procolipase, which is activated by trypsin to form colipase. Exocrine pancreatic secretions also contain sodium bicarbonate in an aqueous solution. Acetylcholine and cholecystokinin stimulate the release of digestive enzymes, while secretin stimulates the release of large quantities of bicarbonate, secreted by epithelial cells of the pancreatic ductulus. Sodium bicarbonate neutralizes gastric secretions in the duodenum.

The pancreas has a high reserve capacity, so signs of maldigestion do not occur until 90% of the exocrine pancreatic function is lost. In rare cases of EPI, there may be selective deficiency of individual pancreatic enzymes. Isolated lipase deficiency has been described in dog. Acute pancreatitis may occur as a mild, oedematous form or a more severe, haemorrhagic form, associated with bleeding from within or around the pancreas. The associated inflammation permits digestive enzymes from the pancreas to spill into the abdominal cavity, inducing digestion of the pancreas itself, and secondary damage to the liver, bile ducts, gall bladder, and intestines. A few dogs who recover from an acute episode of acute pancreatitis may continue to have recurrent bouts of the disease during convalescence, which are called chronic or relapsing pancreatitis. Untreated pancreatitis in dogs can become chronic leading to weight loss and lethargy.

The persistence of undigested food within the small intestine in absence of pancreatic enzymes results in bacterial overgrowth, further compromising intestinal function. EPI is complicated by enteric microbiota dysbiosis (EMD), which is suspected to be contributing to the persistence of clinical signs. Pancreatic fibrosis may affect the islet cells of the endocrine pancreas and the acinar cells of the exocrine pancreas.

Animals with EPI also show evidence of malabsorption, the pathogenetic basis of which is less well understood. The nutrients remaining in the intestinal lumen cause loose, voluminous faeces and steatorrhea. The lack of nutrients also causes weight loss and may lead to vitamin deficiencies. The destruction of pancreatic tissue may not be limited to the acinar cells in animals with EPI, caused by chronic pancreatitis, and concurrent diabetes mellitus may be noticed. In rare cases, painful adhesions between the abdominal organs may occur due to pancreatitis.

Clinical Signs

The clinical signs of pancreatitis are often variable, and the severity of the disease depends on the volumes of enzymes that are prematurely activated. Affected dogs may display a praying position. In this position, dogs raise their hind end in the air and bow their head down towards their front legs. This position is believed to relieve abdominal discomfort of the affected dogs because of the inflammatory processes. Dogs with pancreatitis often experience significant dehydration die to vomiting. The most common clinical signs include nausea, vomiting, fever, lethargy, abdominal pain, diarrhoea, and decreased appetite. Death may occur in cases with severe attack, intense depression, and acute shock

Clinical signs most commonly reported are polyphagia, weight loss, and loose stools. Vomiting and anorexia are observed in some patients and may be a sign of concurrent conditions rather than EPI itself. The faces are most commonly pale, loose, and voluminous, and may be malodorous. In rare cases, watery diarrhoea may be seen.

Clinical signs associated with EPI are caused by defective digestion and absorption of dietary macromolecules, such as protein, fat, and starch. Malnutrition and the catabolic state resulting from a negative energy balance cause the considerable weight loss and polyphagia that are observed in many affected patients, especially dogs. The presence of undigested and unabsorbed macromolecules in the intestinal lumen causes osmotic diarrhoea, which contributes to poor stool quality in dogs and cats with EPI.

Clinical signs most commonly reported are polyphagia, weight loss, and loose stools. Vomiting and anorexia are observed in some patients and may be a sign of concurrent conditions rather than EPI itself. The faeces are most commonly pale, loose, and voluminous and may be malodorous. In rare cases, watery diarrhoea may be seen.

Dogs with EPI present with signs of maldigestion, primarily weight loss despite an increased appetite and diarrhoea or loosely formed faeces. Faeces are usually yellow or grey. In most cases, faecal consistency is loosely formed, but dogs may experience severe watery diarrhoea initially. The diarrhoea is usually accompanied by steatorrhea, flatulence, and borborygmi. Some dogs with EPI also experience vomiting. Along with the weight loss, these dogs may have a poor coat. They may also seem nervous, aggressive, or irritable as a result of abdominal discomfort.

Clinical Pathology

Low serum concentrations of TLI are consistent with a diagnosis of EPI, and most laboratories also utilize a diagnostic grey zone in which repeat testing is recommended. While there is some evidence that there is a Increased TLI concentrations can be seen in animals with pancreatitis, which may complicate with diagnosis of EPI secondary to CP. Although serum trypsinogen is cleared by glomerular filtration, renal insufficiency has to be marked to lead to increased serum TLI concentrations. Measurement of pancreatic lipase immunoreactivity concurrent to TLI may assist in identification of concurrent pancreatitis.

EPI rarely affects serum chemistry profile and complete blood count results. Amylase and lipase activities are not useful in diagnosing EPI. Occasionally, a serum chemistry profile may reveal hypocholesterolaemia due to fat maldigestion. Elevation of alanine transaminase (ALT) activity has been documented uncommonly, although the cause of this is unknown. One theory is that an increased uptake of hepatotoxins occurs through the compromised small intestinal wall

Diagnosis

The diagnosis of EPI in dogs is facilitated by species-specific immunoassays for serum trypsinogen (trypsin-like immunoreactivity (TLI), the concentration of which directly reflects the mass of functioning pancreatic acinar tissue. Diagnosis is generally made with laboratory testing, revealing elevations in white blood cells and/or pancreatic enzymes. In recent years, newer pancreatic tests have become available to diagnose pancreatitis more accurately. The SPEC-CPL test, which stands for Specific Canine Pancreatic Lipase test, can be run in the clinic quickly, providing a positive or negative result.

Ultrasound studies often diagnose pancreatic inflammation or local peritonitis caused by this condition. Unfortunately, some dogs with pancreatitis, especially chronic pancreatitis, will escape detection with many of these tests. Consequently, the diagnosis of pancreatitis may be tentative or presumptive in some cases and is based solely on clinical signs and medical history.

The results of a minimum database (CBC, serum chemistry, and urinalysis) are typically unremarkable in animals with uncomplicated EPI. Abnormalities that may be present typically reflect the presence of concurrent disease, which requires a separate diagnostic approach. Abdominal ultrasonography is commonly performed in dogs with clinical signs consistent with EPI or small intestinal disease. Mean pancreatic thickness may be reduced, but EPI cannot be ruled out on the basis of a normal pancreatic ultrasonographic appearance or size. Other abnormalities that may be seen include pancreatic duct dilation.

Trypsin-like immunoreactivity measures serum trypsin and trypsinogen concentrations and is considered the test of choice for EPI. Prior to the advent and validation of the species-specific TLI assays, several methods are used to diagnose EPI, including gelatine film digestion tests, the N-benzoyl-L-tyrosyl-p-aminobenzoic acid test, faecal proteolytic activity, observation of lipid droplets and starch in faeces, and fat absorption tests. These methods were methodologically complex, unreliable, lacked specificity, or some combination, thereof. Faecal proteolytic activity has been used in the diagnosis of EPI in cats. However, a minimum of at least 3 faecal samples must be collected to acquire reliable results. For these reasons, the TLI should still be considered the diagnostic test of choice for the diagnosis of EPI in cats and dogs. The pancreatic acinar cells are replaced by adipose tissue (arrow) and with minimal inflammation or fibrosis.

Immunohistochemical analysis of pancreatic biopsy samples from dogs with subclinical EPI reveals a predominance of intra-acinar CD4+ and CD8+ T lymphocytes, supporting an immune-mediated cause. Other histologic characteristics include piecemeal tissue destruction, large groups of lymphocytes that resemble lymphoid follicle germinal centres, and necrotic and apoptotic acinar cell death

Prognosis

The prognosis depends on the severity of the disease and the response to initial therapy. Most dogs recover without any long-term consequences. Dogs with depression and shock have a very guarded prognosis. Most mild forms of pancreatitis have a good prognosis with aggressive treatment. Dogs not treated may progress to the haemorrhagic form and suffer severe consequences, including sudden death.

Treatment

Successful pancreatitis management will depend on early diagnosis and prompt treatment. Allowing the pancreas to rest and heal itself may work in cases of mild and oedematous pancreatitis. Vomiting dogs are fasted until the vomiting subsides. Food can be withheld for a few days. Dogs, who are not vomiting, are advised to feed a low-fat, highly digestible diet during recovery.

Analgesics are given to control the intense pain, and intravenous fluids are administered to maintain normal fluid and electrolyte balance and perfuse the pancreatic tissues. Many cases will also require anti-inflammatory drugs. Antibiotics are administered to deal with concurrent infection. Intensive care with aggressive doses of intravenous fluids and medications are given to counteract shock in haemorrhagic pancreatitis with the signs of systemic shock.

A lack of proper food digestion may follow the destruction of a significant number of acinar cells that produce digestive enzyme. This is known as exocrine pancreatic insufficiency (EPI) and can be treated with daily administration of an enzyme replacement powder.

Several different formulations of pancreatic enzymes exist, including enteric-coated and noncoated formulations in addition to raw pancreas. Enteric coatings are designed to protect the pancreatic enzymes as they pass through the stomach, but studies suggest no difference in efficacy between coated and noncoated preparations in dogs. Noncoated preparations are therefore often initiated as first-line therapy due to cost, but enteric-coated enzymes are considered by the authors in refractory cases when increasing doses of non–enteric-coated enzymes are ineffective. It is also important to note that enteric coatings are often formulated for the human gastrointestinal tract, and it cannot be assumed that their properties are well suited for use in dogs or cats. Potential adverse effects of PERT include oral bleeding. Dose reduction is recommended in these cases. Fresh raw pancreas contains digestive enzymes and may be fed to animals with EPI and may be inexpensive (when available). Porcine pancreas is commonly utilized and is effective. Raw products may be contaminated with bacterial pathogens and other infectious agents that could have negative impacts on the animal or their owner. They should be used as a last resort. Standard starting doses for PERT are guided by the manufacturer but typically start around 1 teaspoon/cup of food (or per 10 kg of body weight) for dogs per meal.68 If an inadequate clinical response is noted, the dose can be increased by half a teaspoon/cup of food, up to a maximum of 2 teaspoons of PERT/cup of food. It is the authors' belief that doses > 2 teaspoons of high-quality pancreatic enzyme/cup of food are not necessary in the vast majority of cases. Other causes of poor treatment response should be investigated in these cases. Any changes in stool consistency should result in discontinuation of treatments.

Dietary Modification

Dietary fat and fibre content are important in the management of EPI. Historically, low-fat diets were recommended but can exacerbate malnutrition due to poor energy density, and they have not been associated with improved outcome. Low-fat diets may also worsen fat-soluble vitamin deficiencies and are generally avoided, unless refractory steatorrhea is present. Highly digestible, low-residue diets are often utilized, but dietary response appears to vary per animal. High-fibre diets are also generally avoided due to low energy density and concerns that fibre may interfere with lipase activity and nutrient assimilation.

Cobalamin supplementation: Hypocobalaminemia is prevalent in dogs (> 60%) with EPI. Hypocobalaminemia likely occurs as a result of increased uptake of cobalamin by the enteric bacteria and due to reduced concentrations of intrinsic factor, which is needed for cobalamin absorption. Hypocobalaminemia is an independent risk factor for decreased survival in dogs, and supplementation with cobalamin improves treatment outcome. Ideally, serum cobalamin concentrations would be monitored during therapy to determine whether supplementation was adequate to normalize serum cobalamin concentrations. Both oral and parenteral cobalamin supplementations are effective at normalizing serum cobalamin concentrations.

Acid-suppressant medications have been suggested to reduce gastric pH and improve the amount of intact enzyme that makes it through the stomach and into the small intestine. This is of particular importance for noncoated preparations and raw pancreas. Use of H2-receptor antagonists, however, has not been shown to improve outcome in dogs with EPI.

Most dogs and cats with exocrine pancreatic insufficiency can be successfully treated by supplementation with pancreatic enzymes. Powder is more effective than tablets, capsules, and especially enteric-coated products. Initially, 1 teaspoon/10 kg should be given with each meal for dogs and 1 teaspoon/cat with each meal for cats. Once the clinical signs have completely resolved, the dose can be slowly decreased until the lowest effective dose has been reached.

Fresh pancreas may be a viable alternative to the use of powder. Raw pancreas can be kept frozen for several months without loss of enzymatic activity. Preincubation of the food with pancreatic enzymes or supplementation with bile salts is not necessary. Even though pancreatic enzyme supplementation decreases the clinical signs in almost all animals, nutrient absorption, especially that of fats, is not normalized. Feeding low-fat diets to accommodate impaired fat digestion has been suggested, but this may further decrease fat assimilation and lead to deficiencies of fat-soluble vitamins and/or essential fatty acids. Some types of dietary fibre interfere with pancreatic enzyme activity, and a diet low in insoluble or nonfermentable fibre should be fed. In contrast, fermentable fibre can be given.

Enzyme supplementation alone may not lead to complete resolution of clinical signs. Cobalamin deficiency should be considered as a possible cause. Cobalamin absorption depends on adequate synthesis and secretion of intrinsic factor. In both dogs and cats, the majority of intrinsic factor is synthesized and secreted by the exocrine pancreas, and >80% of dogs and almost all cats with EPI are cobalamin deficient. Also, cobalamin deficiency was the only independent risk factor for poor outcome. Thus, serum cobalamin and folate concentrations should be routinely evaluated in small animals with suspected EPI. Dogs and cats with cobalamin deficiency, suggested by a severely decreased serum cobalamin concentration, should be supplemented with cobalamin, either parenterally or orally. Other hypovitaminoses have also been reported in animals with EPI. For example, vitamin K deficiency leading to a coagulopathy has been reported in some cats with EPI.

Some animals may not respond to enzyme supplementation and cobalamin therapy, and likely have concurrent small-intestinal disease. Animals with EPI commonly have concurrent small-intestinal dysbiosis and may benefit from antibiotic therapy. Inflammatory bowel disease also occurs in some animals with EPI. A proton pump inhibitor can also be tried in dogs that do not respond to therapy.

34

Diseases of Urinary System

The primary functions of the urinary system comprise of excretion of waste products of metabolism, maintenance of a constant extracellular environment through retention and excretion of water and electrolytes, regulation of acid-base balance, synthesis of the hormone like erythropoietin from its precursor erythropoietinogen that is responsible for haematopoiesis, and the enzyme renin that regulates blood pressure and sodium reabsorption, metabolism of vitamin D to its active form, 1,25-dihydroxy cholecalciferol, and excretion of drugs and toxic compounds. The diseases of urinary system disturb the one or more of these functions depending upon the severity of the disease and extent of damages. Common problems of the urinary system and kidneys in animals include urinary tract infection, urinary crystals and bladder stones, feline urological syndrome (FUS), urinary tract obstruction, incontinence, kidney disease and failure, and cancer.

Etiopathogenesis

The principal aetiologies linked to diseases of urinary system include stones, crystals, or debris, infection or inflammation of organs of urinary system, ingestion of toxic substances, trauma, cancer, stress, spinal cord abnormalities, congenital abnormalities and prostate diseases.

The common clinical manifestations include inability to urinate or passing only small amounts of urine, bloody or clouded urine, high rise in body temperature, loss of bladder control, increased frequency of urination or amount of urine, straining or crying out in pain during urination, soiling in inappropriate places, frequent licking of urinary opening, strong odour of urine, lethargy, vomiting, weight loss and anorexia.

Diagnosis and Differential Diagnosis

Many abnormalities of the urinary system can be diagnosed from signalment, history, physical examination, serum biochemical profile, urinalysis, and aerobic bacterial culture of urine. In making a diagnosis, anamnesis should include information regarding changes in water consumption, frequency of urination, urine volume, its appearance and behaviour of the patients. It is also important to obtain information about historical and current drug administration, appetite, diet, changes in body weight, and previous illnesses or injuries. A complete physical examination including palpation of the bladder and examination of external genitalia is performed to detect disorders of the urinary system. The urethra in both the sexes and the prostate in males are examined in dogs. The kidneys are generally easier to palpate in cats than dogs. A full neurological examination should be performed on all patients with micturition disorders.

Additional diagnostic tests such as complete blood cell counts (CBC), blood gas analysis for acid-base status, blood pressure, urine protein-creatinine ratio, symmetric dimethyl arginine (SDMA) test, abdominal radiography, abdominal ultrasonography, contrast studies of the upper and lower urinary tract, cystoscopic examination of the urinary bladder, and renal biopsy provide valuable information.

Urinalysis: It is one of the most important diagnostic tests for evaluating urinary tract disorders. Urine is sampled by one of the four methods such as spontaneous micturition, manual compression of the urinary bladder, catheterization, and cystocentesis with each method having advantages and disadvantages. Urinalysis results should include method used for its collection, specific gravity, colour, turbidity, pH, glucose, ketones, bilirubin, occult blood, protein, and leukocytes. Urine specific gravity is obtained using a refractometer. The urine sediment is examined microscopically for quantifying RBCs, WBCs, epithelial cells, renal casts, bacteria, yeast, parasitic ova, fat, sperm, and crystals. The analysis of urine samples should not be delayed to avoid artifacts such as changes in urine pH, formation of crystals, etc. The protein in urine should be evaluated in reference to its specific gravity. The protein content in a concentrated urine sample might not be noteworthy. However, the same amount in a dilute urine sample indicates a problem.

A urine protein to creatinine ratio from single urine sample or from a pooled 24-hour urine sample is required to quantify the amount of protein in urine. Urine protein to creatinine ratio must be interpreted in the context of other information from the urinalysis. Inflammation and haematuria falsely increase urine protein to creatinine ratio, although haematuria generally has minimal effects.

Bacterial Culture of Urine: The method of urine sample collection is important for interpreting the results. Urine samples for bacterial culture can be obtained by the same methods as used to obtain samples for urinalysis. However, cystocentesis is the preferred method for the bacteriological examination, as it collects sterile samples. A quantitative urine culture is performed, if urine samples are collected by methods other than cystocentesis.

All urinary tract infections are not associated with an inflammatory response. In addition, bacterial counts more 10,000 rods/mL and 100,000 cocci/mL of urine are required to find bacteria consistently in a urine sample using light microscopy. Approximately, 25%–30% of all dogs with urinary tract infections have urine bacterial counts below these thresholds, so urine culture is important to establish infections.

The bacterial colony counts of ≥100,000 CFUs/mL in urine of dogs or ≥10,000 CFUs/mL in cats indicate the presence of clinically relevant bacteriuria consistent with UTI, if the sample is collected by spontaneous micturition or manual compression. Colony counts of 10,000 to 90,000 CFUs/mL in dogs and 1,000–10,000 CFUs/mL in cats suggest urinary tract infections.

Serum Biochemical Profile: The measurement of serum biochemical profile such as BUN, creatinine, calcium, phosphorus, bicarbonate, and serum electrolytes, is useful for diagnosis of many urinary tract disorders and provides an indication of glomerular filtration rate (GFR). The increases in serum BUN and creatinine concentrations suggest renal dysfunction. However, these values are also influenced by nonrenal factors like dehydration. The blood urea nitrogen (BUN) concentration can also be influenced by diet and gastrointestinal bleeding. Serum creatinine concentration is falsely decreased in patients with severe muscle wasting and increased with severe muscle damage. The BUN and serum creatinine concentrations increase as GFR decreases, and this relationship is not linear. Large changes in GFR early in renal diseases lead to only small increases in BUN and serum creatinine concentrations, whereas small changes in GFR in advanced renal disease can be associated with large changes in BUN and serum creatinine concentrations.

Symmetric dimethyl arginine (SDMA) is a sensitive renal biomarker, whose concentrations increase earlier than creatinine with decreases in the glomerular filtration rate. It is considered an important early biomarker of kidney dysfunction in humans and small animals. SDMA concentration is not affected by the factors that affect serum creatinine and BUN concentrations. SDMA level can detect changes in GFR earlier than either serum creatinine or BUN concentration. It may increase with as little as 25% loss of kidney function, versus creatinine, which does not increase until 75% loss of kidney function. Therefore, SDMA is more sensitive than creatinine as an indicator of renal function. The SDMA test has largely replaced iohexol clearance test to detect renal dysfunction. The iohexol clearance test is performed by determining the appropriate dose of iohexol calculated as per the patient's body weight and administering the calculated dose through intravenous route. Blood samples are then collected at 2, 3, and 4 hours after administration to determine iohexol clearance, which is then used to measure glomerular filtration rate. Monitoring iohexol plasma clearance is considered as a useful, reliable, and sensitive tool to establish glomerular filtration rate (GFR) and early stages of kidney disease in both human and veterinary medicine. The assessment of GFR based on iohexol plasma clearance needs repeated blood sampling over different time periods, which is not easily attainable in a clinical setting. More sensitive methods to detect renal dysfunction include plasma clearance tests like inulin clearance, radionuclide techniques, endogenous creatinine clearance, and exogenous creatinine clearance. However, these tests are impractical to perform routinely in clinical practice.

Radiographic imaging, ultrasonographic examination, and cystoscopic examination of the bladder provide additional valuable information depending on the cause of the urinary tract disorder. Renal biopsies can miss appreciable lesions as all portions of the kidney are not equally affected always and thus, it is rarely used to assess the renal dysfunction.

Blood gas analysis or serum bicarbonate concentrations provide useful information on acid-base status, especially for patients with renal dysfunction. Metabolic acidosis is a common problem with chronic kidney disease and can result in protein catabolism. Urinary neutrophil gelatinase-associated lipocalin (NGAL) and cystatin C (Cys C) have shown promise as biomarkers for early detection of renal injury in cats and dogs.

Diseases of Kidney

Nephrosis: It includes degenerative and inflammatory lesions primarily affecting renal tubules. Nephrosis can occur as a sequalae to renal ischaemia and following affection of kidney. Nephrosis is the most common cause of acute renal failure, resulting in uraemia.

Renal Ischaemia: Reduced blood flow through kidney usually results from general circulatory failure. There is transitory oliguria followed by anuria and uraemia, if circulatory failure is not corrected. Ischaemia may be acute or chronic. Acute renal ischaemia occurs due to shock, dehydration, haemorrhagic anaemia, acute heart failure, embolism of renal artery, particularly in horse, and extreme ruminal distension in cattle. Chronic renal ischaemia is seen due to chronic heart failure.

Clinical Signs and Clinical Pathology:

There is oliguria, azotemia and acute renal failure. Analyses of urine as well as blood urea nitrogen and creatinine level are most commonly used for diagnosis. Serum biochemistry of collected sample is used to monitor the response to therapy. Proteinuria is an early indication of renal parenchyma damage. Passage of large volume of urine of low specific gravity after a period of oliguria is usually a diagnostic indication of normal glomerular and tubular function. Diagnosis is made based on symptoms and laboratory findings.

The treatment is aimed at correcting fluid, electrolyte and acid-base imbalances as soon as possible. In case of renal damage, supportive treatment like antibiotic (ceftriaxone, cefotaxim) as well as anti-haemorrhagic drug should be administered.

Diseases of Urinary Tract: Diseases of bladder and urethra are most commonly reported in farm animals than diseases of kidney. In general, renal tubules maintain homeostasis, while the glomerulus controls excretion of metabolic end products.

35

Manifestations and Principles of Treatment of Urinary System Diseases

The major manifestations of urinary tract diseases include abnormalities of urine composition and volume, pain and dysuria, rupture of renal pelvis, bladder and urethra and defect of nervous control during urination. Abnormalities in urine are described below.

Proteinuria: Protein may be present in urine in different conditions like haemoglobinuria, myoglobinuria and haematuria and also in urinary tract infections.

Cast and cells: The urine from animals with renal diseases contains casts and cells.

Haematuria: Blood in urine can result from prerenal causes when vascular damage occurs such as trauma to kidney and septicaemia. Post renal haematuria occurs, particularly in urolithiasis and cystitis.

Haemoglobinuria: Presence of haemoglobin in urine in called haemoglobinuria. False haemoglobinuria occurs in haematuria, when erythrocytes are lysed in urine releasing haemoglobin. Red blood cells can be detected only by microscopically in such cases, by examining the urine sediments for cellular debris. True haemoglobinuria causes a deep red to brown discolouration of urine and gives a positive reaction to biochemical test.

Myoglobinuria: Presence of myoglobin in urine is an indication of severe muscle damage. It is called azoturia in horses. Myoglobin molecule is much smaller than haemoglobin and passes more readily through glomeruli, so a detectable dark brown staining of urine occurs without very high plasma level of myoglobin.

Pyuria: Leucocytes or pus in urine indicates inflammatory exudates at some point in urinary tract, usually renal part or urinary bladder. Pyuria is usually accompanied with presence of bacteria in urine.

Crystalluria: Crystals in urine of herbivore animals have no special significance, unless they occur in very large numbers, and are associated with clinical signs of irritation of urinary tract. Calcium carbonate and triple phosphate crystals are commonly present in normal urine of herbivores. If they occur in large number, it suggests that urine is concentrated and possibility of development of urolithiasis in future.

Glycosuria and ketonuria: Glycosuria in combination with ketonuria occurs only in diabetes mellitus, a rare disease of large animals. Glycosuria may occur in enterotoxaemia due to infection with *Clostridium perfringens* type-D. It may occur after parenteral treatment with dextrose solution and adrenocorticotropic hormone. It also occurs in acute tubular nephrosis. Ketonuria is a more common finding in ruminants during starvation and in acetonemia, in pregnancy toxaemia in ewes and does. A low level of ketonuria is normally present in dairy cows.

Urine flow: The increased daily flow should be differentiated from increased frequency without increased flow. Normal urine production is highly variable in large animals. In healthy adult horse, normal urine production ranges from 15-48 ml/kg/day. In neonatal foals, it is 150 ml/kg/day. Polyuria occurs when there is an increase in volume of urine production. It can also result from extra-renal causes, when horse habitually drinks excess quantity of water. The urine osmolality is determined in polyuria. A urine sample with a constant specific gravity of 1.008-1.012 is produced, then the presence of renal disease should be considered in cases of polyurea. If urea and creatinine are within normal limits, a water deprivation test may be performed in horses to access their ability to produce concentrated urine. Animals unable to conserve water due to renal disease can be rapidly dehydrated, and develop pre-renal uraemia. The specific gravity of urine below 1.008 indicates production of dilute urine. Low specific gravity may occur in diabetes insipidus, following excessive water intake or fluid infusion, or following diuretic administration. Neonatal animals on fluid administration often produce dilute urine.

Oliguria and Anuria: The reduction in daily output and complete absence of urine are called oliguria and anuria, respectively. They occur in same conditions, but vary only in degree. In dehydrated animals, urine flow naturally decreases to conserve water as plasma osmotic pressure increases. Congestive heart failure and peripheral circulatory failure may cause a reduction in renal blood flow, leading to oliguria. Anuria occurs commonly in complete urethral obstruction. Oliguria occurs in terminal stages of all forms of nephritis. Anuria and oliguria lead to retention of solutes and disturbance in acid base balance that supports in pathogenesis of uraemia.

Pain and Dysuria: Both the abnormal pain and dysuria are the expression of discomfort, caused by urinary tract diseases. Acute abdominal pain rarely occurs in urinary tract disease, and it is usually associated with sudden distension of renal pelvis or ureter or infarction of kidney. None of these conditions are common in animals, but occasionally noticed with pyelonephritis. Painful or difficult urination occurs in cystitis and urethritis, and is manifested by frequent passage of scanty urine. Grunting may occur with painful urination, and animal may remain in typical posture after urination is completed.

Clinical feature of acute and chronic renal failure: Signs of urinary tract disease vary with rate of development and stage of the disease. In most cases, clinical signs are those of the initiating cause. Acute renal failure causes signs of uraemia. The most common signs in horses are depression, colic and diarrhoea with oliguria or polyuria. Cattle show similar signs frequently, besides being recumbent. In chronic renal disease of all species, there is severe loss of body weight, anorexia, polyuria, polydipsia and ventral edema.

Uraemia: The clinical signs of uraemia are anuria or oliguria. Oliguria is more common unless there is complete obstruction of urinary tract. Chronic renal disease is usually manifested by polyuria, but oliguria appears in terminal stages, when clinical uraemia develops. Uraemic animals are depressed and anorectic with muscular weakness and tremor. In chronic condition, there is poor body condition, loss of protein via urine, dehydration etc. The rate and depth of respiration is usually increased, but it is not dyspnoeic. In terminal stages, it may become periodic in character. The heart rate is markedly increased due to dehydration and myocardial ischemia. An ammoniacal or uriniferous smell of the breath is rarely noticed. Animals become recumbent and comatose in terminal stage. The temperature falls below normal preceding to death.

Principles of Treatment of Urinary Tract Diseases

Diseases of the urinary system can result from a variety of pathological processes. The appropriate treatment depends on the location, severity, and cause of the problem. Suitable diagnostic samples should be collected for diagnosis and treatment of the renal affections. Some diagnostic tests have potential to cause substantial harm, and some treatments can make diagnosis of specific conditions more difficult. Nonspecific and supportive care such as monitoring the patient, administering fluids, treating acidosis, managing nutrition with an appropriate therapeutic diet, should be instituted, in absence of diagnosis of the specific cause.

Treatment of acute renal failure in all species is aimed at removing the primary cause and restoring the normal fluid balance by correcting dehydration, acid base disorders and electrolyte abnormalities. Prognosis for acute renal failure depends on initiating cause and severity of lesion. The animal may survive on its remaining functional renal tissue, provided the acute disease process can be stopped. The urination resumes with the correction of fluid deficit. The rate of fluid administration should be monitored in cases of continuing anuria or oliguria to prevent over-hydration. Diuretic should be given to restore the normal urine flow if patient has oliguria after correction of fluid volume deficit. Frusemide @1-2 mg/

kg every 2 hour or mannitol @0.25-2 gm/kg in 20% solution can be used. Diuretic should not be used until dehydration is corrected. The resulting dialysis will increase the maintenance fluid requirements after urine flow is restored. Animals that remain anuric have a grave prognosis and can only be managed with peritoneal or vascular dialysis. The treatment of chronic renal failure depends on the stage of the disease and cost of the animals. Selection of antibiotics for treatment of urinary tract infection should be based on quantitative urine culture. The ideal antibiotic for treatment of urinary tract infection should have antibacterial against causal bacteria, be excreted and concentrated in kidney and urine, be active at pH of the urine, has low toxicity, be easily administered and low cost. It should not have harmful interaction with other concurrently given drugs. Antibiotic therapy for lower urinary tract infection should continue for at least 7 days. For lower urinary tract infection, 3-4 weeks of treatment is often necessary. Success of therapy is evaluated by repeating urine culture 7-10 days after last treatment.

36

Cystitis and Pyelonephritis

Cystitis, inflammation of bladder, is usually caused by bacterial infection, characterized by symptoms of frequent and painful urination, presence of blood, inflammatory cells and bacteria in urine. Pyelonephritis *is* inflammation of the pelvis and renal parenchyma, usually resulting from infection ascending from the lower urinary tract, and characterized clinically by pyuria, suppurative nephritis, cystitis and ureteritis. Hematogenous pyelonephritis is extremely uncommon. Ascending pyelonephritis is characterized by inflammation, necrosis, and eventually deformity of the calyces, in association with tubulointerstitial inflammation and necrosis. The location of the lesions distinguishes it from other forms of nephritis. It is usually accompanied by ureteritis and cystitis. The inflammation and necrosis predominate in acute pyelonephritis, with the pelvis and medulla being more severely affected than the cortex. Fibrosis replaces inflammation in chronic pyelonephritis. In both the cases, the disease can be asymmetrical, unilateral, or bilateral. Chronic pyelonephritis produces irregular contracture of the kidneys with pelvic deformities. Pyelonephrosis denotes severe suppuration of the kidney in the presence of complete or near complete ureteral obstruction. The infected hydronephrotic kidney is converted to a sack of pus. Suppuration may extend through the renal capsule during the course of pyelonephritis to produce a perinephric abscess.

Etiology

Cystic calculus, difficulty in parturition or dystocia, contaminated parturition, late pregnancy, paralysis of bladder, particularly when horses graze on Sudan grasses, is the causes of cystitis. The causative bacteria frequently gain entry to bladder, but are usually removed by the flushing action of voided urine before they invade the mucosa. Mucosal injuries facilitate invasion, and stagnation of urine is the most significant predisposing cause. The causes of pyelonephritis include bacterial infection of lower urinary tract, septicaemia in cattle caused by *Pseudomonas* spp. Specific pyelonephritis is caused by *Corynebacterium renale* in cattle and *Eubacterium suis* in pigs.

In cattle, bacterial pyelonephritis has been attributed to ascending infection of the urinary tract by *Escherichia coli* or *Corynebacterium renale*. At least, three *C. renale* serotypes exist as normal flora of the caudal portion of the reproductive tract of female cattle and the sheath of male cattle. Unlike most other gram-positive organisms, *C. renale* possesses pili that promote attachment and colonization in the urinary tract mucosa. Physical or chemical damages to the mucosa in the lower portion of the urinary tract such as dystocia or puerperal infection, bladder paralysis, or catheterization, predispose the cow to pyelonephritis due to ascending infection of *C. renale* from the urinary bladder to the ureters and kidneys. *C. renale* causes a humoral antibody response when renal infection develops, which is not seen when infection is limited to the bladder. Pyelonephritis as a result of *E. coli* infection has a similar pathogenesis to that caused by *C. renale* in that ascending infection from the lower urinary tract occurs following damage to the caudal portion of the reproductive tract but the disease is not as fulminant.

Trueperella pyogenes, *Corynebacterium cystitidis*, *Corynebacterium pilosum*, *Staphylococcus*, *Streptococcus*, *Enterococcus*, *Klebsiella* and *Pseudomonas* species are potential pathogens but those have been isolated infrequently as the cause of pyelonephritis. In calves, septic nephritis and renal abscessation may occur from umbilical artery infection.

Clinical Signs

Acute primary pyelonephritis causes anorexia, a precipitous decrease in milk production and fever with body temperature rising to 103.5° to 105.5°F. Some cows with acute pyelonephritis have colic, manifested by kicking at the abdomen, and restlessness. Renal or ureteral inflammation are usually associated with signs of colic and urinary obstruction caused by blood clots, blocking urine outflow from kidney, ureter, bladder or urethra may also contribute to colic.

Cystitis as a concurrent or precursor disease to pyelonephritis is clinically manifested by twitching of the tail, as an indication of general uneasiness and stress. Stranguria, polyuria, an arched stance, gross haematuria, urinary blood clots or fibrin, or pyuria are also observed in some patients with *C. renale* infection. Acute pyelonephritis should be considered for differential diagnosis from acute colic in post-parturient cattle. Consequently, mandatory components of the physical examination of any sick cows with signs of colic should invariably include per rectal palpation of left kidney and both right and left ureters.

Chronic pyelonephritis is manifested by weight loss, poor hair coat, anorexia, poor production, diarrhoea, polyuria, anaemia, stranguria, and gross urinary abnormalities. Lordosis and stretching out may be apparent in some cows affected with chronic pyelonephritis because of renal pain.

Latent or subclinical pyelonephritis may exist in cattle with multiple health problems, especially during the first few months of lactation. Cattle with abomasal displacement, metritis, mastitis, or cattle that had dystocia may develop pyelonephritis. However, signs of pyelonephritis are masked by more obvious signs of the other systems. Thorough screening of urine and subsequent urinalysis and ultrasound examination confirms the pyelonephritis. The specific clinical signs of pyelonephritis are minimal in more chronic cases.

Diagnosis

The diagnosis of pyelonephritis can be made from clinical signs, per rectal palpation findings, and urinalysis. Fever is usually present in acute pyelonephritis but may be absent in chronic cases. The abnormalities in urinalysis such as haematuria, pyuria, proteinuria, and bacteriuria may be present in both cystitis and pyelonephritis. However, cystitis does not usually lead to systemic illness, and the ureters would not be enlarged as determined by per rectal palpation. Vaginal palpation remains an essential aid to diagnosis because it allows detection of unilateral or bilateral ureteral enlargement that is too subtle to be detected per rectum.

Infection of the left kidney is more common. Rectal palpation may therefore reveal enlargement of the left kidneys in unilateral left kidney infection or in cases of bilateral disease. Normal lobulations of the kidney may be lost, the kidney may feel "mushy," and there may be a pronounced arterial pulsation. Rectal palpation of the right kidney is hardly possible, unless it is seriously enlarged. Ultrasonography is another helpful ancillary aid to diagnosis. It may reveal valuable prognostic information. Dilatation of the ureter on the affected side is the most common ultrasound finding followed by cystic changes within the kidney.

Other laboratory tests may be performed where a definitive diagnosis has not been arrived. Hypoalbuminemia is present in most pyelonephritis patients, and it is more severe in chronic pyelonephritis. Proteinuria appears to be very significant in pyelonephritis. Serum globulin values may be higher (>5.0 g/dL), if infection has been chronic. Generally, a period of at least 6-14 days of renal infection is necessary to elevate globulin values. Adult cattle tend to have higher globulin levels than calves with chronic infection.

Gross examination of the urine has diagnostic significance in acute cases with apparent presence of fibrin, blood clots, and pus in the voided urine. Some cows with acute pyelonephritis will have severe renal haemorrhage that may obstruct the ureter or urethra, thus leading to intermittent or continuous urinary blockage. Sometimes, blood clots may be so substantial to fill and occlude the bladder and urethra. On a rare occasion, inspissated pus may totally obstruct urine flow on one side leading to hydronephrosis and appreciable renal enlargement. Urinalysis confirms the presence of RBCs, WBCs, protein, and bacteria. Routine use of multiple test reagent strips to screen urine during the routine physical examination is an excellent method to detect pyelonephritis and other urinary tract diseases.

Urine culture is the most important laboratory aid as it allows identification of the causative organisms and more importantly, permits antimicrobial susceptibility testing. Previous treatment of the animal with antibiotics may interfere with *in vitro* growth. Therefore, antibiotics should be discontinued for a minimum of 24 to 48 hours before undertaking the urine culture. Urine for culture should be obtained using catheterization or a midstream sample is collected to avoid contamination. Colony counts of $>10^3$/mL in a catheterized sample or $>10^4$/mL on a midstream voided sample should be viewed as cut-offs for significance and are often necessary to determine the infectious organism(s). Concurrent azotemia is the cause for prognostic concern, and it may indicate prerenal conditions such as dehydration, bilateral pyelonephritis with subsequent renal failure, or postrenal urinary obstruction.

Postrenal obstruction becomes usually obvious after the physical and rectal examinations. Prerenal azotemia should be suspected, if the animal is very dehydrated but is capable of concentrating urine to a specific gravity > 1.022. Prerenal azotemia also responds to rehydration using oral or intravenous fluids. Most cattle with pyelonephritis that also are azotemic have bilateral disease and renal failure. These are usually chronic infections and have elevated globulin levels, hypoalbuminemia, with inability to concentrate urine, and they may have electrolyte abnormalities such as hypochloraemia, hyponatremia, hypokalemia, and hypocalcaemia. Therefore, cattle with bilateral pyelonephritis and azotemia have a guarded prognosis.

Anaemia may be suspected based on physical examination, and it is confirmed by haematological examination. Anaemia may develop from blood loss alone in acute cases or more commonly by blood loss from the urinary tract coupled with reduced erythropoiesis subsequent to renal parenchymal damage and inflammation in chronic cases.

Pathophysiology

The pathogenesis of pyelonephritis begins with localization of infection in the lower urinary tract. Organisms involved in urinary tract infection are usually endogenous bacteria of the bowel and skin, such as *Escherichia coli*, *staphylococci*, *streptococci*, *Enterobacter*, *Proteus*, and *Pseudomonas*, and more specific urinary pathogens, such as *Corynebacterium renale*, *C. cystitidis*, and *C. pilosum* in cattle, and *Actinobaculum* (*Eubacterium*) *suis* in pigs. Mycoplasmas are rarely involved in cattle urinary tract infections. Infection can be mixed. *C. renale* is an obligate parasite of urinary tract mucosa. Fungal infections are rare causes of pyelonephritis in dogs.

The virulence of invading bacteria in the urinary tract is enhanced by the presence of pili on their surface that assist in adhesion of *C. renale* to urinary epithelium in a pH-dependent manner. The type of pili expressed by the bacteria can also be important later in the course of pyelonephritis. Type 1 fimbriate *E. coli* induce greater activation of neutrophils and hence more renal scarring than do non-fimbriate or P-fimbriate *E. coli.*

The shedding of mature epithelial cells with attached bacteria is one of the urinary tract defences against bacterial infection and colonization. Normal voiding of urine helps maintain the sterility of the bladder. Once bacteria enter the bladder, they grow well in urine with low osmolality or alkaline pH. Stasis of urine is an important predisposing factor in the pathogenesis of cystitis and of pyelonephritis. Urinary obstruction can be caused by ureteral anomalies in young animals, kinked ureters in pigs, pregnancy, urolithiasis, and prostatic hypertrophy. Females are predisposed to urinary tract infection because of their short urethras, urethral trauma, and possibly because of hormonal effects. Clinically, urinary infection is indicated by bloody or cloudy urine, pyuria, and/or bacteriuria.

The vesicoureteral reflux is most significant mechanism in causing renal infection following an established infection in the bladder. This retrograde flow of urine up the ureters and into the kidney *(intrarenal reflux)* during micturition may carry bacteria as far as the urinary space of glomeruli, especially if there is urinary obstruction. It can also occur during manual compression of the bladder in dogs and cats for collection of urine samples. Swine have a long intravesical ureter, 5 mm at birth, 36 mm at maturity, that should prevent reflux in mature animals. Transient vesicoureteral reflux is common in puppies and is related to the immaturity of the ureterovesical junction, wherein pressure in the bladder overcomes the vesicoureteral valve. Reflux should decrease with age because the intravesical portion of the ureter lengthens and has a more oblique entry through the bladder wall. Vesicoureteral reflux of sterile urine does little renal damage. Cystitis and mucosal edema may impede the valve function, leading to reflux in adult dogs. Importantly, vesicoureteral reflux is a clinical diagnosis based on contrast radiographs or advanced imaging. Positioning of the patient, especially dogs, may affect the clinician's ability to detect this phenomenon, wherein dogs in lateral recumbency may have reflux that is not repeatable when they are in a supine position. Movement of the urinary bladder with kinking of the ureters is the likely cause of this discrepancy.

Cystitis can alter normal ureteral peristalsis, perhaps causing reversed peristaltic waves. Hence persistent pyelonephritis may result from vesicoureteral reflux, and possibly from reversed peristalsis in animals with cystitis. Progression of pyelonephritis probably depends upon persistence of bacterial infection or bacterial antigens.

The medulla of the kidney is most susceptible to infection for a number of reasons. It is relatively hypoxic because of the low haematocrit in vasa recta. Hypertonicity depresses the phagocytic activity of leukocytes, and ammonia may interfere with activation of complement. In pigs, the renal poles are more susceptible to intrarenal reflux and infection because the collecting ducts serving those lobes do not collapse as readily as do those of the central lobes when exposed to the increased pelvic pressure. Invasion of the renal papilla probably progresses by way of the collecting ducts, as is suggested by the early development of lines of suppuration along the straight tubules and the presence of bacterial colonies within tubules. There can also be direct invasion across the eroded pelvic urothelium.

Pyelonephritis can be unilateral or bilateral. However, unilateral pyelonephritis can be clinically silent. The acute disease typically begins with necrosis and inflammation of papilla or renal crest, called "necrotizing papillitis" in an irregular pattern. Bacteria may be abundant in the collecting ducts. Associated wedge-shaped areas of parenchyma are swollen, dark red, and firm. As the hyperaemia subsides, suppurative tubulointerstitial nephritis and tubular necrosis develop in radially distributed wedges. Tubules are obliterated by the inflammation, and neutrophils cross the tubular basement membrane to enter the tubular lumen resulting in tubulitis. The leukocytic casts formed in the tubules may be found on urinalysis. Tubular dilation and obstruction are common. Glomeruli, although initially uninvolved, may eventually be obliterated. As the process becomes chronic, mononuclear cells replace neutrophils, and fibrosis develops. Contraction of the scars results in deep, flat cortical depressions. The scars in pyelonephritis extend from capsule to pelvis. They can be distinguished from those of other nephritides and infarcts by the presence of fibrosis and deformation of the renal papilla. The calyces and pelvis are dilated and often contain exudate and debris. Papillary defects may be subtle in mild cases. Generalized involvement of the kidney will produce a firm pale shrunken kidney with an irregular surface, and this will be difficult to differentiate from other end-stage kidneys of other aetiologies.

Clinical Signs

The first clinical sign observed in a case of bovine cystitis or pyelonephritis may be the passage of blood-stained urine in an otherwise healthy cow. As the infection proceeds up the ureters, causing inflammation and subsequent involvement of the kidney, the animal exhibits discomfort, manifested by frequent attempts to urinate, anorexia, a slight fever, loss of production, colic with restlessness, tail switching, polyuria, haematuria, or pyuria. In chronic cases, the animal may show colic, diarrhoea, polyuria, polydipsia, stranguria, and anaemia. As the disease progresses, the bladder becomes thickened and inflamed. The ureters become thickened and dilated with a purulent exudate. The involved kidneys develop multiple small abscesses on the surface that may extend into the cortex and medulla.

Cystitis is manifested by clinical signs of frequent, scanty and painful urination, sometimes associated with grunting. Acute case of cystitis is associated with abdominal pain, kicking at belly. Acute retention of urine may develop due to blockage of the urethra with pus or blood. The clinical signs are less marked in chronic case of cystitis. Frequent urinations and thickened bladder wall on rectal examination, no palpable abnormality may be detected in acute cases of cystitis but pain may be noticed.

In Pyelonephritis, renal pelvis and medulla are affected initially. Later, infection may extend to cortex. Pyelonephritis causes systemic signs of toxaemia and fever. Extensive uraemia may develop with bilateral renal involvement. Pyelonephritis is always accompanied by pyuria and haematuria due to inflammatory lesions of the ureter and bladder. In sows, there may be an initial period during which vaginal discharge is noticed, but most affected animals die without showing prior symptoms.

Clinical pathology: Erythrocytes, leucocytes and cell debris are present in urine on microscopic examination. Urine culture is necessary to identify the causative bacteria. Blood and pus in urine is a typical finding in acute cases, and urine may have a strong ammonia smell. In less severe cases, no abnormality is noticed on gross inspection. Microscopic examination of urine sediments reveals erythrocytes, leucocytes and desquamated epithelial cells. Quantitative bacterial culture is carried out to confirm the diagnosis and to institute therapeutic measures.

Diagnosis

The presence of blood, protein, and leukocytes in a free-catch urine sample gives an indication of cystitis or pyelonephritis. Catheterization of the bladder, and culture and susceptibility testing of bacterial isolates confirms the disease. The diagnosis of bovine cystitis and pyelonephritis is based on clinical signs of haematuria, a history of recent parturition, palpation of the left kidney for enlargement, loss of lobulation, and pain. Ultrasonographic inspection of the kidneys, ureters, and bladder, endoscopic inspection of the bladder for detection of cystitis, microscopic examination of the urine for WBCs and bacteria, dipstick screening for proteinuria and haematuria, and quantitative urine culture to identify the organism establishes the disease.

The presence of pus and blood in urine may suggest cystitis as well as pyelonephritis. Laboratory examinations confirm pyelonephritis.

Cystitis syndromes resemble with pyelonephritis and cystic urolithiasis. Pyelonephritis is commonly accompanied with bladder involvement, which may be determined by per rectal examination in large animals. However, prognosis in pyelonephritis is less favourable than cystitis. Thickening of bladder wall may suggest cystitis. Presence of calculi in bladder can usually be detected by per rectal examination, radiographic examination or by ultrasonography. In small animals, urethral obstruction may also cause frequent attempts of urination, but urine flow is greatly restricted, urine comes only drop by drop and distended bladder can be felt on per rectal examination.

Treatment

The causative organisms, their susceptibility to antimicrobial agents, the existence of bilateral versus unilateral disease, and the presence or absence of azotemia constitute the major input for a making decision and chalk out case management strategies. After the organism is identified and antimicrobial susceptibility determined, an antimicrobial agent is selected that maintains high concentrations in urine, is not nephrotoxic, and is approved for use in cattle. Penicillin, because of its excretion through urinary route, attains a much higher concentration in urine than in plasma, potentially making the drug effective in vivo against *E. coli* despite their penicillin resistance *in vitro*. When *C. renale* is identified, penicillin is continued for 3 weeks because the organism is uniformly susceptible to penicillin. If the disease is severe and peracute, penicillin may be administered intravenously for the initial treatment. When *E. coli* is identified, penicillin is continued if objective data (e.g., serial urinalyses, temperature) and subjective data (e.g., appetite, attitude) are returning to normal. If no improvement has occurred during the initial 72 to 96 hours of penicillin therapy in patients with E. coli pyelonephritis, another antibiotic must be chosen based on antimicrobial susceptibility. Other antibiotic options include ampicillin, oxytetracycline, or on very rare occasion, extra-label use of an aminoglycoside but only if azotemia is absent, the cultured organism is not susceptible to approved drugs, and prolonged withdrawal time is properly discussed, agreed upon, and documented with the owner. If a potentially nephrotoxic antibiotic is needed, the frequency of treatment may be prolonged based on creatinine concentration. The serum creatinine level of 1.8 mg/dL suggests an approximate 30% decrease in GFR. The recommended antimicrobial therapy for bovine pyelonephritis, as with any UTI, is long term, at least for 3 weeks. Short-term antimicrobial therapy is often ineffective or only transiently improves the animal clinically before the condition returns and becomes chronic.

Fluid therapy can be helpful during initial treatment to correct prerenal azotemia, flush debris and bacteria from the urinary tract, and help maintain normal renal blood flow; it is especially important, if potentially nephrotoxic antibiotics are being used.

The prognosis for cows with acute pyelonephritis that are treated with long-term antimicrobial therapy is good unless functional or mechanical urogenital abnormalities persist. Cows with severe bilateral pyelonephritis and azotemia have a guarded prognosis. Pyelonephritis secondary to bladder paralysis or rectovaginal fistula after dystocia would have a poorer prognosis because of likely recurrence of UTI. The prognosis for chronic pyelonephritis is guarded because abscesses of the kidney or total loss of the normal, functional kidney parenchyma may occur.

Cows affected with chronic pyelonephritis also have a greater risk of developing a bilateral infection, leading to azotemia and renal failure. Chronically affected cattle also have increased incidence of renal stone formation. Surgical therapy to remove a massively enlarged and infected kidney occasionally is indicated for treatment of chronic unilateral pyelonephritis.

37

Urolithiasis and Gout

Urolithiasis

Urolithiasis is a common sub-clinical problem among ruminants raised in a management system, where the ration is primarily composed of grains or where the animals graze on a certain type pasture, grown in soil rich in silica, oxalates and phosphates. In such situation, 40-60% of animals may form calculi in their urinary tract. However, urolithiasis is an important clinical disease of castrated male ruminants. Calculi cause urinary tract obstruction, usually of urethra, characterized by complete retention of urine, frequent unsuccessful attempt to urinate and distension of bladder resulting in perforation of the urethra and rupture of the bladder. Mortality is high in case of complete urethral obstruction, and the treatment is mostly surgical. So, prevention is essential to limit loss from urolithiasis.

Etiology

Urinary calculi or uroliths are formed, when organic or inorganic urinary solutes are precipitated out of the solution. The precipitates occur as crystals or as amorphous deposits. Several factors affect the rate of urolith formation including some condition that affect the concentration of specific solutes in urine.

Epidemiology

Urolithiasis has been observed in all ruminant species. However, it is of greatest economic importance in castrated animals, fed with heavy concentrate rations, and animals on range pasture in particular problem areas. The presence of pasture plants in these range areas, containing large quantities of oxalate, oestrogens, or silica are the predisposing factors. Cattle grazing on pasture containing plants with high levels of silica develop uroliths in all ages and sexes. Females and bulls usually flush the calculi with the urine flow, but obstructive urolithiasis becomes a problem, primarily in castrated male animals.

Pathophysiology

There are three main groups of factors that contribute to urolithiasis: Those that favour the development of a nidus about which precipitation and concretion can occur, those that facilitate precipitation of solutes on to the nidus surface, and those that favour concretion by cementing precipitated salts to the developing calculus.

Nidus formation: A nidus favours the deposition of crystals about itself. A nidus may be a group of desquamated epithelial cells or necrotic tissue that may be formed as a result of local infection in the urinary tract in occasional cases. Some other factors, such as vitamin A deficiency or the administration of oestrogens, associated with excessive epithelial cells desquamation may be the probable cause of urolith formation, affecting large number of animals at the same time with similar management practices. The use of stilbestrol as a growth promoter increases the incidence of obstructive urolithiasis in treated in withers receiving implants. Diets low in vitamin A have been believed as a cause of urolithiasis but vitamin A deficiency does not appear to be a major causative factor.

Precipitation of solutes: Urine is a highly saturated solution containing a large number of solutes, many of those in higher concentrations than their individual solubilities in a simple solution. Several factors may. The presence of protective colloids that convert urine into a gel is probably the most important factor in preventing precipitation of solutes, explaining why those continue in solution. However, these colloids are efficient up to a concentration, but their capacity to maintain the solution may be overcome by abnormalities in one or more factors. Even in normal animals, Crystals of a number of solutes may be present in the urine intermittently and urine must be considered to be an unstable solution. The ease of calculus formation is influenced by the physical characteristics of urine, the amount of solute to be excreted by the kidney, and the balance between water and solute in urine. The management practices also influence these factors in most cases. The pH of urine affects the solubility of some solutes, mixed phosphate and carbonate calculi being more readily formed in an alkaline than an acid medium. Dietary supplementation with ammonium chloride does not consistently prevent the formation of siliceous calculi. The amount of solute presented to the kidney for excretion is influenced by the diet. The urine of these animals often becomes supersaturated with silicic acid, which promotes the polymerization or precipitation of the silicic acid and calculus formation. Feeding sodium chloride prevents the formation of silica calculi by reducing the concentration of silicic acid in the urine and maintaining it below the saturation concentration. An excessive intake of minerals may occur from highly mineralized artesian water, or from diets containing high concentrations, particularly of phosphates in heavy-concentrate diets. Ingestion of plants with high oxalic acid content can be a risk factor for formation of calcium carbonate calculi in sheep. The process of urinary calculi formation is more complex than a simple dietary excess. Feeding practices can influence the function of the kidney and may contribute to calculus formation.

Clinical Signs Although it is difficult to induce urolithiasis by restricting access to water, concentrated urine is a risk factor for calculus formation. Animals can be forced to produce concentrated urine because of lack of easy access to water, a particular problem in pastured animals, lack of familiarity with water delivery systems and poor quality of available water. Water deprivation of any origin can exacerbate urine solutes concentration. Heavy fluid loss by sweating in hot and arid climates accentuates concentrating process resulting in precipitation of solutes and calculi formation. Mucopolysaccharide fraction of mucoprotein acts as a cementing agent and favours the formation of calculi when precipitates are present. A high dietary intake of phosphate, may be an important cause of urolithiasis in livestock.

The exit of urine is blocked due to presence of urolith in the renal pelvis. Acute distension results in pain manifested by stiffness of gait and pain on pressure over the loins. Calculi in bladder may cause cystitis, characterised by the clinical signs of abdominal pain such as kicking at the belly, repeated twitching of penis, strenuous effort to urinate, grunting resulting in voiding of few drops of blood-stained urine. The urethra and bladder are distended and painful on palpation, as noted during per rectal examination. A heavy precipitate of crystals is often visible on prepucial hairs or inside of the thigh. The passage of a lead wire into the urethra after relaxing the penis with epidural anaesthesia may make it possible to locate the site of obstruction. Cattle with incomplete obstruction frequently pass small amount of blood-stained urine. Occasionally, a small strip of urine is voided followed by a complete blockage, which confuses the diagnosis. If obstruction is not relieved, perforation of urethra or rupture of bladder occurs in 48 hours, facilitating leakage of urine into the connective tissue of the ventral abdominal wall and an obvious fluid swelling, which may spread to thorax. Urine is often infected and severe cellulitis may occur with development of toxaemia. There is an immediate relief from discomfort with rupture of the bladder. Anorexia and depression develop and uraemia appears. Animals may continue in this state for as long as 2-3 days before it dies in coma. In occasional cases, death occurs soon after rupture of bladder due to sever internal haemorrhage.

Clinical Pathology

The urine usually contains erythrocytes, epithelial cells and a higher number of crystals in early stages of calculus presence in kidney or bladder. Bacteria may also be present in secondary invasion of traumatic cystitis, and pyelonephritis is commonly noticed. Blood urea level is increased before rupture of either urethra or bladder.

Diagnosis

Diagnosis can be made from clinical signs. But radiological examination and ultrasonographic examination confirms the disease.

Treatment

Treatment of urolithiasis is primarily surgical. In general, calculi can't be dissolved by medical therapy. It is also difficult to prevent a further increase in size of existing stone or the development of new one, once urolithiasis becomes clinically apparent in animals. However, preventive measures are useful for others exposed to same risk factors. In early stages or in case of incomplete obstruction, treatment with smooth muscle relaxant may be given to relax the urethral obstruction and permit the flushing away of the obstructing calculi with urine flow. Animals treated medically should be closely observed to ensure that complete obstruction doesn't occur. Surgical treatment includes urethrostomy, which is necessary to relieve bladder pressure and for removal of calculi.

Gout

The defective metabolism of uric acid results in gout in animals and birds. Uric acid is a breakdown product of nitrogen, formed due to protein metabolism in the body. Reptiles excrete nitrogen waste as uric acid, which is removed from the circulating blood by the kidneys. Gout is a common problem in many reptiles, including terrestrial turtles. Aquatic turtles, although they are reptiles, excrete the majority of their nitrogenous wastes as urea or ammonia and produce only a small amount of uric acid, so they are not usually affected by gout.

Clinical Significance

Gout is frequently seen in humans and primates. It is a common affliction in reptilian patients, and occurs in any of three forms such as visceral, articular, and periarticular. Uric acid in vertebrates, is derived from both exogenous and endogenous nucleic acids that are degraded in the liver to yield free purine and pyrimidine bases, which are reused by the body. Unused free bases are further degraded and excreted. The pyrimidines are catabolized to the end products, CO_2 and NH_3 that enter the ornithine cycle. In some vertebrates, including primates, the dalmatian dog, birds, squamates, and some chelonians, purine degradation can produce significant quantities of uric acid. These disease entities are significant in reptile pets, as they often reflect deficiencies in captive husbandry and nutrition like dehydration, high-protein diets, imbalances of vitamins and minerals or severe renal dysfunction.

Gout is characterized by joint inflammation, caused by formation and deposits of uric acid crystals are around a joint due to high uric acid levels in the blood, called hyperuricemia. The gout-affected joints become swollen and painful, until they return to normal with reduction of uric acid levels.

Etiology

Increased uric acid production can result from purine digestion and breakdown. Purines occur naturally in the body and in certain foods. High consumption of foods that contain purines has been found to contribute to gout attacks.

The main factors that influence risk of developing gout are diet, genetic factors, and overweight. Other risk factors include gender, age, medical conditions, medications, and lifestyle.

Visceral gout is the abnormal deposition of uric acid crystals in the soft tissues. Gout occurs as a result of poor husbandry, dehydration, high-protein diets, and inappropriate use of aminoglycoside antibiotics

Clinical Signs

Clinical signs are variable and range from lethargy and anorexia to sudden death. If articular gout is present, affected joints are swollen and painful. In visceral gout, clinical signs are related to failure of the affected organ due to deposition of uric acid crystals. Neurologic signs are common, if there is uric acid deposition in the brain.

Diagnosis

Diagnosis is based on physical examination and identification of the uric acid crystals in cytologic or histopathologic examination of tissue samples. Gross lesions may be visible during laparoscopy or laparotomy. Uric acid deposits are sometimes visible on ultrasound or radiographs. Serum uric acid concentrations may be elevated in affected animals. However, elevated serum uric acid concentrations have not been recorded in all reptiles with gout. Kidney biopsy may be helpful in cases of renal gout. The prognosis for long-term survival is grave.

Treatment

Supportive care is provided, especially fluid therapy. Improving temperature, humidity, and access to water is critical. High-protein diets, especially those with high purine contents and nephrotoxic drugs should be avoided. The use of furosemide is avoided because it may inhibit renal excretion of uric acid. Antihyperuricemic drugs such as allopurinol, probenecid, sulfinpyrazone, and colchicine can be used to control, but not to cure gout.

Gout in Poultry

Excretion of metabolic waste products is important in poultry and this function is performed by the kidneys. The function of kidneys is affected by a number of specific diseases and disorders. One of the important disorders associated with kidney damage is Gout in Poultry. In birds uric acid is the end product of nitrogen metabolism. Uric acid is a nitrogenous waste from protein breakdown. In mammals, it is converted to less harmful substance with the help of the enzyme uricase. But in birds this enzyme is absent. Hence, uric acid is the final excretory product. Uric acid is produced mainly in the liver and is excreted by the kidneys. High blood levels of uric acid favour its precipitation in tissues. Uric acid is not toxic but precipitated crystals can cause mechanical damage to tissues like kidneys, heart, lungs, intestines and also in the joints. These crystals severely damage body tissues. So Gout is a condition in which kidney function decreases to a point where uric acid accumulates in the blood and body fluids. Avian gout is a metabolic condition where abnormal accumulation of white chalky uric acid or urates occurs in soft tissues of various organs of body. Gout is commonly observed in chicken as they are uricotelic and lack the enzyme uricase. In gout, blood levels of uric acid can be as high as 44mg/100ml as compared to 5-7mg/100ml in a normal bird.

There are two major forms of gout in poultry which are differentiated by the sites of uric acid deposition- visceral and articular gout. In both forms, deposits consist of needle shaped crystals called tophi. Articular gout is considered to be the chronic form of the disease and is less common. Lesions observed are urate deposition around joints, ligaments and tendon sheaths. There is a predilection for peripheral articulation. Clinical signs observed are shifting leg lameness with joints becoming warm, swollen and tender. It is a condition in chicken that has been recognized for more than 30 years. Visceral gout is considered to be the acute form of disease causing huge mortality characterized by the urate deposits on serosal surfaces, most often in the liver, kidney, pericardium, heart and air sacs. Visceral gout is more common in broilers as young as 2-3 days old. In layers, pullets above 14 weeks are more likely to be affected. Whenever there is kidney damage, excretion of uric acid gets affected and uric acid starts accumulating in the blood and later in tissues.

Cause of Gout in Poultry

The causes of gout are many as kidney damage occurs due to multietiological factors. These causes can be broadly categorized as:

Nutritional & Metabolic Causes, Infectious Causes and other Causes of Gout in Poultry.

1. Excess dietary calcium with low available phosphorus results in precipitation of calcium-sodium-urate crystals. High levels of vitamin D3 can also increase calcium absorption from the gut which can favour formation and deposition of urate crystals.
2. Excessive use of sodium bicarbonate when used to combat heat stress to improve egg shell quality in layers. This alkalinity of urine favours kidney stone formation.
3. Prolonged vitamin A deficiency causes sloughing of tubular epithelium and subsequent blockade resulting in accumulation of urates in the kidney.

However, incidence of gout due to vitamin A deficiency is least under field conditions.

4. Gout due to sodium intoxication is seen in younger birds when the sodium levels exceed 0.4% in water and 0.8% in feed. This generally happens when fish meal is used in the diet (even with normal salt content), since fish meal is rich in salt content. Total content of sodium chloride in feed should not exceed 0.3%.
5. Feed containing more than 30% of protein causes uric acid production which in turn creates an excretory load on kidneys. At the same time the presence of sulphates decreases calcium resorption causing excessive calcium secretion through urine. This condition favours gout.
6. Water deprivation leads to concentration of uric acid and other minerals in the blood and later in the kidneys. Water deprivation especially in the summer is dangerous. This can happen during transportation of birds or due to blockage of nipples, inadequate number of waterers, extra height of water lines, overcrowding, water withholding for long durations during vaccination etc.
7. Hard water with higher salt content is also a load on the kidneys.

There are various chemicals and toxins are involved in kidney damage as;

1. Mycotoxins: Mycotoxins are the most common cause of kidney damage and among mycotoxins citrinin, ochratoxin and oosporin are important. The combination of ochratoxins with aflatoxin is found to be more dangerous. Because of kidney damage uric acid excretion is reduced resulting in accumulation of uric acid in the body.
2. Antibiotics: Certain antibiotics like gentamycin, sulphonamides and nitrofurosones are known to cause renal damage especially in young chicks. The drugs which get excreted through the kidneys have their own imbalancing effect on pH and renal metabolism.
3. Disinfectants: Disinfectants like phenol and cresol if used erroneously cause residual toxicity.
4. Chemicals: Chemicals like copper sulphate used in water results in water refusal, dehydration and gout.

Sign & Lesions of Gout in Poultry

Gout is characterized by depression, dehydration and sometimes with greenish diarrhea. Affected chicks appear dull with ruffled feathers and moist vent. Mortality among young chicks is high. There is irregular and excessive enlargement of kidney lobules and cutting open the kidney reveals urate crystals chalky white deposition of urate crystals is seen all over the visceral organs like the heart, liver and kidney under the skin etc.

Severe Visceral Gout –Note the white chalky deposits around the heart (in the pericardium), on all major abdominal organs, including liver, gizzard and intestines, and even in the tissues of the thigh.

Treatment of Gout in Poultry

Individual cases of gout may be ignored. In acute cases of gout mortality following prescription would be beneficial.

1. Provide plenty of water and adequate drinkers and Aquacure must be added in water.
2. Avoid a diet higher in protein than the recommended level as per the age and breed. Provide low protein diet for 3-5 days based on need depending on severity of gout.
3. Review IB vaccination programme. In the areas where IB is endemic it is advisable to vaccinate with nephrotropic strain at around 4 days. Day one beak dip vaccination has proved to be beneficial in broilers.
4. Use of urine acidifier.
5. Ensure adequate levels of A, D3, K and B complex vitamins.
6. Excessive use of sodium bicarbonate i.e. more than 2kg/ton should be avoided.
7. Use of electrolytes through water may assist in controlling mortality.

Prevention of Gout in Poultry

1. Scientifically balanced feed in respect of Calcium-phosphorus ration depending on the type of ration, required level of sodium, chloride and other ions, vitamin A, D3 and other essential vitamins.
2. Conventional sources of protein.
3. Analyse the feed for mycotoxin content and if found positive change the feed or use suitable toxin binders.
4. Judicious use of drugs such as antibiotics, sulpha drugs and anticoccidials to avoid kidney damage.
5. Fresh water should be accessible to birds all the time.

38

Glomerulonephritis and Interstitial Nephritis

Introduction

The inflammation of kidneys, called nephritis, may involve the glomerulus, tubule, or the interstitial renal tissue. The inflammation involving the glomeruli is called glomerulonephritis. Glomerulonephritis can occur as a primary disease or as a component of diseases affecting several other body functions as seen in equine infectious anaemia and chronic swine fever. The primary and secondary glomerulonephritis are not the common causes of clinical disease in farm animals. Proliferative glomerulonephritis is relatively more common cause of chronic renal failure in horse. The kidney diseases involving the kidney structures outside the glomerulus are broadly referred to as tubulo-interstitial disease. Interstitial Nephritis is rarely recognized as a cause of clinical disease in farm animals. It may be diffuse or have a focal distribution. In calves, focal interstitial nephritis (white spotted kidney) occurs as a sequalae to bacteraemia. Diffused interstitial nephritis occurs in pigs, following infection with *Leptospira* sp. The disease becomes clinically important due to destruction of nephrons. Horses with chronic interstitial nephritis have the clinical syndrome of chronic renal failure coupled with uraemia.

Etiology

Interstitial nephritis is a disease that causes inflammation and swelling around glomerulus, called the tubules. Acute interstitial nephritis, that lasts for a short time, is the most common type of interstitial nephritis. Chronic interstitial nephritis lasts for a longer period ranging from weeks, months to years. Acute interstitial nephritis is usually caused by an allergic reaction or other type of reaction to a medicine or drug. The symptoms of acute interstitial nephritis appear quickly. There are more than 100 different medicines that might cause acute interstitial nephritis. The medicines that are most likely to cause acute interstitial nephritis include antibiotics, such as penicillin, nonsteroidal anti-inflammatory drugs (NSAIDs), medicines used to treat excess stomach acid called proton pump inhibitors (PPIs). Chronic interstitial nephritis is usually caused by another health problems and include autoimmune diseases such as lupus, low levels of potassium in the blood, too high calcium levels and sarcoidosis, a disease that causes inflammation, most commonly in the lungs and lymph nodes. Besides, some infections cause interstitial nephritis.

The most frequent causes of acute intestinal nephritis can be found in one of three general categories: drug-induced, infection-associated, and cases associated with immune or neoplastic disorders.

Drug-induced acute intestinal nephritis: The list of drugs implicated in causing acute intestinal nephritis continues to expand. Drugs are more frequently recognized as etiologic factors in acute intestinal nephritis because of the increased frequency with which drugs are used, the increased use of renal biopsy, and the characteristic clinical presentation. Some classes of medication often are associated with certain acute intestinal nephritis. Development of drug-induced acute intestinal nephritis is not dose-related. Acute intestinal nephritis may become clinically evident after two weeks or longer following start of the medication.

Acute intestinal nephritis is associated with primary renal infections such as acute bacterial pyelonephritis, renal tuberculosis, and fungal nephritis. Systemic infections can cause direct injury because of pathologic processes in the kidney, or can be associated with indirect injury caused by medications used in the treatment of infections. Acute intestinal nephritis may be caused by local or systemic autoimmune processes. Several types of glomerulonephritis result in interstitial inflammation, possibly associated with anti-tubular basement membrane autoantibodies.

Epidemiology

Acute interstitial nephritis is an important cause of acute renal failure resulting from immune-mediated tubulointerstitial injury, initiated by medications, infection, and other causes. Acute interstitial nephritis may be implicated in up to 15 percent of patients hospitalized for acute renal failure.

Pathophysiology

Immune system plays a major role in the pathogenesis of glomerular lesions. Glomerular injury can be initiated by an immune response, wherein antibody is directed against intrinsic glomerular antigen or foreign antibody present in glomerulus. More commonly, deposit of circulating antigen-antibody complex in the glomerulus resulting in set up the inflammation process, severely affecting the glomerular filtration process. The inflammatory damage to the glomerulus alters the selective permeability of filtration system, allowing plasma protein, particularly albumin to pass into glomerular filtrate. Interstitial nephritis lowers the ability of the kidneys to clean the blood and produce urine. Usually, it is caused by a reaction to a medicine and the problem resolves after the suspected medicine is stopped.

There is strong evidence that acute intestinal nephritis is immunologically mediated. The precise disease mechanism is unclear, but antigen-driven immunopathology is the key mechanism. The presence of helper-inducer and suppressor-cytotoxic T lymphocytes in the inflammatory infiltrate suggests that T-cell mediated hypersensitivity reactions and cytotoxic T-cell injury are involved in pathogenesis of acute intestinal nephritis. In some cases, humoral mechanisms are involved with complement proteins, immunoglobulins, and anti-TBM antibodies found in the interstitium.

Clinical Signs and Clinical Pathology

Clinical features of acute intestinal nephritis are essentially those of acute renal failure from any cause. Apart from a history of new illness or medication exposure, there are no specific history, physical examination or laboratory findings that distinguish acute interstitial nephritis from other causes of acute renal failure. Classic findings of fever, rash, and arthralgias may be absent in up to two thirds of patients.

Acute intestinal nephritis typically presents with nonspecific symptoms of acute renal failure, including oliguria, malaise, anorexia, nausea and vomiting, with acute or subacute onset. The clinical presentation can range from asymptomatic elevation in creatinine or blood urea nitrogen (BUN) or abnormal urinary sediment, to generalized hypersensitivity syndrome with fever, rash, eosinophilia, and oliguric renal failure. A relatively rapid decrease in renal function, as measured by elevated creatinine and BUN, is typical.

The classic triad of low-grade fever, skin rash, and arthralgias was primarily described with methicillin-induced acute interstitial nephritis, but it is present only about one third of the time. This triad is present in only 5 percent of cases of acute intestinal nephritis. Each component of the triad is present in 70 to 100 percent, 30 to 50 percent, and 15 to 20 percent of acute intestinal nephritis cases, respectively. Other associated symptoms may include flank watching, gross haematuria, or other clinical findings associated with an underlying disease process.

Tachycardia, conjunctival edema, nystagmus, walking in circle, and convulsion are commonly noticed clinical signs. Kidneys are enlarged and tender. Severe proteinuria and low plasma albumin are observed. Blood urea nitrogen (BUN) is increased with hypophosphatemia and hypocalcaemia. The common symptom of interstitial nephritis is urinating less than normal, blood in your urine, fever, rash, swelling of limbs and other body parts, weight gain and high blood pressure.

Pathology

The acute intestinal nephritis is characterised by the infiltration of inflammatory cells within the renal interstitium, associated with edema, sparing the glomeruli and blood vessels. Interstitial fibrosis is initially sparse, but develops later in the course of the illness. Fibrotic lesions may be diffuse or patchy, beginning deep in the renal cortex, most prominently at the medullo-cortical junction. The inflammatory infiltrate is typically composed of mononuclear cells and T lymphocytes, with a variable number of plasma cells and eosinophils. Eosinophils may be totally absent from the infiltrate or may concentrate in small foci, forming eosinophilic microabscesses.

Nonsteroidal anti-inflammatory drugs (NSAIDs) are commonly associated with glomerular involvement producing no or minimal changes. Peritubular infiltration and occasional invasion of lymphocytes beneath the TBM may occur with mild to extensive tubular damage, which may be difficult to distinguish from acute tubular necrosis (ATN). In chronic interstitial nephritis, the cellular infiltrate is largely replaced by interstitial fibrosis. The third pathologic category is granuloma formation with epithelioid giant cells usually found in acute intestinal nephritis, secondary to tuberculosis, or sarcoidosis.

Diagnosis

Diagnostic studies such as urine eosinophils and renal gallium 67 scanning provide suggestive evidence, but they are unable to reliably confirm or exclude the diagnosis of acute interstitial nephritis. Gallium 67 scanning is considered as a useful test to diagnose acute intestinal nephritis.

Renal biopsy in cases of acute intestinal nephritis is the gold standard for diagnosis, but it may not be required in mild cases or when clinical improvement is rapid after removal of an offending agent or medication. The time until removal of such agents, and renal biopsy findings, provide the best prognostic information for return to baseline renal function. Renal biopsy is the only definitive method of establishing the diagnosis of acute intestinal nephritis. This step is usually undertaken when the diagnosis is unclear and there are no contradictions for the procedure, or when the patient does not improve clinically following discontinuation of the medication suspected as the cause of acute intestinal nephritis and renal failure. Other laboratory features are used to provide suggestive evidence of acute intestinal nephritis, to guide conservative management, or to permit empiric treatment with steroids. Unfortunately, none of these tests have sufficient predictive value to be diagnostically reliable.

Renal ultrasonography may demonstrate kidneys that are normal to enlarged in size, with increased cortical echogenicity, but there are no ultrasonographic findings that will reliably confirm or exclude acute intestinal nephritis versus other causes of acute renal failure.

Urine Eosinophils: Urine eosinophils are frequently tested to provide confirmatory evidence of acute intestinal nephritis, though the typical constellation of fever, rash, arthralgias, eosinophiluria, and renal insufficiency rarely presents completely. However, other conditions such as cystitis, prostatitis, and pyelonephritis can also be associated with eosinophiluria.

The challenges with clinical diagnosis of acute intestinal nephritis due to lack of clinical features and lack of accuracy of existing tests, there has been significant interest in identifying non-invasive biomarkers for this disease. Monocyte chemotactic protein-1 (chemokine CCL-2) and neutrophil gelatinase associated lipocalin (NGAL) rises in patients with acute intestinal nephritis than in controls. Acute intestinal nephritis is associated with rise in urine cytokines interleukin-9 and tumour necrosis factor

Treatment

The interstitial nephritis can cause kidney damage or kidney failure, if left untreated. The cessation or change of the incriminating medicine alleviates acute interstitial nephritis. Often, this restores kidney function quickly and the symptoms go away. However, acute interstitial nephritis can cause permanent kidney damage, if it is not found early enough. The severe loss of kidney function due to acute interstitial nephritis may require dialysis for a short time until kidney function improves. Treatment consists of addressing the cause, such as by removing an offending drug. There is no clear evidence that corticosteroids help. Nutrition therapy consists of adequate fluid intake, which can require several litres of extra fluid.

The health problems responsible for chronic interstitial nephritis are treated with medicines that help control the blood pressure and lower inflammation. Corticosteroids appear to provide some benefit in terms of clinical improvement and return of renal function.

39

Nephrosis

Degenerative and inflammatory lesions primarily affecting renal tubules can occur as a squeal to renal ischemia and following affection of kidney. Nephrosis is the most common cause of acute renal failure, resulting in acute uraemia or may occur in terminal stages of chronic renal disease.

Renal Ischemia

Reduced blood flow through kidney usually results from general circulatory failure. There is transitory oliguria followed by anuria and uraemia, if circulatory failure is not corrected.

Etiology

It may be acute or chronic. Acute renal ischaemia occurs due to shock, dehydration, hemorrhagic anaemia, acute heart failure, embolism of renal artery particularly in horse, extreme ruminal distension in cattle. Chronic renal ischaemia is seen due to chronic heart failure.

Clinical Signs

Oliguria, and azotemia are the characteristic clinical signs in acute renal failure. Clinical signs of nephrosis in animals include fluid retention like puffy eyelids, swelling in the legs, or fluid in the abdomen, weight loss with gaunt and emaciated appearance, increased urination, frothy or discoloured urine, loss of appetite or vomiting, shortness of breath or increased respiratory effort, pale gums, ulcers in the mouth, or diarrhoea. Nephrosis in dogs is manifested by increased thirst, strained urination, and Chronic kidney failure. Nephrosis in cats is manifested by increased thirst and urination, intermittent vomiting, lethargy, anorexia and weight loss. Nephrosis is a severe form of kidney disease that can cause fluid retention and swelling. Fluid can accumulate in the abdomen, around the lungs, or in the eyelids.

Clinical Pathology

Analysis of urine and blood urea nitrogen and creatinine level is most commonly used for the diagnosis. Serum biochemistry of collected sample may also be used to monitor the response to therapy. Proteinuria is an early indication of renal parenchyma damage. Passage of large volume of urine of low specific gravity after a period of oliguria is usually a good indication of normal glomerular and tubular function. Diagnosis is made based on symptoms and laboratory findings.

Treatment

Treatment must be done for correcting fluid, electrolyte and acid base disturbance as soon as possible. In case of renal damage, supportive treatment like antibiotic (ceftriaxone, cefotaxim) as well as anti-haemorrhagic drug should be given.

40

Renal Insufficiency and Failure, Renal Secondary Hyperparathyroidism

Introduction

The functions of kidneys are to excrete the end products of tissue metabolism except CO_2 and to maintain homeostasis with respect to fluid, solutes and acid base balance by selective excretion of these substances, by varying volume of water and the concentration of solutes in the urine. In general, renal tubules maintain homeostasis, while the glomerulus controls excretion of metabolic end products. Therefore, knowledge on physiology of blood filtration, urinary secretion and excretion helps to understand the urinary tract diseases.

Renal function depends on the integrity of the individual nephron, the structural and functional unit of kidney. Insufficiency can occur from an abnormality in the rate of renal blood flow, glomerular filtration rate and efficiency of the tubular re-absorption. Circulatory emergency may lead to marked reduction in glomerular filtration, but they are extra renal in origin and cannot be considered as true cause of renal insufficiency. Diseases of kidney, and in some instances of ureter, bladder and urethra reduce the efficiency of the renal function, renal insufficiency. It is associated with disturbances in protein, acid base, solute and water homeostasis and excretion of metabolic end products. Renal failure follows, when kidneys can no longer regulate body fluid and solute composition. A prolonged circulatory disruption may cause renal ischaemia and renal insufficiency. The progress and end result of any type of renal disease tends to be very comparable. The glomerulonephritis is associated with primary reduction in glomeruli function but it is mostly associated with secondary involvement of tubules. It is imperative to consider renal blood flow, glomerular and tubular functions independently, while assessing the kidney function clinically. The clinical and clinico-pathological signs of renal dysfunction often depend on anatomical location of lesion and on the functional imbalance between glomeruli and tubules. Renal dysfunction tends to be a dynamic process, so the degree of dysfunction varies with time. If renal dysfunction is so severe that the animal existence is not possible, it is said to be in a state of renal failure and clinical syndrome of uraemia develops.

The diseases of kidney are less commonly diagnosed in farm animals than that of bladder and urethra. The major manifestations of urinary tract diseases are abnormalities of urine volume and composition, pain and dysuria, rupture of renal pelvis, bladder and urethra and defect in nervous control of urination.

Clinical Features of Acute and Chronic Renal Failure

Signs of urinary tract disease vary with the rate of development and stage of the disease. In most cases, clinical signs are those of the initiating cause. Acute renal failure causes signs of uraemia. In horses, most common signs are depression, colic and diarrhoea with oliguria or polyuria. Cattle show similar signs besides being recumbent frequently. Severe loss of body weight, anorexia, polyuria, polydipsia and ventral edema are the common signs in chronic renal disease of all species.

Uraemia

serum calcium and elevated phosphate due to failure of kidneys to excrete phosphate, increases the secretion of the parathyroid hormone. Prolonged stimulation results in parathyroid hyperplasia. Secondary hyperparathyroidism also arises in vitamin D-deficient rickets, malabsorption, and pseudohypoparathyroidism.

Pathophysiology

Hypocalcaemia is the most important stimulus for increased secretion of PTH from parathyroid glands in SHPT. The increased stimulation also results in hyperplasia of the parathyroid glad. Increased PTH level results in increased calcium and phosphate absorption from the intestine. Parathyroid hormone (PTH) acts as a stimulus for increased osteoclast activity, which results in calcium and phosphorus resorption from the bone. PTH activates vitamin D in the kidneys to its active form. Fibroblast growth factor 23 (FGF-23) is secreted by osteocytes that plays an important role in phosphorus homeostasis by increasing phosphorus clearance in the renal tubules. FGF-23 also inhibits 1, alpha-hydroxylase activity. FGF-23 is not known to modulate PTH secretion directly but may do so indirectly through regulating phosphate and vitamin D metabolism.

In chronic kidney disease, reduced glomerular filtration rate (GFR) leads to increased secretion of PTH. Decreasing GFR leads to decreased phosphate clearance and hyperphosphatemia, which then stimulates the parathyroid glands to secrete PTH. Hyperphosphatemia causes hypocalcaemia as phosphorus forms complexes with calcium The clinical signs of uraemia are anuria or oliguria. Oliguria is more common unless there is complete obstruction of urinary tract. Chronic renal disease is usually manifested by polyuria, but oliguria appears in terminal stages, when clinical uraemia develops. Uremic animals are depressed and anorectic with muscular weakness and tremor. In chronic conditions, there is poor body condition, loss of protein via urine, dehydration etc. The rate and depth of respiration is usually increased, but it is not dyspnoeic. The respiration may become periodic in character in terminal stages. Heart rate is markedly increased due to dehydration and myocardial ischemia. An ammoniacal or uriniferous smell of the breath is sometimes noticed. The affected animals become recumbent and comatose in terminal stage. The temperature falls below normal followed by death.

The treatment of urinary tract diseases in all species is aimed at removing the primary cause and restoring the normal fluid balance by correcting dehydration, acid base and electrolyte disorders. Prognosis for acute renal failure depends on originating cause and the severity of lesions. The animal may survive on its remaining functional renal tissue, provided acute disease process is stopped. The patient should urinate following correction of fluid deficit. If anuria or oliguria continues, the rate of fluid administration should be monitored to prevent overhydration. Diuretic should only be given after restoration of normal urine flow, if the patient has oliguria even after correction of fluid volume deficit. Frusemide @1-2 mg/kg body weight every 2 hr or mannitol @0.25-2 gm/kg as 20% solution can be used. Diuretics should only be used until dehydration is corrected. After urine flow is restored, the resulting dialysis will increase the maintenance fluid requirements. Animals those remain anuric have a grave prognosis and can only be managed with peritoneal or vascular dialysis. The treatment of chronic renal failure depends on the stage of disease and cost of the animals. Selection of antibiotics for treatment of the urinary tract infection should be based on quantitative urine culture and antibiotic sensitivity test. The ideal antibiotic for treatment of urinary tract infection should have antibacterial action against causal bacteria, be excreted and concentrated in kidney and urine, be active at pH of the urine, has low toxicity, be easily administered and of low cost. It should not have harmful interaction with other concurrently given drugs. Antibiotic therapy for lower urinary tract infection should continue for at least 7 days. Antibiotic treatment is often necessary for 3 to 4 weeks for lower urinary tract infections. The success of a therapy is evaluated by repeating urine culture, 7-10 days after last treatment.

Secondary Hyperparathyroidism

Hyperparathyroidism can be primary, caused by growths or enlarged glands, or secondary due to kidney diseases or low calcium levels. Hyperparathyroidism is called secondary, when the glands are normal but extra parathyroid conditions, like kidney failure, may lower calcium levels and cause the body to react by producing more parathyroid hormone. It is called Tertiary hyperparathyroidism is when long-standing secondary hyperparathyroidism starts to behave like primary hyperparathyroidism.

Secondary hyperparathyroidism (SHPT) is a condition where the parathyroid glands produce too much parathyroid hormone (PTH) in response to low blood calcium levels. There are several conditions that lower blood calcium level including chronic kidney disease (CKD) or vitamin D deficiency. The defect in the activation of vitamin D in the kidneys due to chronic kidney disease (CKD) leads to hypocalcaemia and hyperphosphatemia, resulting in a compensatory increase in parathyroid gland cellularity and production of parathyroid hormone, causing secondary hyperparathyroidism (SHP). It occurs due to another primary disease associated with low calcium levels in the body over a protracted period. It is commonly seen in kidney disease and after certain intestinal surgeries. The increased parathyroid hormone levels to keep the calcium level up in the standard range are the pathogenesis of the disease. Hyperparathyroidism secondary to chronic renal failure is a condition characterized by inappropriate secretion of parathyroid hormone (PTH), secondary to a phosphocalcic metabolic disorder.

Secondary hyperparathyroidism (SHPT) is a state associated with an increased secretion of PTH due to parathyroid hyperplasia caused by triggers such as hypocalcaemia, hyperphosphatemia, or decreased active vitamin D. The increased PTH secretion, in turn, causes increased calcium in the blood by acting on bones, intestines, and kidneys. The parathyroid hormone (PTH), secreted by parathyroid glands, located close to thyroid gland, plays a role in calcium and skeletal metabolism. Hypocalcaemia and hyperphosphatemia are the important triggers for PTH secretion. On the other hand, the secretion of PTH is reduced by 1, 25 (OH) vitamin D. A prolonged secondary hyperparathyroidism is often associated with disturbances of bone turnover, as well as visceral and vascular calcifications, which are responsible for cardiovascular morbidity and mortality.

Etiology: Secondary hyperparathyroidism is frequently linked to vitamin D deficiency and chronic kidney disease (CKD), where the kidney cannot convert vitamin D into the physiologically active 1, 25-cholecalciferol. Reduced intestinal absorption of calcium resulting in a low, thereby stimulates FGF-23 and increased PTH secretion.

Clinical signs: The symptoms of secondary hyperthyroidism include bone and joint pain, weak or broken bones, kidney stones, frequent urination, abdominal pain, looking tired or weak easily, and loss of appetite. Extraosseous manifestations are also seen. Calcifications affect the arterial walls, viscera, periarticular tissue, cutaneous tissue, and the cornea and conjunctiva of eyes. There may be muscle weakness, red-eye syndrome, and intense pruritus. Pruritus is especially seen in advanced renal disease and is due to the deposition of calcium and phosphorus on the skin. Calcifications can occur in the heart, myocardium, aortic, and mitral valves leading to increased cardiovascular events such as ischemia, left ventricular dysfunction, congestive heart failure, arrhythmias, and death.

Calciphylaxis can be seen as ulceration of the skin, resulting from arterial obstruction, with cutaneous necrosis of the extremities. Calciphylaxis is caused by high levels of PTH, calcium, and phosphorus induced by high levels of calcium in dialysate and phosphate binders. In calciphylaxis, there is calcification of small arterioles and venules with severe intimal hyperplasia and can be complicated by thrombosis leading to painful skin necrosis. Increased risk of infections, sepsis, and ischemia contributes to increased mortality risk in calciphylaxis. Other manifestations in SHPT include neurological problems and malnutrition.

Clinical Pathology: Evaluation in secondary hyperthyroidism consists of monitoring the laboratory values of PTH, calcium, phosphorus, vitamin D levels, and renal function. There are certain striking radiological features in renal osteodystrophy. Osteosclerosis increases bone density, especially in the axial skeleton. However, bones are structurally weak and prone to stress fractures in SHPT. Osteomalacia, softening of the bone due to poor mineralization of the newly formed osteoid, is characterized by the presence of looser zones on imaging. The brown tumour in SHPT is a lytic bone lesion caused by increased osteoclastic activity and proliferation of fibroblasts. On a standard radiograph, it presents itself as a well-defined hypodense lesion. These tumours are located more commonly in hands, feet, facial bones, and skull. Brown tumour represents the terminal stage of bone mineral disorder in secondary hyperparathyroidism. Brown tumours can be misdiagnosed as neoplasms on imaging.

Osteitis fibrosa cystica produced by subperiosteal resorption, is seen on standard radiographs, especially at the distal phalanges, clavicles, distal ulna, and skull. Cortical thinning in long bones, bone cysts, and densification of the trabecular bone are also observed. Periarticular, vascular, and more rarely, visceral metastatic calcifications can also be found.

Histopathology: The parathyroid glands in secondary hyperparathyroidism are characterized by asymmetric enlargement and nodular hyperplasia. In chronic renal failure, there is an increase in the ability of parathyroid cells to synthesize and secrete hormone, which is responsible for an increase in serum PTH concentration at first. This results in hyperplasia of the gland linked to both cell hypertrophy and increased cell proliferation. As renal function deteriorates, the expression of calcium, vitamin D, and FGF23 receptors gradually decreases in the parathyroid gland, which leads to parathyroid gland hyperplasia. Parathyroid hyperplasia is classified into four categories: diffuse hyperplasia, diffuse and multinodular hyperplasia, multinodular hyperplasia, and simple nodular hyperplasia.

Renal osteodystrophy comprises of a group of bone mineral disorders. There are five types based on histological classification. Mild disease is a state of slightly increased bone remodelling and usually seen in early or treated SHPT. Osteitis fibrosa cystica is a state of high bone turnover from increased osteoclast activity and bone destruction and resorption caused by the high PTH levels. It is characterized by peri-trabecular fibrosis.

Osteomalacia is the softening of the bone due to inadequate osteoid or insufficient mineralization of the osteoid, depending on the rate of bone remodelling. The problems with mineralization arise due to abnormal calcium and phosphorus metabolism. Mixed disease has features of both osteomalacia and osteitis fibrosa and is associated with aluminium deposition.

Diagnosis: Both forms of hyperparathyroidism are diagnosed by blood tests, revealing a high level of PTH, Primary hyperparathyroidism indicates high calcium, and secondary hyperparathyroidism shows normal circulating calcium but high phosphorous.

Differential Diagnosis: Primary and tertiary hyperparathyroidism are differentiated from secondary hyperparathyroidism. In primary hyperparathyroidism, there is increased secretion of PTH, which leads to increased calcium and phosphate levels. In secondary hyperparathyroidism, there is hypocalcaemia and hyperphosphatemia, which leads to increased PTH levels. In tertiary hyperparathyroidism, the PTH levels are excessively high, and the calcium and phosphorus levels are high as well.

Treatment: Management of SHPT targets abnormal phosphocalcic metabolism. Maintaining the serum calcium and phosphorus levels within the normal range along with control of PTH and vitamin D levels is the key in management of secondary hyperparathyroidism.

Phosphate binders, vitamin D, and calcimimetics have been suggested in the management of calcium and phosphate levels in human patients with CKD. In addition, patients are advised to restrict dietary intake of phosphorus by limiting phosphate-rich foods such as soda beverages, meat, cheese, and dietary products. Phosphate binders can be calcium-containing or calcium-free. Calcium-containing phosphate binders are known to increase vascular and soft tissue calcification and are associated with lower survival as compared to calcium-free phosphate binders. Vitamin D metabolites include cholecalciferol (vitamin D3) and ergocalciferol (vitamin D2). Vitamin D analogs such as calcitriol, paricalcitol, alfa calcidol, and doxercalciferol are grouped under vitamin D receptor activators (VDRA) based on their site of action. Vitamin D could have a possible survival benefit in CKD patients. Calcimimetics, agents that increase the sensitivity of calcium-sensing receptors (CaSR) in the parathyroid gland, decreases PTH production. However, they do not increase the levels of calcium or phosphorus. Cinacalcet, along with low dose vitamin D, minimizes the risk of calcification, while offering the benefits of lowering PTH.

Surgical Treatment: Parathyroidectomy is a surgical modality available if medical therapy is unsuccessful or refractory. Other indications include calciphylaxis, refractory pruritus, severe hypercalcemia with serum calcium greater than 10.2 mg/dL, or hyperphosphatemia with serum phosphorus greater than 5.5 mg/dL, hyporesponsive anaemia to erythropoietin, PTH levels more than 800 pg/mL for more than 6 months despite medical therapy and extra-skeletal calcification. Assessment of parathyroid mass is an important factor in predicting the response to medical management. Glands larger than 1 cm in size or greater than 500 mm^3 on ultrasound represent the glandular hyperplasia and are usually refractory to medical treatment. Sometimes, in total parathyroidectomy, small amounts of the parathyroid gland could be autografted post-surgery.

Secondary hyperparathyroidism can be treated with surgery, but none of the surgical techniques permanently fix the underlying cause, so there's a high risk of the condition returning. Other treatments include renal transplantation, but this often doesn't fully resolve the hyperparathyroidism. Notwithstanding advances in medical treatment, surgical treatment of SHPT is often essential, particularly in refractory cases. Administration and overuse of vitamin D analogs, calcimimetics, and phosphate binders decrease PTH levels, which in turn, leads to a state of low bone turnover with normal mineralization.

41

Feline Lower Urinary Tract Diseases and Feline Urolithiasis

The diseases of the lower urinary tract such as bladder and urethra of cats are collectively labelled as Feline lower urinary tract disease (FLUTD). Although problems may occur at any age, those are usually realized in middle-aged, overweight cats that get little exercise and have little or no outdoor access. Changes in routine, moving to a new home, or living in a multi-cat household are the potential stressors, enhancing the risk of FLUTD. The partial or complete obstruction of urethra resulting in little or no urine passage are included under FLUTD that makes the cat increasingly distressed.

Etiology

Some of the conditions that are included FLUTD include urinary tract infection (UTI), urolithiasis, Feline Idiopathic Cystitis (FIC), cancer of bladder or urethra.

Urinary Tract Infection (UTI): This is the most common cause of symptoms referrable to urinary tract diseases in cats those are more common in females. UTIs are caused by colonization of bacteria in the urethra or bladder. Risk factors for urinary tract infections include obesity, poor grooming, other diseases such as diabetes and kidney disease, or anatomical changes in the bladder or urethra. The infection of the urinary tract with bacteria, fungi, parasites, or viruses can cause signs of FLUTD. Although bacterial infections are more common than other types of urinary tract infections, later are still relatively uncommon in cats. The urinary tract infection is a common cause of FLUTD in cats, especially females, older than 10 years of age. Cats in this age group are at risk of other health issues, like kidney disease and diabetes that cause changes to the urine that support infection. In younger cats, urinary tract infections are an uncommon cause of FLUTD because the acidity and concentration of their urine prevents infection.

The bacterial urinary tract infections in cats have been categorized as sporadic bacterial cystitis, recurrent bacterial cystitis, pyelonephritis, bacterial prostatitis, and subclinical bacteriuria. A new separate category is UTIs in dogs and cats receiving urological surgery, minimally invasive urological procedures and urologic implants. Primary bacterial infections are rare in cats, although secondary infections can occur as complicating factors. Bacterial infections are more common in cats over ten years of age. Many older cats will have both bacterial infections and bladder stones.

Urolithiasis (Bladder Stones): Bladder stones may develop in cats in urinary tract infections, or result from factors such as diet and genetics. Those are formed in the bladder, urethra, or at both the locations, causing irritation or blockage to urine flow. Urinary calculi or bladder stones is responsible for approximately 20% of all cases FLUTD in cats under ten years of age These stones can be made of different minerals, most commonly struvite (magnesium ammonium phosphate) or calcium oxalate. Urethral obstruction is the most serious cause of FLUTD, and the impact may range from being very uncomfortable to life-threatening. Cats with bladder stones are at risk for urethral obstruction, blocking the exit for urine, which can cause death within hours.

Feline Idiopathic Cystitis (FIC): Feline idiopathic cystitis (FIC), also termed interstitial cystitis or Pandora syndrome, is the most common cause of FLUTD in cats in the age group of less than 10 years. The exact cause is not known, so the term, idiopathic. This complex disease involves both the urinary system and other body systems. Chronic anxiety from perceived threats, such as conflict with another cat in the household, is believed to play a major role. Affected cats, particularly males may also have urethral obstruction. Cats with FIC experience inflammation of the bladder that is not caused by bacteria or stones. Unfortunately, FIC is just as severe as other forms of FLUTD and can still cause urethral obstruction in male cats. Feline idiopathic cystitis is a more complicated diagnosis.

Cancer of bladder or urethra: The cancer of the bladder or urethra is a rare cause of FLUTD, but it is a serious condition. A full workup is important to rule out cancer in a cat with FLUTD, even if UTI, bladder stones, or FIC is suspected. FLUTD may be caused less commonly by tumours of the urinary tract, anatomical abnormalities, or injury to the urinary tract or spinal cord.

Anatomical changes, obesity, diet, and genetics can increase the risk of urinary tract infections and bladder stones in cats. There is a linking between FLUTD symptoms and stress or perceiving a threat in the environment. Common things that you not see as stressors might, such as Certain scents, noises and even furniture changes may cause a lot of stress for your cat. The source of stress may include stray cats outside, new babies, losing a family member, and so on. It is obscure how these stress events and situations cause physical changes in the bladder of a cat.

There are many potential causes of FLUTD. The average age of onset is four years old. As mentioned, about half of affected cats experience severe bladder and/or urethra inflammation without an identifiable cause (idiopathic). These idiopathic cases must be differentiated from other potential causes so as to provide appropriate treatments.

Epidemiology

Urethral obstruction is most often found in male cats, one to 10 years of age. The neutered or intact male cats are at a greater risk than females because of longer and narrower urethra. The common causes of obstruction include urethral plugs, a soft material containing minerals, cells and mucus-like protein, and urinary stones. The urine starts to back up into the kidneys, once the urethra is completely blocked., The kidneys fail without treatment, toxins build up in the blood, and electrolyte imbalances occur, leading to death in less than 24 to 48 hours.

Female cats are more likely to be affected subclinical bacteriuria. UTIs are significantly more common in spayed female cats. The most common systemic comorbidities in affected cats are CKD and endocrine diseases such as diabetes mellitus and hyperthyroidism.

Bacterial UTIs occur much less frequently in cats than in dogs, with only 1–2% of cats suffering from UTIs in their lifetime. In cats with clinical signs of FLUTD, such as pollakiuria, macroscopic haematuria, stranguria, periuria or urethral obstruction, the proportion of cases with a UTI ranges from 2–19%. UTI is less common in younger cats. However, it is an important cause of FLUTD in cats older than 10 years, affecting 40–45% of the population.

Subclinical bacteriuria: Clinical signs of FLUTD are observed in 65% of cats with a positive urine culture, and 35% cats show sub-clinical bacteriuria. The impaired local defence mechanisms predispose to the presence of positive urine cultures. Factors associated with an increased risk of positive urine cultures are urinary incontinence, transurethral procedures and urogenital surgery, gastrointestinal disease, decreased urine specific gravity (USG) and decreased body weight. However, a low urine specific gravity does not appear to be a consistent risk factor for positive urine cultures.

Perineal urethrostomy (PU) is a well-recognised risk factor for UTI, with 22-53% of affected cats having undergone the procedure.

Nephroliths can predispose cats to pyelonephritis in addition to obstructing the urinary outflow and injuring the renal parenchyma. There is little evidence to suggest that feline immunodeficiency virus or feline leukaemia virus infection and immuno-suppressive drug treatment are predisposing conditions for significant bacteriuria.

Pathophysiology

UTIs have traditionally been classified as uncomplicated or complicated. An uncomplicated UTI was defined as a sporadic bacterial infection of the bladder in an otherwise healthy individual with normal urinary tract anatomy and function. UTIs in animals with any anatomic and/or functional abnormalities of the urinary tract, animals with co-morbidities that predispose to persistent infection, recurrent infection or treatment failure, or animals experiencing more than three episodes within a 12-months period were categorised as complicated.

Development of UTI and subclinical bacteriuria is multifactorial and depends on the interplay between bacterial virulence and alterations in host defences. The healthy feline urinary tract is a remarkably hostile environment that is not conducive to bacterial migration and colonisation owing to various host defence mechanisms. A transient or permanent breach in host defence mechanisms leads to developments of urinary tract infections, allowing virulent microbes to adhere, multiply and persist within the urinary tract. The major host defences against bacterial colonisation include frequent and complete voiding of an adequate urine volume, presence of a normal resident microflora, a physiological urinary tract anatomy, antimicrobial characteristics of the urine and systemic immunocompetence.

Bacterial virulence and fitness factors enable pathogens to colonise and to invade the urinary tract. Virulence factors determine not only the severity of a UTI, but also the infection site. Virulence factors are clustered on pathogenicity-associated islands, which can be easily spread among bacterial populations by horizontal gene transfer. The virulence of uropathogenic *E coli* (UPEC) is enhanced by adhesins (eg. type 1 and P fimbriae), iron acquisition systems and toxins (eg, haemolysin). P fimbriae are bacterial adhesion molecules found in most (80%) pathogens causing pyelonephritis, in 22% of cystitis strains, and in only 15% of asymptomatic strains in people. Bacterial invasion into epithelial cells often triggers the cell to undergo apoptosis and exfoliation. Some *E coli* strains in humans and mice are able to invade deeper tissues, persisting intracellularly, and may also form intracellular biofilm. Thus, those pathogens cannot be isolated from urine and can escape antimicrobial treatment.

Emphysematous, encrusting and polypoid cystitis are rare forms of complicated lower UTIs. Emphysematous cystitis is characterised by gas accumulation within the bladder wall and lumen, secondary to infection with glucose-fermenting bacteria, mainly *E coli.* Animals with diabetes mellitus are predisposed because of their high urinary glucose levels. The urea-splitting bacteria, such as *Corynebacterium urealyticum*, may lead to mineral precipitation, resulting in bladder wall encrustation. Polypoid cystitis refers to mass-like proliferations or diffuse thickening of the bladder mucosa induced by chronic inflammation, and is most commonly associated with infections due to *Proteus* species. Although adequate antimicrobial treatment can lead to complete resolution, surgical intervention may be required.

Significant bacteriuria is most commonly caused by pathogens from the host's own enteric or distal urogenital flora. The bacteria ascend from the distal urethra into the normally sterile proximal urethra, urinary bladder and upper urinary tract. *Escherichia coli* clones responsible for urinary tract infections can frequently be isolated from the faecal samples of the same animal. Haematogenous spread is rare but can cause UTIs, particularly in cases of renal abscesses.

Most UTIs are caused by single bacterial pathogen (>85%), while infections with multiple bacterial species are more common in cats with indwelling urinary catheters or other local comorbidities. *E coli* has been most commonly isolated pathogen in feline urine.

The cause FLUTD is diagnosed as feline idiopathic cystitis (FIC), if all the known causes of the disease have been eliminated. Idiopathic is a term that means, the exact cause is unknown.

Clinical Signs

Typical signs in cats with FLUTD are associated with inflammation and irritation of the lower urinary tract, and include pollakiuria, increased urination frequency, dysuria, difficulty in urinating, straining in the litter box for a long time, passing only small quantities of urine, and blood in the urine, called haematuria. Urine that is cloudy smells foul. The affected cats urinate in unusual places. Other clinical signs include excessive grooming or licking in the genital region.

Complete urinary tract obstruction, mostly in male cats, results in the inability to urinate, straining to urinate persistently without producing any urine due to urinary tract obstruction, necessitating immediate veterinary care as it is very painful for the cat, and can be a life-threatening complication, if untreated.

The most common symptoms of FLUTD in cat include inappropriate urination, and passing urine outside the litter box. Inappropriate urination often occurs on clothes, bedding, and rugs. Some cats may choose one location, but others may urinate throughout the home. Other symptoms include vocalizing while urinating and/or the presence of blood in the urine. The principal manifestations of FLUTD include crying out while urinating, urinating frequently small amount of urine, urinating outside the litter box, licking the genital area, straining in the litter box, and passage of blood in the urine. Cats affected with uroliths strain to urinate and produce little or no urine. Although it may seem like they are constipated and trying to pass stool, straining in the litter box is more often a sign of blocked urethra.

In most cases, episodes of FIC are short-lasting, and cats without urethral obstruction often improve within a week, even without treatment. However, a large proportion of these cats will have at least one other episode within one or two years. Less commonly, FIC can persist, never totally going away. Stress or anxiety can make episodes worse, and can increase the risk and severity of feline idiopathic cystitis.

Diagnosis

The initial diagnosis of FLUTD is based on identifying signs of lower urinary tract inflammation. The clinical signs displayed by the cat are often characteristic of FLUTD. A urinalysis will confirm the presence of inflammation or infection.

When clinical signs are persistent or recurrent, various tests may be required to differentiate idiopathic FLUTD from the other known causes of urinary tract inflammation. These diagnostic tests include laboratory analysis of a urine sample, bacterial culture of a urine sample, analysis of blood samples to look for other evidence of urinary tract disease or another systemic disease, such as kidney disease. Radiographs (X-rays) and/or ultrasound examination of the bladder and urethra may be carried out for diagnosis of the underlying causes.

The diagnosis of specific problem or cause underlying an episode of FLUTD can be difficult. A thorough physical examination, a detail history-taking, and laboratory tests of urine and blood samples for pH, specific gravity, and the presence of crystals, and indicators of inflammation and infection, bacterial culture, abdominal X-rays or ultrasound can help to arrive at a diagnosis. A thorough diagnostic evaluation may not identify the specific cause. Some cats can have multiple episodes of FLUTD, and the specific cause underlying each episode may be different, with more than one cause possible per episode, necessitating a thorough evaluation of each episode. Urine samples in suspected cases of urinary tract infection, are subjected to bacterial culture to identify the infecting organism, and antibiotic susceptibility testing to determine the best drug for treatment.

There is no specific diagnostic test for FIC. The disease is commonly diagnosed by ruling out other causes of FLUTD and examining the inside of the urethra and bladder with a special device, called cystoscope.

Distinguishing between bacterial cystitis and idiopathic cystitis with concurrent clinical or subclinical bacteriuria, is challenging, as clinical signs and urinalysis results may be identical. Optimal treatment of subclinical bacteriuria requires clarification as there is currently no evidence that demonstrates a beneficial effect of routine treatment.

Urinary tract infection (UTI) refers to the adherence, multiplication and persistence of an infectious agent within the urogenital system that causes an associated inflammatory response and clinical signs. In the vast majority of UTIs, bacteria are the infecting organisms. UTIs due to parasitic, fungal or viral infections are fewer than 1%.

In addition to cats with typical clinical signs of feline lower urinary tract disease (FLUTD) or upper UTI, many cats have asymptomatic or subclinical bacteriuria. In human medicine, asymptomatic bacteriuria is defined as the isolation of a specified number of bacteria in a urine specimen from a patient without symptoms referable to UTI. In veterinary medicine, use of the term subclinical UTI has been suggested for these cases, where clinical signs might be too subtle to be detected. The terms asymptomatic bacteriuria in human beings and subclinical bacteriuria in dogs and cats, are also used if there is evidence of inflammation such as pyuria, and haematuria in the urine sediment.

Treatment

There is no universal treatment for FLUTD. Treatment depends on the underlying cause. Each case has to be investigated to determine the underlying cause, and then the treatment has to be tailored to the individual cat. Sometimes, despite appropriate tests and treatment, clinical signs may still recur, requiring further diagnostic testing. Bacterial infections of the lower urinary tract usually respond well to antibiotic therapy.

Initially, a cat with uncomplicated FLUTD may be treated symptomatically with pain medication, dietary modification, or with antibiotics in certain circumstances. However, if the symptoms do not respond to this treatment or there is a recurrence of the clinical signs, additional diagnostic tests may be required to identify the underlying cause of the FLUTD. There also is no specific treatment or cure, which can make FIC frustrating to manage for all involved.

Management of recurrent UTIs remains a challenge as evidence for most alternatives used for prevention in cats is mainly anecdotal, and no preventive treatment modality is currently recommended. Many treatments have been recommended for managing cats with FLUTD. The recommendations should be based on results of randomized and well-controlled scientific studies performed in clinical patients with the spontaneously occurring disease of interest. In the absence of such studies, one is left to practical relevance.

UTIs are one of the most important indications for antimicrobial use in veterinary medicine and contribute to the development of antimicrobial resistance. Adherence to treatment guidelines and confinement to a few first-line antimicrobial agents is imperative to avoid further deterioration of the antimicrobial resistance situation. The decision to treat with antimicrobials should be based on the presence of clinical signs, concurrent diseases, and the results of urine culture and susceptibility testing.

Treatment of encrusting cystitis can be challenging and long-term antimicrobial therapy, urine acidification and surgical debridement of plaques are often required.

Additional supportive treatment is instituted as per the abnormalities detected during the diagnostic evaluation, and may include fluid or electrolyte therapy, and medications to manage pain, nausea, and anxiety.

Treatment for struvite stones typically involves feeding the cat with a special stone-dissolving diet. The surgical removal is generally recommended, if the diet changes are not implementable. Fortuitously, minimally invasive procedures can be used to remove urinary stones, with little to no recovery time for cats and less discomfort than with the traditional, more invasive surgical options of cystotomy or urethrotomy.

Urethral obstruction is an emergency and requires immediate veterinary care. Dislodging the obstruction may be tried by flushing sterile solution through a catheter, placed up and into the urethra. The catheter may be left in place for a few days once the obstruction is removed, to allow swelling to go down. Dehydration and electrolyte imbalances are treated with intravenous fluid therapy. Pain medication, a stone-dissolving diet, and other medications are used to make the cat more comfortable.

The current goals of treating cats with FIC are to decrease the severity and frequency of FIC episodes. A key aspect of this involves making changes to the environment to increase their sense of control and reduce their sense of threat, a strategy known as "multimodal environmental modification (MEMO)" Diet changes and certain medications also may be helpful.

The treatment of choice for calcium oxalate urolithiasis is urolith removal, followed by methods to prevent urolith recurrence. General goals of preventive management are to decrease urine saturation with calcium oxalate, increase the concentration or activity of calcium oxalate inhibitors in urine, and promote production of urine that is more dilute. At present, the standard of care for preventing calcium oxalate urolith recurrence is to feed a moist therapeutic food and encourage water intake

Control

It is impossible to completely prevent lower urinary tract diseases. FLUTD is more common in cats that have low water consumption and in cats that are inactive and obese. All these factors may relate, at least in part, to the frequency with which a cat urinates. The lower urinary tract disease can be prevented by providing safe places to rest and sleep, an environment that meets the cat's sensory needs such as taste, sight, smell, sound, touch, access to clean, fresh water at all times, an adequate number of litter boxes, usually one more than the number of cats in the household, with the type of litter and litter box size and type as per the preference of the cat. Safe access to litter boxes, such as in quiet areas of the house, clean litter boxes, with waste ideally removed twice a day and litter changed once a week or more, daily opportunities for play, a healthy body weight, a positive, consistent, and predictable an environment are necessary to prevent FLUTD.

The provision of separate areas for sleeping, eating, water, litter box use, scratching, climbing, and play are required for multiple cats in a house to minimize conflict between cats and allow each cat to be separate from the others.

Index

E

F

G

H

I

T

U

V

W

Z